Alexey Bezryadin

Superconductivity in Nanowires

Related Titles

Vedmedenko, E.

Competing Interactions and Patterns in Nanoworld

2007
ISBN: 978-3-527-40484-1

Wilkening, G., Koenders, L.

Nanoscale Calibration Standards and Methods
Dimensional and Related Measurements in the Micro- and Nanometer Range

2005
ISBN: 978-3-527-40502-2

Waser, R. (ed.)

Nanoelectronics and Information Technology

2012
ISBN: 978-3-527-40927-3

Reich, S., Thomsen, C., Maultzsch, J.

Carbon Nanotubes
Basic Concepts and Physical Properties

2004
ISBN: 978-3-527-40386-8

Alexey Bezryadin

Superconductivity in Nanowires

Fabrication and Quantum Transport

WILEY-VCH

WILEY-VCH Verlag GmbH & Co. KGaA

The Author

Prof. Alexey Bezryadin
Department of Physics
University of Illinois
1110 West Green Street
Champaign IL 61801
USA

All books published by **Wiley-VCH** are carefully produced. Nevertheless, authors, editors, and publisher do not warrant the information contained in these books, including this book, to be free of errors. Readers are advised to keep in mind that statements, data, illustrations, procedural details or other items may inadvertently be inaccurate.

Library of Congress Card No.:
applied for

British Library Cataloguing-in-Publication Data:
A catalogue record for this book is available from the British Library.

Bibliographic information published by the Deutsche Nationalbibliothek
The Deutsche Nationalbibliothek lists this publication in the Deutsche Nationalbibliografie; detailed bibliographic data are available on the Internet at http://dnb.d-nb.de.

© 2013 WILEY-VCH Verlag GmbH & Co. KGaA, Boschstr. 12, 69469 Weinheim, Germany

All rights reserved (including those of translation into other languages). No part of this book may be reproduced in any form – by photoprinting, microfilm, or any other means – nor transmitted or translated into a machine language without written permission from the publishers. Registered names, trademarks, etc. used in this book, even when not specifically marked as such, are not to be considered unprotected by law.

Print ISBN 978-3-527-40832-0
ePDF ISBN 978-3-527-65196-2
ePub ISBN 978-3-527-65195-5
mobi ISBN 978-3-527-65194-8
oBook ISBN 978-3-527-65193-1

Composition le-tex publishing services GmbH, Leipzig
Printing and Binding Markono Print Media Pte Ltd, Singapore
Cover Design Adam-Design, Weinheim

Printed in Singapore

Printed on acid-free paper

Contents

Preface – Superconductivity and Little's Phase Slips in Nanowires IX

Abbreviations – Short List XI

Notations – Short List XIII

Color Plates XV

1 Introduction 1

2 Selected Theoretical Topics Relevant to Superconducting Nanowires 15
2.1 Free or Usable Energy of Superconducting Condensates 15
2.2 Helmholtz and Gibbs Free Energies 18
2.3 Fluctuation Probabilities 23
2.4 Work Performed by a Current Source on the Condensate in a Thin Wire 27
2.5 Helmholtz Energy of Superconducting Wires 29
2.6 Gibbs Energy of Superconducting Wires 31
2.7 Relationship between Gibbs and Helmholtz Energy Densities 35
2.8 Relationship between Thermal Fluctuations and Usable Energy 36
2.9 Calculus of Variations 38
2.10 Ginzburg–Landau Equations 39
2.11 Little–Parks Effect 46
2.12 Kinetic Inductance and the CPR of a Thin Wire 50
2.13 Drude Formula and the Density of States 51

3 Stewart–McCumber Model 53
3.1 Kinetic Inductance and the Amplitude of Small Oscillations 60
3.2 Mechanical Analogy for the Stewart–McCumber Model 62
3.2.1 Defining the Supercurrent Through Helmholtz Free Energy 65
3.2.2 Cubic Potential 66
3.2.3 Thermal Escape from the Cubic Potential 67
3.3 Macroscopic Quantum Phenomena in the Stewart–McCumber Model 68
3.3.1 MQT in a Cubic Potential at High Bias Currents 71

3.4	Schmid–Bulgadaev Quantum Phase Transition in Shunted Junctions 74
3.5	Stewart–McCumber Model with Normalized Variables 76
4	**Fabrication of Nanowires Using Molecular Templates** *79*
4.1	Choice of Templating Molecules 86
4.2	DNA Molecules as Templates 86
4.3	Significance of the So-Called "White Spots" 88
5	**Experimental Methods** *91*
5.1	Sample Installation 91
5.2	Electronic Transport Measurements 95
6	**Resistance of Nanowires Made of Superconducting Materials** *101*
6.1	Basic Properties 101
6.2	Little's Phase Slips 105
6.3	Little's Fit 108
6.4	LAMH Model of Phase Slippage at Low Bias Currents 115
6.5	Comparing LAMH and Little's Fit 122
7	**Golubev and Zaikin Theory of Thermally Activated Phase Slips** *125*
8	**Stochastic Premature Switching and Kurkijärvi Theory** *131*
8.1	Stochastic Switching Revealed by V–I Characteristics 131
8.2	"Geiger Counter" for Little's Phase Slips 135
8.3	Measuring Switching Current Distributions 139
8.4	Kurkijärvi–Fulton–Dunkleberger (KFD) Transformation 143
8.5	Examples of Applying KFD Transformations 148
8.6	Inverse KFD Transformation 152
8.7	Universal 3/2 Power Law for Phase Slip Barrier 153
8.8	Rate of Thermally Activated Phase Slips at High Currents 157
8.9	Kurkijärvi Dispersion Power Laws of 2/3 and 1/3 160
9	**Macroscopic Quantum Tunneling in Thin Wires** *163*
9.1	Giordano Model of Quantum Phase Slips (QPS) in Thin Wires 165
9.2	Experimental Tests of the Giordano Model 175
9.3	Golubev and Zaikin QPS Theory 183
9.4	Khlebnikov Theory 185
9.5	Spheres of Influence of QPS and TAPS Regimes 187
9.6	Kurkijärvi–Garg Model 189
9.7	Theorem: Inverse Relationship between Dispersion and the Slope of the Switching Rate Curve 195
10	**Superconductor–Insulator Transition (SIT) in Thin and Short Wires** *197*
10.1	Simple Model of SIT in Thin Wires 207

11 Bardeen Formula for the Temperature Dependence of the Critical Current *213*

Appendix A Superconductivity in MoGe Alloys *215*

Appendix B Variance and the Variance Estimator *217*

Appendix C Problems and Solutions *223*

References *241*

Index *247*

Preface – Superconductivity and Little's Phase Slips in Nanowires

This book presents an account of fabrication techniques, charge transport experiments, and theoretical models related to superconducting nanowires, that is, thin wires made of superconducting materials and having dimensions at the nanometer scale. A wire is classified as one-dimensional (1D) if its diameter d is smaller than the superconducting coherence length ξ. In such wires, the superconducting condensate wavefunction only depends on the position along the wire, while it is independent of the position within the wire cross section. A nanowire is classified as quasi-one-dimensional (quasi-1D) if $d < \pi\sqrt{2}\xi$. Such condition ensures that vortices are not energetically stable in the wire. Therefore, the order parameter is approximately constant within the wire cross section. Since ξ diverges at the critical temperature T_c, it is not too hard to make a wire which is quasi-1D near T_c. It is more challenging to fabricate nanowires which are 1D or quasi-1D even in the limit of zero temperature. Recent perfection of nanotechnology allows us to make wire as thin as 5 nm, which can be regarded as quasi-1D even at zero temperature. This book provides a detailed analysis of the transport properties of such ultrathin wires. Among other things, the small size of the wires causes them to behave as quantum objects. Due to the Heisenberg uncertainty principle, such wires might not be true superconductors. Their state becomes uncertain because the wires exist in a quantum superposition of normal and superconducting states. As the admixture of the normal phase increases, nominally superconducting nanowires can lose their ability to carry supercurrent and can become either normal or, what it more typical, slightly insulating.

Even if quantum fluctuations are not sufficiently strong to destroy superconductivity, the ohmic resistance R of a thin wire is always greater than zero, provided that the temperature, T, is above absolute zero. Due to thermal fluctuations, the phase of the superconducting condensate wavefunction jumps irreversibly, leading to a decay of the supercurrent. Such jumps are called thermally activated phase slips (TAPS) or Little's phase slips (LPS). A detailed discussion of LPS is one of the main topics of this book.

Nanowires made of superconducting materials can generally be classified into three different categories: superconducting or S-type wires, normal or N-type wires, and insulating or I-type wires. Assume that $R(T) \to R(0)$ as $T \to 0$, where T is the temperature of the nanowire and $R(T)$ is its electric resistance as a function of temperature. Also, assume that R_n denotes the resistance of the nanowire in the normal state, that is, at $T > T_c$. Then, $R(0) = 0$ (at $T = 0$) defines the S-type wires, $0 < R(0) < R_n$ corresponds to the N-type, and $R(0) > R_n$ characterizes, by convention, I-type wires. There is no direct way to determine the type of the wire since the condition $T = 0$ is impossible to satisfy experimentally. Thus, the classification is usually done using indirect approaches. For example, if the resistance is observed to increase with cooling, then the wire is classified as the I-type. On the other hand, if the ohmic resistance decreases exponentially with cooling, then the wire should be classified as the S-type. The S-type classification also requires that the wire exhibits a finite supercurrent at sufficiently low temperatures. In the book, we summarize the properties of the two most common types of nanowires, namely, the S-type and I-type wires, and analyze the quantum phase transition between these two phases.

The book is intended for undergraduate and beginning graduate students interested in nanoscale superconductivity. The readers are assumed to have some familiarity with the basics of quantum mechanics and superconductivity. Recommended introductory books on general aspects of superconductivity are *Introduction to Superconductivity* by M. Tinkham [1] and *Superconductivity of Metals and Alloys* by P.G. de Gennes [2]. The present monograph is focused on selected topics related to thin superconducting nanowires.

It is my pleasure to acknowledge my indebtedness to the graduate students and postdocs I have supervised since 2000. Of special importance is the advice and the helpful discussions with Liliya Simpson, Anthony T. Bollinger, Ulas C. Coskun, Celia Elliott, David S. Hopkins, Robert C. Dinsmore III, Mitrabhanu Sahu, Sergei Khlebnikov, Thomas Aref, Matthew W. Brenner, Paul Goldbart, Andrey Rogachev, Andrey Belkin, Andrey Zaikin, Konstantin Arutyunov, and Victor Vakaryuk. The work was supported by the DOE grant DE-FG02-07ER46453 and by the NSF grant DMR 10-05645.

Campaign, August 2012 *Alexey Bezryadin*

Abbreviations – Short List

BCS	Bardeen, Copper, and Schrieffer theory of superconductivity.
BCS condensate	a coherent quantum states of a macroscopic number of electrons appearing in the BCS theory.
CP	Cooper pair.
CPR	Current–phase relationship.
GL	Ginzburg and Landau phenomenological theory.
GLAG	Ginzburg, Landau, Abrikosov, and Gor'kov theory, which is a microscopic theory derived from BCS, but similar in form to the GL theory.
GZ	Golubev and Zaikin theory (either of TAPS or of QPS).
I-type wire	an insulating nanowire, i.e., a nanowire made of a superconducting metal, but, due to various reasons, having its zero-temperature resistance larger than its normal-state resistance.
JJ	Josepshon junction.
KG	Kurkijärvi–Garg model.
LAMH	Langer, Ambegaokar, McCumber, and Halperin combined model.
LPS	Little's phase slip.
MT	molecular templating.
MQT	Macroscopic quantum tunneling.
N-type wire	A normal nanowire, that is, a nanowire made of a superconducting metal, but, due to various reasons, having its zero-temperature resistance larger than zero but lower or equal compared to its normal-state resistance, which is measured at $T > T_c$.
QPS	quantum phase slips, usually in quasi-1D superconducting wires. QPS is a particular case of MQT.
SB	Schmid–Bulgadaev quantum phase transition.
SIS	Superconductor–insulator–superconductor junction.
SNS	Superconductor–normal metal–superconductor junction.
SIT	Superconductor to insulator quantum phase transition. The meaning of it is that on the S-side the wire has zero resistance

	at zero temperature, while on the I-side of the transition the wire acts as an insulator or a bad conductor.
SNT	superconductor–normal quantum phase transition.
SRT	"superconducting" to "resistive" quantum phase transition. The meaning of the phase transition is that in the S-phase the wire has zero resistance at zero temperature, while in the R-phase the wire is resistive. SNT and SIT are examples or particular cases of the more general SRT.
S-type wire	A nanowire characterized by zero resistance at zero temperature.
SM	Stewart and McCumber model, predicting the washboard potential.
TAPS	Thermally activated phase slips, that is, Little's phase slips energized by thermal fluctuations.

Notations – Short List

A_{cs}	Cross section area of a nanowire, taken perpendicular to the wire.
a_G	Giordano constant, used as a fitting parameter in the QPS rate, under the exponent.
Δ	Superconducting energy gap.
e	Charge of one electron.
$E_J = \hbar I_c / 2e$	Josephson energy.
ΔF_{LAMH}	Barrier for a thermally activated phase slip, as it appears in the LAMH model.
ΔF_{GZ}	Barrier for a thermally activated phase slips, due to Golubev and Zaikin.
h	Planck's constant.
$\hat{H}$	Hamiltonian in the Schrödinger equation describing the phase particle.
$I_0 = \dfrac{4ek_B T}{h} \approx 13\,\text{nA/K}$	Nonlinearity onset current.
I_s	Supercurrent.
$I_s = I_s(\phi)$	Supercurrent, which is a function of the phase difference ϕ.
I_c	Critical current, defined as the maximum of the CPR function $I_s(\phi)$.
$i_b = I/I_c$	Normalized bias current.
$F_b = -dU_b/x_p$	The force acting on the phase particle in the SM model occurring due to the bias current.
F_ϕ	Phase difference force acting on the phase particle in the SM model.
l_e	Electronic elastic mean free path.
m	Mass of one electron.
m_p	Mass of the phase particle describing JJ in SM model.
n	Total electronic density.
$N_0 = \dfrac{m^2 v_F}{2\pi^2 \hbar^3}$	Density of states at the Fermi level, per one spin projection.

$R_q = h/4e^2 \approx 6.5\,\text{k}\Omega$	Quantum resistance or von Klitzing constant for pairs.
S	Entropy.
S_{mqt}	Action of a macroscopic quantum tunneling event.
T	Temperature.
ϕ or $\Delta\phi$	Phase difference.
$\phi(r,t)$	Local phase of the order parameter.
Φ_0	Magnetic flux quantum.
Ω_{mqt}	Corresponding attempt frequency of a macroscopic quantum tunneling event.
ω_p	Plasma frequency.
$\omega_p(0)$	Plasma frequency of a superconducting device at zero bias current.
$\Psi_p = \Psi_p(x_p)$	Wave function of the phase particle, representing the phase difference on a JJ.
ψ_p	Time-independent wave function of the phase particle.
U_{wb}	Washboard potential in which the phase particle moves, within Stewart–McCumber model.
U_b	Potential corresponding to the bias current force F_b acting on the phase particle and generated by the bias current.
U_ϕ	Potential energy corresponding to the phase force F_ϕ acting on the phase particle.
v_I	Sweep speed in the switching statistics experiments.
V	Voltage.
V_s	Volume of sample or system.
x_p	Coordinate of the effective phase particle, $x_p = \phi$, in the SM model.
$\xi_0 = \hbar v_F / \pi \Delta$	Coherence length in pure metals without disorder at zero temperature.
$Z_0 = 377\,\Omega$	Impedance of vacuum.

Color Plates

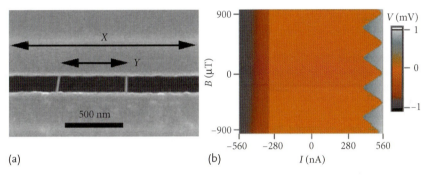

Figure 2.1 (a) An SEM micrograph of a sample with two parallel wires [105, 110, 111] made using DNA molecules as templates. The same device is illustrated on the cover of the book. The width of the electrodes is X and the distance between the wires is Y. (b) Color-coded plot of the voltage as a function of the bias current and the magnetic field, applied perpendicular to the thin film electrodes. The Little–Parks oscillation, induced by the magnetic field, is visible [105, 110, 111] as periodically occuring gray triangles which correspond to the normal state.

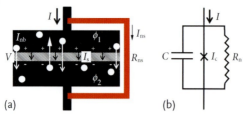

Figure 3.1 Stewart–McCumber model for Josephson junctions. (a) A schematic cross-section of a superconductor-insulator-superconductor (SIS) sandwich junction. It includes two superconducting electrodes (black) separated by a very thin insulating oxide layer (dashed region). The bias current, I, is shown by the large black arrow. The current through the junction contains the supercurrent I_s (small black arrows) and the bogoliubon current I_{nb} (white arrows). The bogoliubons are shown as white circles. The optional Ohmic normal shunt, shown in red, has a resistance R_{ns} and can carry a current of normal electrons I_{ns}. Moreover, the electrodes separated by the oxide form a capacitor which becomes charged if the currents through the junction and the shunt do not altogether amount to the bias current. (b) An equivalent electrical circuit for a JJ, called the Stewart–McCumber (SM) model [95, 96] or the resistively and capacitively shunted junction (RCSJ) model. The model is based on the current conservation law $I = I_n + I_s + \dot{Q}$. Here, $I_s = I_c \sin\phi$ is the supercurrent and I_n is the total dissipative current, which includes the current of bogoliubons through the junction and the current of normal electrons in the shunt, that is, $I_n = I_{nb} + I_{ns}$ (see part (a) for the notations). The displacement current $\dot{Q}$ is the current that does not flow through the junctions, but produces extra charge accumulated on the plates of the capacitor, that is, on the electrodes forming the junction. One of the electrodes gets charge Q and the other always gets charge $-Q$. The total normal current resistance R_n is defined as a parallel connection of the shunt R_{ns} and the quasiparticle tunnel resistance R_{nb} as $R_n^{-1} = R_{nb}^{-1} + R_{ns}^{-1}$. The normal resistance can also be defined such that the total normal current is $I_n = V/R_n$ and the Joule heating in the entire system is $P_J = V^2/R_n = I_n^2 R_n$. Here, the voltage V is, as usual, the difference between the electric potentials of the electrodes forming the junction. The potential as well as the superconducting phases ϕ_1 and ϕ_2 are assumed constant within each of the electrodes. The phase difference is simply $\phi = \phi_1 - \phi_2$, where ϕ_1 is the phase of the order parameter in the top electrode and ϕ_2 is the phase in the bottom electrode.

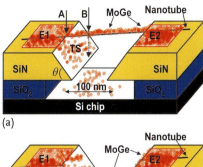

(a)

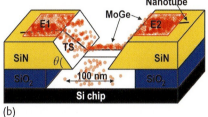

(b)

Figure 4.1 Schematic (not to scale) of molecular templating (MT) using a carbon nanotube [100]. (a) The templating nanotube is shown as the solid black line. The tube is coated with atoms of the desired metal (red disks). The atoms, deposited by DC magnetron sputtering, stick to the nanotube and form a continuous nanowire. The smallest achievable diameter of the wire is comparable to the diameter of the nanotube, which is ~ 2 nm. The wire forms naturally, perfectly connected to the electrodes which are superconducting films, marked "E1" and "E2," deposited simultaneously with the wire at the sides of the trench. The electrodes are shaped by photolithography. The segment of the nanowire located between the marks "A" and "B" is suspended over the tilted side (TS) of the trench. This is why the region AB appears brighter on scanning electron microscope (SEM) images. Such "white spots" (see Figure 4.3) at the ends of the wire are the indicators that the wire is straight and is well-connected to the electrodes. (b) A similar schematic, but having the molecule diving down into the trench. Such samples do not show "white spots" in SEM images. Such an arrangement is typical for DNA molecules which are flexible on the length scale of the trench width. The van der Waals force is sufficient to force the molecule to stick to the surface over the maximum possible length. Thus, the molecule crosses the trench at points where the gap is the narrowest, which, in this geometry, happens at the level of the bottom surface of the SiN film.

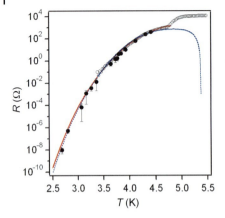

Figure 6.5 Resistance versus temperature plot obtained on a narrow Dayem bridge of width less than $4.4\xi(T)$ (sample B2 of [135]). Open circles show the linear resistance of the sample measured using a bias current lower than the nonlinearity onset current $I_0 = \frac{4ek_B T}{h} = 13$ nA/K and much lower than the critical current of the bridge. The filled black circles are obtained indirectly, by conducting differential resistance measurements at high bias currents and then by performing an extrapolation procedure in order to obtain the linear or low-bias resistance. The solid (red) and the dashed (blue) curves give the best fits generated by the Little (6.6) and the LAMH (6.30) formulas correspondingly. The critical temperature of the bridge was used as a fitting parameter. The best fits were obtained by taking $T_c = 4.81$ K for the Little fit and $T_c = 5.38 K$ for the LAMH fit. The first value, corresponding to the Little's fit, is realistic because the critical temperature of the film is 4.91 K and the bridge is expected to have the same critical temperature or slightly lower. The value of T_c required for the LAMH fit is significantly larger than the T_c of the electrodes, which is physically unrealistic because the T_c of the MoGe nanoscale samples of any sort always decreases as the sample dimensions are reduced.

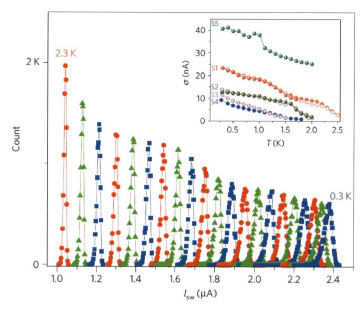

Figure 8.4 Examples of distributions of the switching current corresponding to different temperatures [131], all taken on the same sample (S1). The horizontal axis is divided into consequent segments, 4 nA each, which are called bins. The total number of measurements of the I_{sw} is $N = 10\,000$ at each temperature. The vertical axis corresponds to the number of counts in each bin. In other words, the "Count"-axis gives the number of switching events for each bias current bin. Note that the current bins are not explicitly marked on the horizontal I_{sw}-axis. Inset: The standard deviation σ vs. T for five different nanowires, including sample S1. For samples S1 and S2, the measurements have been repeated (and the results are shown by different symbols) to verify the reproducibility of the temperature dependence $\sigma(T)$. In these examples, it is observed that $d\sigma/dT < 0$, which is unusual. This is explained in terms of multiple phase slips [131]. The detailed analysis of [131] showed that below about 1 K, the switching events are caused by individual phase slips and each phase slip causes a switching. Therefore, the fact that at $T < 1$ K, no decrease of σ is seen as the temperature is decreased provides evidence that the phase slips in these samples occur by quantum tunneling, that is, they are quantum phase slips (QPS). Some increase of σ at $T < 1$ K could be due to the fact that the barrier for phase slips increases slightly with cooling. Note that if the switching at low temperatures would be caused by thermally activated phase slips (TAPS) then σ would decrease with cooling as $\sigma \sim T^{2/3}$. This is not observed even at the lowest temperature tested. Therefore it is concluded that QPS is the dominant mechanism causing the switching events.

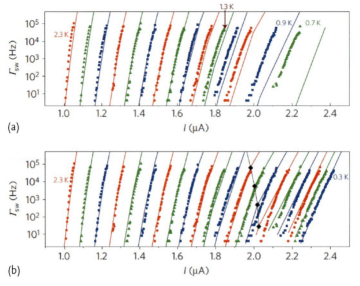

Figure 8.5 Switching rates plotted versus bias current for different temperatures for the sample S1 [131]. The rates are computed from the data of Figure 8.4 using the KFD transformation (8.19). (a) Switching rates for the temperatures between 2.3 K (left-most) and 0.7 K (right-most), for Sample 1. The rates are shown for all temperatures between 2.3 and 1.1 K, with a step of 0.1 K, and, also, for $T = 0.9$ K and $T = 0.7$ K. The symbols are the experimentally obtained rates and the solid curves (using the same colors) are the fits to the overheating model that only includes TAPS-initiated switching events and assumes that the QPS rate is zero. The fits agree well with the data down to $T \approx 1.3$ K. The data and the curve corresponding to $T = 1.3$ K are indicated by an arrow. At low temperatures, such a TAPS-only model does not agree with the data since it predicts switching rates much lower than the rates obtained experimentally. For example, for $T = 0.7$ K, the TAPS-only predicted rate is about 10 000 times lower than the measured one. All the fitting parameters, namely, $T_c = 3.872$ K, $\xi(0) = 5.038$ nm, $I_c = 2917$ nA, are kept the same for all temperatures. (b) The same switching rates for S1 as in (a), extended down to lower temperatures. The rates are plotted for the temperatures between 2.3 K (left-most) and 0.3 K (right-most), the step being 0.1 K. The symbols are the experimentally obtained rates and the solid curves (using matching colors) are the fits to the overheating model that incorporates Γ_{TAPS} as well as Γ_{QPS}, as in (8.20). The single-phase-slip switching occurs to the right of the black curve indicated by the black diamonds. These best fits are obtained assuming that the effective quantum temperature is $T_q = 0.726$ K $+ \Delta T_q$, where the temperature dependent correction is $\Delta T_q = 0.4 T$ and T is the temperature. The physical reason for the correction is not well understood yet. It could be either due to the temperature dependence of the superconducting energy gap or due to a survival of multiphase-slip switching even at low temperatures. At $T < 0.7$ K the QPS rate is much larger than the TAPS rate. The QPS based model agrees with the data very well, thus providing a solid evidence that the switching is caused by QPS at $T < 0.7$ K.

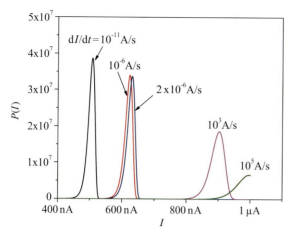

Figure 8.6 Probability distribution of the switching current calculated using the Kurkijärvi theory [179]. Here, P is defined such that $P(I)dI$ is the probability that the switching event occurs in the interval from I to $I+dI$. The critical current in this example is 1 µA. The curves are given for various sweep speeds $v_I \equiv dI/dt$, which are indicated near corresponding curves. The sweep speed for the blue curve is twice as large than for the red one, illustrating that doubling the sweep speed only causes a minor change of the distribution curve. In practice, the sweep speed is of the order of 10^{-6} A/s. It is clear that the mean switching current is smaller than the critical current by many widths of the distribution curve. This is because the attempt frequency is usually very high, such as $\Omega = 10^{12}$ Hz in this example.

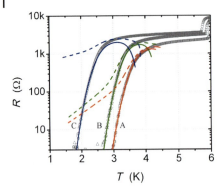

Figure 9.3 Resistance versus temperature curves of three representative $Mo_{79}Ge_{21}$ nanowires [132]. The data are shown by symbols. The continuous curves are the best fits to the LAMH model, generated using $C_T = 29$, 31, and 16 for the samples A, B, and C, correspondingly. The dashed curves represent fits to the Giordano model which are generated using the same parameter plus two additional Giordano parameters $a_G = 1.3$ and $B_G = 7.2$. Apparently, the model predicts a strong contribution of QPS, which is in disagreement with experimental curves. An explanation for the unobservable low rate of QPS is the low normal resistance of all three wires, which satisfy $R_n < R_q$, as is evident from the plots. According to Khlebnikov and Pryadko [46], coupling a nanowire to an impedance that is lower than $R_q = h/4e^2 \approx 6.5\,k\Omega$ makes the wire a true superconductor, characterized by zero rate of QPS. Note that the Giordano model predicts a finite QPS rate at $T = 0$, that is, according to the Giordano model the wires should be classified as normal or resistive, not as superconducting.

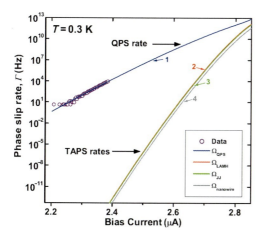

Figure 9.4 A comparison of various models predicting the rate of Little's phase slips to the experimentally obtained rate (circles) plotted versus bias current. This is a high bias current result which appeared in [131]. The blue curve is the fit to the Giordano model of QPS, which agrees with the data well. The other color curves represent TAPS models computed using different choices of the attempt frequency. It is illustrated that the choice of the attempted frequency has a very minor impact on the fitting curve. Predicted TAPS rates are many orders lower than the experimental switching rates. Thus QPS is confirmed.

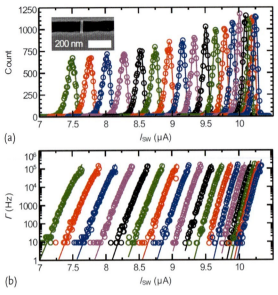

Figure 9.6 Measurements and analysis of switching current fluctuations [115, 116] in a MoGe nanowire. (a) Examples of switching current distributions (circles) measured at various temperatures ranging from 2 K (the left-most curve) to 0.3 K (the right-most distribution curve); the step is 0.1 K. The fits, shown as solid curves of the same color, are generated using the inverse KFD transformation (8.24) applied to the switching rate fits shown in (b). Insert: SEM image of a nanowire which was annealed in vacuum using high current pulses [114] and thus was crystallized (see the discussion about such Joule-heat-treated wires in the text). (b) Switching rates obtained from the distribution plotted in (a) using the discrete KFD transformation (8.19). The experimental switching rate is shown by circles. The solid curves of the same color are the fits to the KD model, namely, to (9.24). The exponent is $b = 3/2$ for all curves. The fits are done using two fitting parameters, namely, the attempt frequency Ω and the escape temperature T_{esc}. Both of these parameters are analyzed in the next figure.

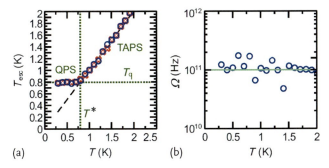

Figure 9.7 Plots of the fitting parameters used to generate fits in Figure 9.6b (see [115, 116]). (a) The fitting parameter T_{esc}, which defines the escape rate in (9.24), is plotted versus temperature (blue circles). The crossover temperature is marked T^* and the quantum fluctuations temperature is T_q. They are approximately equal, as expected from the KG model. (b) The attempt frequency Ω, which is another adjustable parameter in (9.24), is plotted as a function of temperature. The horizontal green line represents a rough theoretical estimation which is explained in the text.

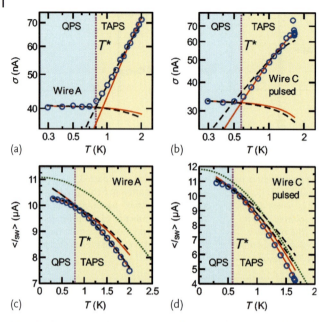

Figure 9.8 Mean switching current (c, d) and its dispersion (a, b) are plotted versus temperature [115, 116]. The plots (a) and (b) are in the log–log format. (a) and (c) – Sample A, unpulsed. (b) and (d) – Sample C, voltage-pulsed and crystallized before I_{sw} and σ were measured. In panels (a) and (b), the fits are generated by (9.25) in which the temperature dependence of the critical current, I_c, is computed according to the Bardeen formula. The fits shown by dashed black curves correspond to $b = 3/2$, while the red line fits represent the case $b = 5/4$. The two almost horizontal curves (solid and dashed) in (a) and (b), fitting well the low-temperature part, correspond to the QPS-dominated regime, meaning $T_{esc} = T_q$. They are computed assuming $T_q = 0.8\,\text{K}$ for sample A, and $T_q = 0.6\,\text{K}$ for sample C. The two other curves (solid and dashed), which fit the high-temperature part of the data well, represent TAPS according to (9.25), in which $T_{esc} = T$. In (c) unpulsed and (d) pulsed, the mean switching current is plotted, having the fits generated by (9.26), according to the same rules as in (a) and (b). The crossover temperature T^* is shown by the vertical dotted lines. The T^* is defined as the temperature below which $\sigma = \text{const}$. The green dotted curves are $I_c(T)$ theoretical curves generated by the Bardeen formula (Section 11). The most important observation here is that σ does saturate below T^*, while $\langle I_{sw} \rangle$ keeps growing with cooling below T^*. This is exactly what the KG model predicts, judging by the fits. Yet, this would not be the case if the saturation of sigma would be due to noise-generated heating of the electrons in the nanowire.

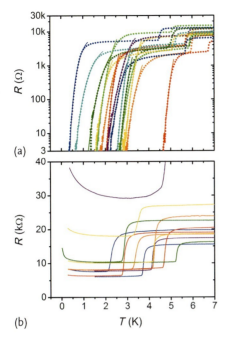

Figure 10.2 Resistance versus temperature curves for a large ensemble of nanowires [130]. All wires are made using the same fabrication process, namely, molecular templating. There are two geometric parameters in which the wires differ. First, the length is different and varies between 29 and 490 nm. Second, the nominal thickness of the wires is also different and varies from 5 to 10 nm. The width of the templating molecule changes between the samples also. The wires have been sorted into two qualitatively distinct classes. Those whose resistance drops with cooling are placed on the top panel (a), and those having the resistance increasing with cooling are collected at the bottom panel (b). The top-panel samples are S-type. All of the curves are in excellent agreement with the Little's fit $R(T) = R_N \exp(-\Delta F/k_B T)$ (solid curves). Each of the bottom-panel wires shows all the characteristics with the I-type ($dR/dT < 0$, $d^2R/dT^2 > 0$). Normal wires are not observed among short (less than $\approx$ 500 nm) and homogeneous wires, at least in the experiments in which the wires are made of the superconducting amorphous alloy of $Mo_{79}Ge_{21}$.

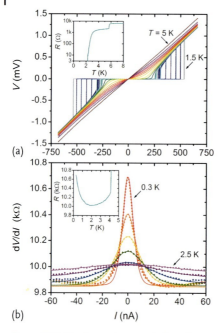

Figure 10.3 The figure shows typical nonlinear transport measurements for a representative (a) superconducting and a representative (b) insulating sample. (a) Voltage–current characteristics of a wire classified as S-type. The wire length is $L = 61$ nm and $T_c = 3.3$ K. The curves are taken at $T = 1.5, 1.6, 1.7, 1.8, 1.9, 2.0, 2.1, 2.2, 2.3, 2.4, 2.6, 2.8, 3.0, 3.25, 3.5, 4.0$, and 5.0 K. The key fact illustrated by these curves is that S-samples always become more resistive (i.e., less superconducting) as the bias current is increased. Mathematically speaking, as the current is increased, the $V(I)$ curves exhibit nonlinearity and the curvature, d^2V/dI^2, is zero or positive at sufficiently low bias. The insert shows the zero-bias resistance measured at low bias, about 10 nA or less. (b) The differential resistance is plotted versus bias current for an I-type sample. The wire length is $L = 140$ nm. The curves are taken at $T = 0.3, 0.5, 0.75, 1.0, 1.5, 2.0$, and 2.5 K. The fits, shown by solid curves, represent the weak Coulomb blockade theory, due to Golubev and Zaikin [190]. The key fact is that the differential resistance, dV/dI, decreases with increasing the bias current I, which means that the second derivative d^2V/dI^2 is negative. The insert shows the linear resistance, which always increases with cooling for I-samples after it reaches its minimum around 2 K.

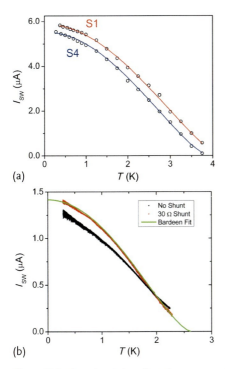

Figure 11.1 Experimental confirmation of Bardeen formula for the depairing current in thin superconducting filaments (nanowires) [197]. (a) The mean value of the switching current is plotted versus temperature for two samples (S1 and S4), which were shunted using 5 and 10 Ω resistors respectively. The shunting is found to suppress the fluctuations of the switching current and push the switching current up to the depairing current. The solid lines are fits to the Bardeen formula for the depairing current, (11.1). The fitting parameters used to generate the fits are $I_c(0) = 5.88$ and 5.53 μA and $T_c = 4.2$ and 3.9 K for S1 and S4, correspondingly. (b) Mean switching current I_{sw} is plotted versus T for sample S5, for the unshunted bare wire (black) and for the same wire being shunted using a 30 Ω normal resistor (the red data points). The Bardeen fit (yellow) is in excellent agreement with the shunted wire date, but not with the unshunted case data. The fitting parameters were $I_c(0) = 1.415$ μA and $T_c = 2.62$ K.

1
Introduction

In 1913, the Nobel Prize in Physics was awarded to Heike Kamerlingh Onnes "for his investigations on the properties of matter at low temperatures which led, *inter alia*, to the production of liquid helium." His crowning achievement was the liquefaction of helium in 1908, which pioneered a new era in low-temperature physics and enabled him to discover superconductivity [3] in 1911.

The theory of superconductivity was developed by Bardeen, Cooper, and Schrieffer (BCS) in 1957 [4]. "In the competitive world of theoretical physics, the BCS theory was the triumphant solution of a long-standing riddle. Between 1911 and 1957, all the best theorists in the world ... had tried and failed to explain superconductivity." [5]. The path to the development of the theory of superconductivity was cleared by the pioneering work of Bardeen and Pines [6], who examined the superconducting isotope effect, took into account the electron–phonon interactions, and determined that electrons could overcome the Coulomb repulsion and attract each other. This weak attraction between the electrons inside the superconducting material is the key to explaining the condensation of electrons and their superconductivity.

In traditional superconductors, the electron–electron attraction, which translates into a small but noticeable reduction of the total energy, occurs between two electrons having opposite wavevectors and opposite spins. At low temperatures, because of the electron–electron attraction, the electrons form a condensate, which is a collective bound state having zero entropy and a reduced total energy, compared with the zero-temperature Fermi distribution energy. This reduction of energy is achieved by allowing, even at zero temperature, for some fraction of the electrons in the superconducting material to have momenta larger than the Fermi momentum, thus maximizing their interaction potential.

Starting from the BCS theory, it was rigorously derived [7] that the condensate can be described by a collective wavefunction, also called superconducting order parameter. The order parameter provides the most complete description of the ensemble of superconducting electrons and depends on three spatial coordinates only, as

$$\psi(\mathbf{r}, t) = \left(\frac{n_{\text{cond}}(\mathbf{r}, t)}{2} \right)^{1/2} \exp[i \phi(\mathbf{r}, t)] \qquad (1.1)$$

Superconductivity in Nanowires, First Edition. Alexey Bezryadin.
© 2013 WILEY-VCH Verlag GmbH & Co. KGaA. Published 2013 by WILEY-VCH Verlag GmbH & Co. KGaA.

The normalization of the wavefunction is chosen such that the square of the absolute value of the wavefunction equals half the local density of the condensed electrons $n_{\text{cond}}(r, t)$.[1] Like the single-particle wavefunction, the collective condensate wavefunction is a complex function of the radius-vector r and time t. The fact that the effective wavefunction of many electrons only depends on one radius-vector r reflects how all electrons behave "coherently," that is, as a single particle.

If the gradient of the phase of the order parameter is not equal to zero (i.e., if $\nabla\phi(r, t) \neq 0$), then the condensate "flows," that is, it carries a nonzero electrical current, called "supercurrent." (Remember that the gradient operator ∇ is a vector having three components given by the spatial partial derivative operators $\partial/\partial x$, $\partial/\partial y$, $\partial/\partial z$, where x, y, and z are the Cartesian coordinates.) In some sense, the BCS condensate acts as a huge "macromolecule" of electrons. As with actual molecules, one needs to perform a positive work on the condensate to free an electron from such a huge electronic macromolecule. This work is called the superconducting energy gap Δ. The energy of the condensate is reduced due to the attractive interactions between the electrons. The most important property of the BCS condensate is that it can flow through the lattice of positively charged ions without friction. This happens because slowing the entire bound state of all electrons is much more difficult than slowing down single unbound electrons, which exist in normal (i.e., nonsuperconducting) metals.

It is frequently stated that electrons in a superconductor form "Cooper pairs" (CP) or "BCS pairs," which, by virtue of being bosons, are able to condense at low temperatures, thus forming a superfluid bosonic state. Qualitative statements of this sort are difficult to prove or disprove in physics because they do not carry any precise meaning, unless accompanied by corresponding formulas or graphs. The view that a superconducting state is a condensate of bosons (CPs) should be considered incorrect though because there are no bound electronic pairs in a conventional superconductor. The BCS state is a collective condensed state of a macroscopic number of fermions (electrons), not bosons. All condensed electrons participate equally in the condensate. The term "condensate" represents all electrons participating in a collective ground state, in which all the electrons behave quantum-coherently. Even if a localization phenomenon occurs in disordered superconducting films and wires, it is still more appropriate to speak about localized condensate "droplets" or condensate "lakes" than about localized Cooper pairs because two electrons do not form a superconducting, BCS-condensed state.

The Cooper pair density is usually defined as $n_{\text{CP}} \sim N_0 \Delta$, where N_0 is the density of states at the Fermi level and Δ is the energy gap. Here, we use a different convention and define the number of electronic pairs (superpairs) in the condensate as n_s. The two quantities, n_s and n_{CP}, differ strongly. To see this, consider that in clean superconductors (i.e., not having impurities or defects) at zero temperature, $n_s = n/2$, where n is the total density of electrons. Therefore, $n_{\text{CP}}/n_s \sim \Delta/E_F \sim 0.0001$.

---

1) In that respect, the superconducting order parameter is different from the wavefunction of a single electron because the squared absolute value of the single-electron wavefunction equals the probability density of finding the electron at the specified location.

The physical reality of the density of superconducting electrons, which in our notations equals $2n_s$, is asserted by the measurements of the depth of the magnetic field penetration. According to the Meissner effect, superconductors expel magnetic field. Yet the expulsion is not perfect. The field penetrates to a certain depth, called the clean-limit penetration depth, or the London penetration depth, which is expressed as $\lambda_L^2 = mc^2/8\pi n_s e^2$. At zero temperature, the depth is defined through the total density of electrons in the condensate, which, at $T = 0$, equals the total electronic density n. Thus, at $T = 0$, one gets $\lambda_L^2 = mc^2/4\pi n e^2$.

The reason it is more convenient to use the number of electronic pairs in the condensate is the superconducting parity effect. In a series of beautiful experiments, Tuominen and coworkers [8] showed that the number of electrons participating in the BCS condensed state is an even number. If the total number of electrons in a superconducting island is odd, one of the electrons gets expelled from the condensate, causing a significant energy (Δ) increase of the whole system. This "uncondensed" or "unpaired" electron is located, energetically, above the energy gap of the superconductor. These experiments, while revealing the parity effect, were done in a setting resembling a single-electron tunneling transistor [9]. Its advantage is that in such a device, the number of electrons on a metallic Coulomb "island," which is just some small, micrometer-scale, metallic disc, can be controlled precisely, using the gate electrode of the transistor [9].

The frictionless flow of the BCS condensate requires an explanation, or at least a discussion. A frictionless flow is equivalent to a current flow, with zero voltage applied, which continues indefinitely if the system is not perturbed. Such a stable persistent current is called a supercurrent. The existence of a frictionless supercurrent may be justified as follows. First, note that the velocity of the condensate, also called the "superfluid velocity" v_s, is proportional to the phase gradient of its wavefunction, [1], namely,

$$v_s = \left(\frac{\hbar}{2m}\right)\nabla\phi(r,t) \qquad (1.2)$$

where $\hbar = 1.054 \times 10^{-34}$ J s is the reduced Planck's constant, and $m = 9.109 \times 10^{-31}$ kg is the electronic mass. This formula is correct only when the magnetic field is zero everywhere, so the vector potential is put to zero.[2] For now, we assume that *the vector potential is zero* everywhere. Then, the electrical current density carried by the condensate, called the "supercurrent density," can be expressed as

$$j_s = 2en_s v_s = \left(\frac{\hbar e}{m}\right) n_s \nabla\phi(r,t) \qquad (1.3)$$

---

[2] To simplify the discussion, we assume that the magnetic field is negligible everywhere, so the corresponding magnetic vector potential can be chosen as zero, $A = 0$. If the magnetic field is present, the superfluid velocity is proportional to a linear combination of the phase gradient and the vector potential.

where $e = -|e|$ is the electronic charge, and $n_s = |\psi|^2 = n_{\text{cond}}(r, t)/2$ is the mean density of the electrons pairs participating in the BCS condensate.[3] Thus, it is clear that for the supercurrent to remain steady in time (i.e., to have $\partial j_s/\partial t = 0$), it is sufficient to have a constant phase gradient of the corresponding wavefunction (i.e., $\partial \nabla \phi(r, t)/\partial t = 0$) and a constant density of the condensate (i.e., $\partial n_s/\partial t = 0$).

Let us now argue that these two quantities remain fixed in time if no voltage is applied. Assume that at $t = 0$, both n_s and the phase gradient are greater than zero and constant in space, that is, there is a uniform supercurrent flow. Then, the wavefunction can be written as $\psi = \sqrt{n_s} \exp(i k r)$. The phase then is $\phi(r) = k \cdot r$, where the vector k is called the wavevector of the wavefunction. The corresponding superfluid velocity is $v_s = \hbar k/2m$. To show that the resistance of the superconductor is zero, we will argue that the superfluid velocity, the phase gradient, and n_s do not change with time if the electric field is zero.[4]

First, let us consider the superfluid density n_s. Its value is set by the requirement that the corresponding thermodynamic potential is minimized. For example, if the volume and the temperature are fixed and the electric field in the superconductor is zero, then the corresponding thermodynamic potential is Helmholtz free energy, $F = U - TS$, where U is the internal energy, T is the temperature and S is the entropy. Since U and S must be functions of n_s, for F to be constant and remain at its minimum, the density of the condensate n_s must remain constant in time. Small fluctuations near the mean value might be present, but they average to zero and do not cause any change of the mean superfluid density which defines the mean supercurrent. The key fact is that in a superconductor at a temperature below its critical temperature the superfluid density is larger than zero provided that the thermodynamic equilibrium is established.

The phase gradient of the condensate wavefunction also does not change with time if the electric and chemical potentials are constant within the sample. Gor'kov has shown theoretically [10], using his microscopic theory [7, 10], that the phase of the superconductor wavefunction changes in time as

$$\phi(r, t) = \frac{2e\mu(r)t}{\hbar} + \phi(r, 0) \tag{1.4}$$

where $\phi(r, 0)$ is the phase at time zero and $\mu(r)$ is the local value of the electrochemical potential, which is defined by the equation $E_{N+2} - E_N = 2e\mu(r)$. Here, E_N is the energy of the condensate containing N electrons, and E_{N+2} is the energy of the same condensate, after introducing an additional superpair at position r.

---

3) Such normalization is traditionally used to stress the superconducting parity effect, that is, the fact that the number of electrons in a BCS-condensed state is an even number. It is curious to note that the BCS quantum state is such that the number of the pairs is not exactly defined, but it is subject to quantum fluctuations. For a large, macroscopic sample, the *uncertainty* of the number of pairs is by many orders of magnitude smaller than the number itself.

4) Compare this with the time-evolution of wavefunctions of a single electron. Such wavefunctions can change in time because of scattering over impurities or phonons, or other perturbations. Thus, the fact that the condensate is able to maintain a constant nonzero velocity or momentum is not trivial and requires some discussion. For example, it would be interesting to understand why a flowing condensate cannot dissipate its momentum to phonons.

The absolute value of the phase does not have any physical significance since it cannot be measured. On the other hand, the phase difference can be measured. Let us define the phase difference between two points, r_2 and r_1, as $\phi = \phi_2 - \phi_1 = \phi \equiv \phi(r_2, t) - \phi(r_1, t)$. The time-evolution equation (1.4) can be transformed for the phase difference as

$$\phi = \frac{2e\Delta\mu t}{\hbar} + \phi(0)$$

where $\Delta\mu = (\mu(r_2) - \mu(r_1))$ is the difference of electrochemical potentials and $\phi(0) = \phi(r_2, 0) - \phi(r_1, 0)$ is the phase difference at time zero.

The electrochemical potential is the sum of the chemical potential and the local electric potential. Assume that the chemical potential is constant everywhere in the superconducting sample. Then, the difference of electrochemical potentials becomes the difference of electric potentials, which is the voltage V between two points. Therefore, $\Delta\mu = V$, and the time-evolution equation becomes

$$\phi = \frac{2eVt}{\hbar} + \phi(0)$$

Finally, one can differentiate it with respect to time and obtain

$$\hbar\frac{d\phi}{dt} = 2eV \tag{1.5}$$

where V is the voltage between two points specified by the arbitrary chosen radius vectors r_2 and r_1. It was Anderson and Dayem [11] who first introduced this popular presentation of the phase evolution equation, in which the phase difference, ϕ, rather than the local value of the phase itself, $\phi(r, t)$, is used. Since, fundamentally, the time-evolution of the phase of the macroscopic superconducting wavefunction was first derived by Gor'kov (1958) (P.W. Anderson, private communication, 2007), we elect to call (1.5) as the Gor'kov phase-evolution equation. It was also named by various authors as the Gor'kov–Josephson equation [12, 14], or the AC Josephson equation [1, 15], or simply the phase-evolution equation. Fundamentally, it is analogous to the time-dependent Schrödinger equation (see more on this analogy below).

Incidentally, note that (1.5) is the only equation in the field of superconductivity which is exact; all others are only approximate. This is why the Gor'kov equation is used in metrology, in which case the phase rotation is synchronized with the external electromagnetic field of a known frequency f, so $d\phi/dt = 2\pi f$. The factor 2π occurs because as the phase completes one cycle, it changes exactly by 2π. Then, according to (1.5), $2\pi f\hbar = 2eV$. Thus, by measuring voltage, the fundamental constants ratio $\hbar/e$ can be determined as $\hbar/e = V/\pi f$.

If the voltage is zero, the phase difference between any two points on the wire, r_1 and r_2, does not depend on time. Furthermore, if the electric field is zero, $E = 0$, then the phase gradient is also time-independent. Remember that the phase difference and the phase gradient are proportional to each other as $\phi = (r_2 - r_1)\nabla\phi$, assuming that the two points are close to each other. In this notation, the voltage is also zero, $V = (r_2 - r_1)E = 0$. So, if $d\phi/dt = 0$ and $r_2 \neq r_1$, then $d(\nabla\phi)/dt = 0$. (Note that the phase gradient $\nabla\phi$ is a vector.)

Thus, we have argued that the supercurrent is time-independent under zero electric field because the supercurrent is a product of the phase gradient $\nabla\phi$ and the density of the condensate n_s, both of which are time-independent, as was discussed above.

To develop a physical intuition and qualitatively understand the physical origin of the time-evolution equations of the phase of the wavefunction, we note that the phase evolution given by (1.4) and (1.5) is analogous to the evolution in time of the phase of a single quantum particle in the ground state. Below, we develop this analogy. Consider a particle in a ground state with energy E_0. Its complete wavefunction satisfies the time-dependent Schrödinger equation $i\hbar\partial_t \Psi = \hat{H}\Psi$, where, for convenience, we use the notation for the partial time derivative as $\partial_t \Psi \equiv \partial\Psi/\partial t$. However, in the ground state, we can also write the time-independent Schrödinger equation as $\hat{H}\Psi = E_0\Psi$. Combining these two equations, $i\hbar\partial_t \Psi = E_0\Psi$. The solution, that is, the wavefunction of the considered quantum particle in the ground state, is well known, namely, $\Psi(r, t) = \psi(r)\exp(-iE_0 t/\hbar)$, where $\psi(r)$ is the time-independent complex function that defines the spatial distribution of the particle probability amplitude, t is the time, r is the radius-vector of the particle, and the imaginary unit satisfies the equality $(-i)i \equiv 1$. Let us find the phase of this wavefunction. First, remember that any complex number X_c can be presented in the form $X_c = X_a \exp(i\phi_x)$. The real number ϕ_x is called the phase of X_c. The absolute value or the magnitude of X_c is $X_a = \sqrt{X_c^* X_c}$. Accordingly, for the wavefunction in the ground state, the first, time-independent factor can be presented as $\psi(r) = |\psi(r)|\exp(i\phi_0)$. Here, ϕ_0 is the phase at time zero. Thus, the entire wavefunction is $\Psi(r, t) = |\psi(r)|\exp[i(\phi_0 - E_0 t/\hbar)]$. So the phase of the single-particle wavefunction is $\phi = \phi_0 - E_0 t/\hbar$. This expression is analogous to the equation describing the phase of the superconducting condensate, that is, (1.4).

To develop the analogy further, assume that the quantum particle under investigation is a single electron exposed to a spatially constant electric potential μ. Then, the Hamiltonian is $\hat{H} = (-\hbar^2/2m)\nabla^2 + e\mu$. Thus, the ground state energy is $E_0 = e\mu$ and, therefore, the phase of the wavefunction is $\phi = \phi_0 - e\mu t/\hbar$ which is already very similar to (1.4).

To understand the origin of the factor 2 in front of μ in (1.4), remember the parity effect. The BCS condensate always contains an even number of electrons. Each pair has the charge $2e$ and the mass $2m$. Thus, the Hamiltonian for a single pair is $\hat{H} = (-\hbar^2/4m)\nabla^2 + 2e\mu$, the energy of the ground state is $E_0 = 2e\mu$, and, therefore, the phase of the wavefunction depends on time as $\phi = \phi_0 - 2e\mu t/\hbar$. The result is in agreement with (1.4), which follows from the BCS and the Gor'kov theory. Since in a superconductor all pairs behave coherently, one expects that the phase evolution of one pair is the same as the phase evolution of the phase-coherent ensemble of pairs.

An important property of a BCS condensate, either stationary or moving with respect to the crystal lattice, is that its spectrum of excitations is usually "gapped," that is, a finite amount of energy, Δ, is required to create an excited state. Such excited states are called quasiparticles, or Bogoliubov quasiparticles, or bogoliubons (see [1], p. 61). According to the BCS theory, the gap is $\Delta = 1.76 k_B T_c$, where $k_B = 1.38 \times 10^{-23}$ J/K is the Boltzmann constant and T_c is the critical temperature of the superconductor. The T_c is the temperature below which superconductivity develops. For completeness, we should mention that gapless superconductivity is in general also possible [16], so the presence of a gap in the spectrum of excitations is not a necessary condition for zero resistance (for more details, see [1], p. 390).

It is interesting to compare superconductors to semiconductors, in which the spectrum of excitations is also gapped. The difference is that in semiconductors, the gapped state, that is, the state in which the valence band is completely filled and the conduction band is completely empty, is characterized by zero total current. To create a nonzero electrical current in a semiconductor, some number of electrons must be excited from the valence band to the conduction band. Such excited states are not gapped since the electron(s) present in the conduction band can change energy by an infinitesimal amount, for example, under the action of external electric field or impurities. With time, the excited electrons give up their energy to phonons and relax back to the lower-energy valence band. As soon as all excited electrons relax, the electrical current decays to zero. In a superconductor, however, the supercurrent is associated not with excitations, but with the condensate itself. Even in the ground state, the supercurrent can be large. For example, if a superconducting wire loop is exposed to a perpendicular magnetic field, the velocity of the condensate is proportional to the magnetic vector-potential, which, in turn, is proportional to the magnetic flux piercing the loop. Such a magnetically induced supercurrent is called Meissner current. It is possible because all electrons in superconductors behave coherently, as a single quantum particle (single electron). For a single electron, the velocity is proportional to vector-potential, assuming that the phase gradient is zero. The Meissner current does not decay since it is associated with the ground state, that is, the BCS condensate. The ground state cannot relax because there are no states having lower energies. The electrons in a normal metal also participate in persistent currents if a magnetic field is applied. However, these currents are all different, and their signs are different since the electrons are not coherent in a normal metal. Thus, they all add to an extremely small value, of the order of a current of one electron. In a superconductor, a macroscopic number of electrons participate in a collective persistent current. In such cases, the currents of all condensed electrons add up. That is why Meissner currents can be much stronger than persistent currents in normal metals.

The ability of a superconductor to carry a dissipationless current, that is, a current under zero applied voltage, disappears if the superconductor is shaped into a *thin* cylinder or a thin wire, or, in other words, if the superconductor is quasi-one-dimensional (see Figure 1.1). This is because if the diameter of the superconductor is small, the rate of strong thermal fluctuations, which bring short segments of the wire into the normal state, is essentially greater than zero at finite temperatures.

Such fluctuations, first predicted by William Little in 1967 [17] and called Little's phase slips (LPS), occur stochastically at random spots on a superconducting wire and interrupt the dissipationless flow of the condensate. Each such local fluc-

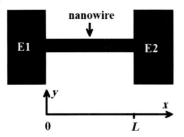

Figure 1.1 Illustration of a typical experimental realization of a transport experiment on a nanowire. The wire is connected at its ends to two macroscopic superconducting electrodes E1 and E2. The wire itself is a thin cylinder having a diameter much smaller than the magnetic field penetration depth, and also smaller than the superconductor's coherence length. It will always be assumed that the x-axis is directed along the wire and the wire starts at $x = 0$ and ends at $x = L$. Since the wire is assumed to be thinner than the coherence length, ξ, it follows that the order parameter is approximately constant within any cross-section of the wire, taken perpendicular to the wire axis. Thus, the assumption that the wire has an exact cylindrical geometry is not essential, that is, the cross-section can be of any shape, without having any qualitative effect on the wire behavior. To qualify as quasi-one-dimensional, the dimensions of the cross-section of the wire must be smaller than $\pi\sqrt{2}\xi$ since, in this case, vortices are not energetically favorable on the wire [18]. The term "nanowire" is usually applied to wires which are much thinner than 1 μm in diameter.

tuation allows the phase difference between the ends of the wire to "slip" by 2π (in other words, to decrease by 2π), causing the supercurrent to diminish. To undergo a phase slip, the free energy of the condensate must increase somewhat to overcome a certain energy barrier (usually denoted ΔF). This barrier equals the condensation energy density multiplied by the volume of the normal region associated with the LPS. As with any barrier crossing process, the LPS are driven by thermal fluctuations at sufficient temperatures. Such phase slips are referred to as thermally activated phase slips (TAPS).

As the temperature is lowered, the rate of TAPS exhibits a rapid decline described by the Arrhenius activation law. The resistance of the wire is linearly proportional to the rate of TAPS. Thus, as the temperature is reduced, the wire resistance drops exponentially, or, to be more precise, according to the Arrhenius law [21], namely, as $R \sim \exp(-\Delta F(T)/k_B T)$, where $\Delta F(T)$ is some effective barrier, which will be discussed in detail below. Such activation dependence of resistance on temperature was confirmed in experiments by Lukens, Warburton, and Webb [22] and Newbower, Beasley, and Tinkham [23]. Although the resistance of a superconducting wire is exponentially low at low temperatures, nevertheless, strictly speaking, it does not become zero at any finite temperature because TAPS has a nonzero probability at any finite temperature. In other words, there is no thermodynamic phase transition in a thin wire. As the temperature is reduced below the thermodynamic critical temperature T_c, the resistance decreases continuously, never reaching zero.

There is no qualitative difference in the state of the wire above T_c and below T_c. Above T_c, superconducting fluctuations occur. Thus, the wire is not completely normal. Below T_c, there are normal-state fluctuations (i.e., the LPS), so the wire is not fully superconducting. Thus, the wire undergoes a crossover from a predominantly normal state above T_c to a predominantly superconducting state below T_c, but not a phase transition. In fact, the crossover does not happen at $T = T_c$, but at a temperature T_1 such that $\Delta F(T_1) \sim k_B T$. It should be emphasized that nothing experimentally noticeable happens with the wire either at $T = T_c$. Thus, when fitting data, the T_c should be treated as a fitting parameter. The parameter T_c controls the behavior of the resistance through the effective barrier ΔF since $\Delta F = 0$ at T_c and increases with cooling. The T_c is not a parameter that is directly measurable in thin wires. This is in contrast with bulk superconductors, in which the T_c is simply the temperature at which the resistance drops to zero.

Generally, one expects that at low temperatures, the thermal activation rate decreases exponentially with cooling while the quantum tunneling rate should remain roughly constant, thus becoming dominant below a certain crossover temperature, typically denoted T^*. Therefore, TAPS, occurring in superconducting wires below T_c, should be succeeded by tunneling of phase slips at sufficiently low temperatures, namely, at $T < T^*$. Such tunneling phase slips are usually called quantum phase slips (QPS) since, qualitatively speaking, they derive their existence from quantum fluctuations and the Heisenberg uncertainty of the energy. If the system undergoing quantum tunneling possesses many internal degrees of freedom which get involved into the tunneling event, then the tunneling is called "macroscopic." For example, tunneling of a condensate involving many electrons or tunneling of a large molecule composed of many atoms would be considered as a macroscopic quantum tunneling (MQT). Thus, tunneling of Little's phase slips is an example of MQT since a large number of electrons occur in the virtual normal core of QPS. According to this classification, QPS is a particular case of MQT.

Macroscopic quantum tunneling is one of the advanced research topics of modern physics, as it belongs to the transitional region between classical and quantum mechanics. Note that at the fundamental level, the relationship between classical and the quantum theories is still not fully understood because of the quantum mechanics' reliance on classical mechanics for its justification. This fact is exemplified by the problem of quantum measurement, which requires the wavefunction to collapse when a quantum system is measured with a classical measuring apparatus. If the system is strictly isolated, such a collapse is difficult to justify.

In older textbooks, such collapse was explained by making an explicit assumption that the measurement apparatus is classical, not quantum, by definition. The statement that the apparatus is classical infers that it cannot exist in a quantum superposition of macroscopically distinct states. For example, a voltmeter cannot, in principle, exist in a superposition of states having different readings, for example, $V = 0$ and $V = 1$ V simultaneously. It must "choose" one particular reading.

Although such an assumption seems reasonable, it remains desirable to formulate quantum mechanics in a self-sufficient manner. Within quantum theory, the system can be in any quantum superposition of allowed states. For example, electrons can be described by extended wavefunctions, meaning that they are not located in any particular point of space, but rather they can occupy many remote points of space simultaneously. And, although counterintuitive, a voltmeter isolated from any interaction with the external world should be able to accept a state of quantum superposition of states having different readings. Such would be a Schrödinger cat state.

Therefore, a search for fundamental physical phenomena causing wavefunctions of large isolated objects to collapse continues [24]. Of course, the puzzle of collapse is a puzzle only insofar as the measurement apparatus is allowed to only interact with the quantum system, but not with the environment. On the other hand, if the measurement apparatus interacts with its environment, say with the rack supporting it or with a physicist looking at it, then quantum theory alone predicts that the wavefunction of the apparatus collapses because of decoherence [25]. Yet, when an isolated system is considered, such as a hypothetical Schrödinger cat [26] or, as a different example, the whole Universe, which is presumed to include everything with which anything can interact, then the expected collapse of the wavefunction remains unjustified theoretically. These fundamental difficulties led to such impressive ideas as the many-worlds interpretation of quantum mechanics by Hugh Everett, which is currently a mainstream interpretation [27].

Initiated by Leggett, the field of macroscopic quantum physics has seen widespread development [28–39]. Definitive experimental evidence that a macroscopic system can behave according to the laws of quantum mechanics has been obtained by Clarke and collaborators [33]. Evidence of MQT was also found in experiments using magnetic nanoparticles, in which case the entire particle reverses its magnetization within a single quantum tunneling event [35]. These experiments quite convincingly demonstrate that rather large systems (large when compared with single atoms) can exist in quantum superpositions of macroscopically distinct states. Recent fundamental recognition [40, 41] of the potential advantages of computational methods based on quantum bits (qubits) has initiated the search for practical implementations of systems which can maintain for a sufficiently long time a quantum superposition of macroscopically distinct states. Such systems can be built and can indeed operate as qubits [42–44]. It was also proposed that superconducting nanowires could be used as active elements in flux qubits, provided that quantum tunneling of the phase difference can occur coherently in nanowires [45, 50, 51]. According to Khlebnikov [45], "the process [of such tunneling in thin superconducting wires] may be suitable for forming quantum superpositions of flux states." Mooij and Nazarov also proposed that QPS could be used to build current standards, and thus could advance the field of exact metrology [51]. Consequently, understanding QPS is an important topic of modern quantum physics.

The search for QPS in superconductor nanowires was first undertaken in experiments by the Mooij group [52], although QPS was not observed. Later, Giordano's experiments [53] gave evidence that QPS might exist. The difficulty of observing QPS is related, in general, to the fact that the tunneling rate is exponentially suppressed, not only by the width and the height of the tunnel barrier and the large effective mass of macroscopic systems, but also by their strong interaction with the environment [30].

Qualitatively speaking, the suppression of the quantum tunneling by the environment can be classified as the quantum Zeno effect. This effect refers to a situation in which an unstable system, if somehow "observed" continuously, can never decay by tunneling. Thus, it is possible to strongly slow down the evolution of the system by continually measuring its state. The quantum Zeno effect is quite general. It refers to a situation in which the Schrödinger-type time-evolution of a quantum system is strongly slowed not only by measurements, but also by quantum decoherence caused by various interactions with the environment. The name originates from Zeno's arrow paradox, which states that an arrow in flight is not observed to move at any single instance, and therefore cannot possibly be moving at all. Of course, arrows can move in space very well. Thus, some sort of paradox is present since the qualitative reasoning leads to a different conclusion. The paradox was resolved by Newton and Leibniz with the invention of calculus, which is a mathematical apparatus allowing exact logical analysis of infinitesimal displacements.

A quantitative description of the environmental effects on a macroscopic quantum system was introduced by Caldeira and Leggett [28, 30, 32]. According to their theoretical approach, the interaction with the environment can be modeled as an interaction with a gapless ensemble of harmonic oscillators. The strength of such an interaction can be characterized by the classical coefficient of viscosity η. The prediction of the theory is that if a system interacts with an environment (or, as is sometimes said, is subjected to "quantum dissipation," or it couples to a "bath of harmonic oscillators"), then its tunneling rate is suppressed by a factor $\exp[-A_{CL}\eta(\Delta q)^2/\hbar]$ relative to the case in which the tunneling system is perfectly isolated from any environment, but tunnels through the same energy barrier. Here, A_{CL} is a numerical factor of order unity, η is the viscosity coefficient defined in the classically accessible region, and Δq is the size of the classically inaccessible region, that is, the tunnel barrier width. The theory is valid only if the distribution of the oscillators representing the environment is gapless; that is, the distribution of the oscillator frequencies reaches zero frequency. So the reservoir of oscillators must be infinite in size. This is the reason why quantum systems coupled to such dissipative reservoirs are able to undergo quantum phase transitions, such as the dissipative Schmid–Bulgadaev transition.

As will be discussed in detail later, for superconducting devices, the effective viscosity that sets the environmental suppression of the rate of QPS depends on the normal conductance of the system. The normal conductance is well defined only if the device is shunted with a macroscopic normal resistor, which, ideally, should not depend on temperature. A normal resistor contains gapless normal electrons which act as an ensemble of harmonic oscillators damping the QPS. In superconducting wires, the damping effect might occur because QPS, like Abrikosov vortices, have normal cores in which the superconducting gap goes to zero.

One of the biggest remaining puzzles is the origin of superconductor-insulator transitions (SIT) in which a nanowire loses its ability to carry any measurable constant supercurrent. A qualitative difference between the superconducting state and a nonsuperconducting state exists only at $T = 0$. At higher temperatures, a nanowire is always resistive because of TAPS. Proving that an SIT does occur as some parameter of the wire is changed is difficult because of the obvious fact that zero temperature is inaccessible experimentally. Thus, conclusions about the occurrence of a quantum transition are usually achieved indirectly. For example, resistance versus temperature, $R(T)$, curves could be extrapolated to zero temperature. To argue that an SIT does exist, it is necessary to show that the sample exhibits at least two qualitatively distinct types of behavior – a superconducting regime and an insulating regime. The transition between the two distinct regimes is usually induced by some control parameter, for example, the wire normal-state resistance R_n or its diameter d.

The SIT in thin wires has been analyzed theoretically by many groups. Andrei Zaikin and collaborators were the first to suggest a model of the SIT in 1D by making a quantum analogue of the well-known Kosterlitz–Thouless transition [168].

If the ensemble of samples studied is such that all samples are qualitatively similar and differ only quantitatively, then the system is said to undergo a crossover, but not a quantum phase transition. For example, suppose a series of experiments on a group of nanowires shows that for all samples as $T \to 0$, then $R(T) \to R_0$, and R_0 is some sample-specific constant, such that $0 < R_0 < \infty$. All samples would then saturate at a constant resistance with cooling. Such results would indicate that there is no SIT in the studied type of samples. On the other hand, a crossover might still be present, if, for example, the experiments show that R_0 gradually changes from $R_0 \ll R_n$ to $R_0 \gg R_n$, as the wire diameter is gradually reduced. The Giordano models of QPS predicts such crossover behaviors [108]. It predicts that any wire has a QPS rate above zero and therefore its resistance is greater than zero at zero temperature, although, within this model, R_0 depends exponentially on the wire diameter. Other quantum models predict that a superconductor-insulator phase transition should occur in thin superconducting wires [168, 194]. Experimental evidence in favor of SIT was published by Bollinger et al. [130].

In many theories of SIT, a quantum tunneling of Little's phase slips is the key factor in determining whether the wire is superconducting or insulating. The basic idea is as follows. If the QPS is suppressed completely (at zero temperature), then the wire stays phase coherent indefinitely and the supercurrent does not decay; thus, the wire is classified as superconducting. On the other hand, if the dissipation and other factors are not sufficiently strong to suppress QPS, the QPS occur and cause the supercurrent to decay, thus making the nanowire resistive (either normal or insulating). Although, in many cases, the experimentally observed transition in thin wires is called SIT, a better name might be SRT, that is, superconductor-resistor transition. The reason that a short wire can be superconducting is related to the net rate of QPS being zero at $T = 0$. The wire can also act as a resistor if the QPS rate is greater than zero. But, it is difficult to prove and/or expect that for a short wire the resistance is infinite. So, the insulating state is usually defined merely by the fact that the resistance increases with cooling. Such behavior, although it resembles insulators in some sense, might better be called a resistive state, not an insulating state. Therefore, in each concrete case of an SIT observation, it is important to explain the meaning of the I-state and the S-state. On the other hand, as was stated above, the meaning of the S-state is always the same within this book – it is a state of zero resistance at $T = 0$.

A qualitatively different approach to SIT is the idea that certain factors, such as enhanced electron–electron repulsion, or unpaired spins, or dangling bonds on the surface of the wire can become more and more influential as the diameter is reduced. As a result, these factors can suppress T_c of the wire to zero, thus leading to an SIT for long wires (in which the normal state is localized and thus insulating) or an SRT for short wires. In latter chapters, we will consider the existing evidence for quantum transitions in thin wires and some of the theoretical models.

2
Selected Theoretical Topics Relevant to Superconducting Nanowires

The microscopic theory of superconductivity was developed by Bardeen, Cooper, and Schrieffer (BCS) for the case of homogeneous superconductors [4]. The theory was then generalized by Bogoliubov [62, 63] to include cases where the density of superconducting electrons varies in space. Such variations frequently occur as a result of various causes, such as a joint boundary with a normal metal or a vortex normal core. Gor'kov has shown theoretically, using Green's functions [7], that in the limit of high temperatures, that is, $T_c - T \ll T_c$, BCS theory can be reduced to a simpler form, known as the Ginzburg–Landau (GL) theory. We will distinguish the phenomenological form of the GL theory [70] and the more advanced form of the theory, augmented by the microscopic derivations by Gor'kov and Abrikosov [7, 157], which is called Ginzburg–Landau–Abrikosov–Gor'kov (GLAG) theory. We begin our discussion of the physics of Little's phase slips in the framework of the GL theory, which is known to give qualitatively correct results even at temperatures significantly lower than T_c, especially in strongly disordered superconductors.

2.1
Free or Usable Energy of Superconducting Condensates

The treatment of the condensate in the GL theory [70] is similar in many ways to the treatment of a single electron in a magnetic field in elementary quantum mechanics. All information about the state of the superconducting condensate is contained in its effective wavefunction, also known as the superconducting order parameter. The condensate effective wavefunction (which will be called, interchangeably, the "wavefunction" or the "order parameter") is a complex function of three spatial coordinates (x, y, z) and time. The wavefunction has the form $\psi(\mathbf{r}, t) = \sqrt{n_s(\mathbf{r}, t)} \exp[i\phi(\mathbf{r}, t)]$, where the radius-vector is defined, as usual [69], as $\mathbf{r} = x\mathbf{e}_x + y\mathbf{e}_y + z\mathbf{e}_z$, and the unit vectors directed along the x-axis, y-axis, and z-axis are denoted as $\mathbf{e}_x, \mathbf{e}_y, \mathbf{e}_z$, respectively. Frequently, the order parameter is written as $\psi(\mathbf{r}, t) = |\psi(\mathbf{r}, t)| \exp[i\phi(\mathbf{r}, t)]$, or simply $\psi = |\psi| \exp(i\phi)$, where the absolute value of the order parameter is $|\psi(\mathbf{r}, t)| = \sqrt{n_s(\mathbf{r}, t)}$ or $n_s = |\psi|^2$. Here and everywhere, $n_s = n_s(\mathbf{r}, t)$ is, by convention, the local density of super-

Superconductivity in Nanowires, First Edition. Alexey Bezryadin.
© 2013 WILEY-VCH Verlag GmbH & Co. KGaA. Published 2013 by WILEY-VCH Verlag GmbH & Co. KGaA.

conducting electrons divided by factor two. The factor two is used to emphasize the superconducting parity effect.

If the temperature is near T_c, Helmholtz free energy density of a superconductor can be written as [1, 7, 70]

$$f(r) = f_n + \alpha|\psi|^2 + \frac{\beta}{2}|\psi|^4 + \frac{1}{4m}\left|\left(\frac{\hbar}{i}\nabla - \frac{2e}{c}A\right)\psi\right|^2 + \frac{h^2}{8\pi} \tag{2.1}$$

where f_n is the free energy density in the normal state, which is assumed constant. In what follows we always assume $f_n = 0$. The expression (2.1) will be taken here without derivation. Its validity is confirmed by the multitude of experiments on superconductors and, from the theoretical perspective, by the Gor'kov theory [7]. To justify this expression, one can say that the order parameter is expected to be small near the critical temperature and so the free energy can be approximated by the Taylor series [70]. The series is written in powers of the superfluid density, $|\psi|^2$, and its gradients. Another way to guess the free energy density expression is to make it such that its variation leads to an equation similar to the Schrödinger equation.

The coefficients of the GL free energy density expression are

$$\alpha(T) = -\frac{2e^2}{mc^2}H_c^2(T)\lambda_{\text{eff}}^2(T) \tag{2.2}$$

$$\beta = \frac{16\pi e^4}{m^2 c^4}H_c^2(T)\lambda_{\text{eff}}^4(T) \tag{2.3}$$

$$\lambda_{\text{eff}}^2 = \frac{mc^2}{8\pi|\psi|^2 e^2} \tag{2.4}$$

Thus, the coefficients of the free energy density are expressed through the material-dependent parameters, namely, the thermodynamic critical field $H_c(T)$ and the magnetic field penetration depth $\lambda_{\text{eff}}(T)$.

The factor $\alpha(T)$ is linear near T_c. It can be expressed as

$$\alpha(T) = \alpha_0(T - T_c) \tag{2.5}$$

where the constant $\alpha_0 = (d\alpha(T)/dT)_{T=T_c}$. This derivative can be found by differentiating (2.2) and then taking $T = T_c$. At $T = T_c$, the coefficient β is positive; it can be taken as a constant near T_c, for the purpose of simplified calculations, that is,

$$\beta = \text{const} \tag{2.6}$$

The Helmholtz free energy of the entire system is

$$F = \int f(r)d^3r \tag{2.7}$$

the integral of the free energy density f over the entire space (since the order parameter is defined only inside the sample, the terms containing the order parameter must be integrated only over the volume of the sample).[1] Note that when discussing superconducting systems, we will use the terms "free energy," "Helmholtz free energy," "Ginzburg–Landau free energy," "GL free energy," or "Helmholtz energy," interchangeably. All of these terms refer to the same physical quantity, namely, (2.7), which expresses the usable energy stored in the superconductor and in the magnetic field created by it.

To see how the free energy density expression can be used, consider the simplest case – the order parameter is constant in space and time and it is a real number. Such a situation occurs in a homogeneous superconductor that is not subjected to any external field or current. Then, the free energy density is $f(\mathbf{r}) = f_n + \alpha \psi^2 + (\beta/2)\psi^4$. The order parameter can be chosen to be real and positive in such a special case, so $\psi = |\psi|$. As will be discussed in detail below, if the temperature is fixed and no external work is done on the superconductor, then the free energy F reaches its absolute minimum in the equilibrium. Since for a constant order parameter $F = f V_s$, the minimum of the free energy density corresponds to the minimum of the free energy. Thus, we will seek to minimize the free energy density. Here V_s is the volume of the superconductor.

In this case, the free energy density is defined by just one adjustable parameter, namely, ψ. If ψ is varied by an infinitesimal amount, $\delta\psi$, the corresponding variation δf of the free energy density is $\delta f = (2\alpha\psi + 2\beta\psi^3)\delta\psi$. If f is the minimum, then any infinitesimal variation of the order parameter $\delta\psi$ does not lead to any change of f in the first-order approximation. Thus, we have to require $\delta f = 0$. Consequently, $2\psi(\alpha + \beta\psi^2)\delta\psi = 0$ for any small values of $\delta\psi$. This is possible if either $\alpha + \beta\psi^2 = 0$ or $\psi = 0$. The choice $\psi = 0$, which is the normal (i.e., nonsuperconducting) state of the system, provides the lowest free energy if $\alpha \geq 0$, which is the case at temperatures above the critical temperature (see (2.5)). Otherwise, the minimum is achieved at $\psi = \sqrt{-\alpha/\beta}$. This equilibrium value of the order parameter in an unperturbed superconductor will be denoted

---

[1] To understand that the integration of the magnetic free energy needs to be done over the entire space consider an example of a superconducting loop. Suppose there is a nonzero supercurrent in the loop. Such a current obviously creates a magnetic field. The total usable energy stored in the loop includes the kinetic energy of the condensate and the energy of the magnetic field. Thus, in computing the free energy of the system, one has to integrate, among other things, the energy of the magnetic field ($h^2/8\pi$) over the volume inside as well as outside the loop. If the loop is exposed to an external homogeneous field, the integral over the entire space would be infinite. This does not mean that one can extract an infinite amount of work from the loop since the maximum work is the difference between the free energy of a state and the minimum free energy of the system. One practical possibility of avoiding dealing with infinities is to assume that the superconducting sample is placed into a large but finite-size solenoid. This is the assumption we will always make if the magnetic field is involved. The integration of the magnetic field energy over the entire space gives a finite number since the field far from the solenoid rapidly approaches zero. Another way to remove the infinite free energy occurring if the field is uniform and occupies the entire space is to redefine the free energy as $F \to F - \int (h_0^2/8\pi) d^3r$ where h_0 is the magnetic field that would be present in each point of space if the superconducting sample would disappear. The integral is again taken over the entire space. This method will not be used in this book.

ψ_0. Thus, we can write $\psi_0 \equiv \sqrt{-\alpha/\beta}$. It is then correct to say that if the superconductor is unperturbed, then equilibrium is achieved at $\psi = \psi_0$. Note that near T_c, the temperature dependence of the equilibrium order parameter can be obtained from (2.5) and (2.6), and equals $\psi_0(T) = \sqrt{a_0/\beta}\sqrt{T_c - T}$. Therefore, the density of condensed electrons increases linearly with cooling, that is, the density of the superpairs changes as $n_s(T) = \psi^2 = (a_0/\beta)(T_c - T)$.

We can now calculate the energy difference between the equilibrium superconducting and the unstable equilibrium normal states. This difference, calculated per unit volume, is called the condensation energy density, f_{ced}. A detailed discussion of f_{ced} can be found in [71]. We define the condensation energy density $f_{\text{ced}}(T) = f_s(T) - f_n(T)$. The physical meaning of f_{ced} is the minimum amount of work (per unit volume of the condensate) that must be performed on a normal metal sample to convert it into its superconducting state at a fixed temperature and fixed phase gradients. At $T < T_c$, we have $f_{\text{ced}} < 0$ because the superconducting state is the equilibrium state of the system at low temperatures. Thus, in principle, if $T < T_c$, a positive work can be extracted from a normal metal if it is allowed to relax into its equilibrium state, which is the superconducting state. And, a positive work must be performed on a superconducting sample to make it normal.

From (2.1), it follows that the condensation energy density is

$$f_{\text{ced}} = f_s - f_n = \alpha\psi_0^2 + \left(\frac{\beta}{2}\right)\psi_0^4 = -\frac{\alpha^2}{2}\beta \sim (T - T_c)^2 \qquad (2.8)$$

This expression is valid only if the phase gradient is zero and the magnetic field is zero. Note also that the thermodynamic critical field is defined as $H_c^2/8\pi = |f_{\text{ced}}|$. Since the f_{ced} is zero at $T = T_c$, the phase transition into the superconducting state, as the temperature is reduced, is a second-order phase transition.

2.2
Helmholtz and Gibbs Free Energies

Thermodynamic systems can be described in terms of thermodynamic potentials. A simple example of a thermodynamic system is a gas of atoms. Here, we provide a brief reminder about two such potentials, namely, Helmholtz and Gibbs free energies. In the following sections, we will generalize these potentials to include superconducting condensates. In the following discussion, we will frequently use a gas of atoms as an example. However, all our conclusions are applicable to solid objects as well, with all their electrons and ions being referred to as particles. What is important is that the thermodynamic system contains a large number of elementary or point-like particles; that is, the energy of each particle only includes its kinetic energy and its potential energy (but no internal thermal energy).

The Helmholtz free energy is defined as $F = U - TS$, where U, T, and S are, as always, the internal energy, the temperature, and the entropy of the thermodynamic system, respectively. The internal energy U of the thermodynamic system includes the kinetic energy of all the elementary particles contained in it and the total potential energy related to mutual interactions between the particles. In this section, we assume that no external fields act on the particles.

The second potential to be discussed is the Gibbs energy $G = F + p V_s$, where p is the pressure and V_s is the volume of the system. The usefulness of the free energy potentials rests in their property to minimize their values in equilibrium. External conditions determine which thermodynamic potential is minimized. Briefly speaking, if $T = $ const and $V_s = $ const, then F is minimized, and if $T = $ const $p = $ const, then G is minimized. Another important property of the potentials is that they define the capacity of the system to perform work. In that respect, they are similar to the potential energy in classical mechanics, which equals the maximum amount of work the system can give out. To be more precise, the maximum work equals the value of the potential minus the value of the potential in equilibrium, which is the absolute minimum of the potential. Thus, the system can perform zero work if it is in equilibrium; only systems in nonequilibrium states can perform work. These properties will be discussed in more detail in subsequent sections. The reason we pay attention to the free energy potentials is that they can be generalized and applied to superconducting systems, and they can be used to compute the order parameter strength as well as probabilities of thermal fluctuations, such as phase slips.

Now, in much greater detail, we discuss Helmholtz energy and, later, return to the discussion of Gibbs energy. Let us first determine the conditions under which Helmholtz free energy approaches its minimum as the system approaches thermodynamic equilibrium. From the first law of thermodynamics, which is the law of total energy conservation, it follows that

$$dU = \delta Q + \delta W \tag{2.9}$$

where dU is the change of the internal energy during the time interval between t and $t + dt$ ($dt > 0$). Here, δQ is the heat entering the system within time dt, and δW is the total work performed on the system during dt. The corresponding increment of Helmholtz free energy is

$$dF = dU - d(TS) = \delta Q + \delta W - SdT - TdS \tag{2.10}$$

Since the heat, as well as the work, cannot be presented as exact differentials of any function, their infinitesimal amounts are denoted by the symbol δ. To understand this, consider that it is not possible to integrate δQ and say how much heat the system contains. The same is true for work: It is not possible to say how much work the system contains (thus, we use δW). Because heat and work "mix up" inside the thermodynamic system, both contribute to its total internal energy change and cannot be separated. The differential operator d will be used to denote exact differentials, which can be integrated and which only depend on the present state of the system, and not on the process or history that led to the present state. For example, one can integrate the change of the internal energy differential dU. The result will be the total energy of the system U, which is a well-defined function of the system's present state, defined by its volume and entropy.

Let us now discuss the relationship between heat influx and entropy change. If the system is in the state of internal equilibrium and no work is done on the system, then the heat entering the system from outside is the only source of its increase in entropy. In this case, the heat influx δQ is related to the entropy increase as $dS = \delta Q/T$, which is the thermodynamics definition of entropy. Yet, if some parts of the system are out of equilibrium, then entropy will increase with time as the system equilibrates, even if $\delta Q = 0$ and $\delta W = 0$. In the general case, the entropy increment is the sum of the increment generated by the heat influx δQ and the additional entropy increment related to the equilibration of the system. Thus, we arrive at the second law of thermodynamics, which can be formulated as the following relationship between δQ and dS (under the condition $\delta W = 0$):

$$TdS \geq \delta Q \tag{2.11}$$

We can now combine (2.10) and (2.11) and thus arrive at

$$dF \leq -SdT + \delta W \tag{2.12}$$

Assume now that the temperature of the system equals the temperature of the environment, which is held constant in time (so $\delta T = 0$). Additionally, we assume $\delta W = 0$ in order to obtain the condition for the direction of change of the free energy in time, namely,

$$dF \leq 0 \tag{2.13}$$

or, since we have assumed $dt > 0$, the same inequality can be written as

$$\frac{dF}{dt} \leq 0 \tag{2.14}$$

The inequality signifies that the free energy of a system decreases in time if the system is initially away from its equilibrium. If the system is in thermal equilibrium with the environment, then it does not change its state in time, $dF = 0$, unless work is done on it, but we have assumed zero work for now. An example of a nonequilibrium state would be, for example, a superconducting film, in which a vortex-antivortex pair is created. The pair experiences an attractive force that could, in principle, be used to derive work. However, if the work is zero, then the pair will collapse and annihilate, the generated heat will go to the environment, thus bringing the free energy F, which measures the ability of the system to do work, to its minimum.

If the process is reversible, the only source of increase in entropy is the external heat, and therefore $TdS = \delta Q$. Then, the change of Helmholtz energy follows from (2.10) and takes the form

$$dF = \delta W - SdT \tag{2.15}$$

If the process is isothermal (T = const and so $\delta T = 0$) and reversible, then $dF = \delta W$; that is, the change of Helmholtz free energy equals the work done on the system.

2.2 Helmholtz and Gibbs Free Energies

Consider again a reversible process. It can be reversed, and the system can perform work, at the expense of its free energy. Thus, the free energy is a measure of the amount of work the system can do, assuming it is in thermal (but not thermodynamic) equilibrium with its constant-temperature environment. Let us emphasize again that if the system is in thermodynamic equilibrium, its ability to do work is zero; it must be in a nonequilibrium state to do work.

If the only work allowed is the one achieved by changing the volume V_s of the system (such is typical for ideal gases, for example), then the work is $\delta W = -p\, dV_s$. Then, (2.15) becomes

$$dF = -S\, dT - p\, dV_s \qquad (2.16)$$

which means that F is a function of two independent, free variables, namely, V_s and T. Therefore, $F = F(T, V_s)$, and the full differential can be written as $dF = (\partial F/\partial V_s)\, dV_s + (\partial F/\partial T)\, dT$. By comparing this general calculus formula to (2.16), we conclude that the pressure is related to F as $p = -\partial F/\partial V_s$.

Equation (2.16) is applicable only if the changes in the system, namely, the changes of the temperature and/or the volume, are done reversibly. In this case, the system acts as a reservoir of usable energy. One can, for example, compress a gas of molecules and thus increase its potential energy. The gas in such a case acts essentially as a spring which can be shortened and thus can accumulate some potential energy. The gas, similar to the spring, can expand and thus can perform usable work on outside objects. In the ideal case, if the process of compression is completely reversible, all work done on the system goes into the change of free energy. In other words, if the initial state of the system has free energy F_i and the final state has free energy F_f, and the work done on the system (as the system transforms from the initial to the final state) is W_{fi}, then $F_f - F_i = W_{fi}$. In the general case, some usable energy can be dissipated into heat, thus reducing the amount of the usable energy in the final state; therefore, $W_{fi} > F_f - F_i$, for irreversible processes.

Gibbs free energy is introduced with the goal of obtaining a thermodynamic potential that would minimize, in a manner analogous to Helmholtz free energy, under constant pressure rather than constant volume.

Gibbs free energy is defined as

$$G = F + p V_s \qquad (2.17)$$

Qualitatively speaking, it is defined like this because the environment puts pressure p on the system, and this pressure tends to reduce the volume. Going against this pressure would increase the energy, just as moving uphill, against gravity, increases the mechanical potential energy of an object. Thus, the term $p V_s$ can be understood as the potential energy of a system of volume V_s in a constant-pressure environment. The term $p V_s$ represents, formally speaking, the amount of energy the system would get from the environment if it were able to shrink its volume from V_s to zero. This additional energy stored by the system by expanding itself against the external pressure can contribute to the maximum work the system can perform. Since the free energy is constructed such that it represents the maximum work the system can perform, it appears reasonable to include the term $p\,V_s$, with a positive sign, into the total calculation of usable or free energy.

Let us analyze the issue using math now. From the definition above, it follows that $dF = dG - d(p\,V_s)$. This expression for dF can be inserted into the inequality of (2.12). The result is $dG - d(p\,V_s) + S\,dT - \delta W \leq 0$. The work on the system is caused by the volume change, $-p\,dV_s$, plus the work performed by other causes, δW_{other}. Therefore, $\delta W = -p\,dV_s + \delta W_{\text{other}}$. With such definitions, the thermodynamic inequality becomes

$$dG \leq -S\,dT + V_s\,dp + \delta W_{\text{other}} \tag{2.18}$$

If we now assume that the process is isothermal ($dT = 0$) and isobaric ($dp = 0$), and also that all forms of work on the system, with the exception of the work caused by the volume change, equal zero (i.e., we assume $\delta W_{\text{other}} = 0$), then

$$dG \leq 0 \tag{2.19}$$

and, correspondingly, the time derivative is

$$\frac{dG}{dt} \leq 0 \tag{2.20}$$

Thus, the Gibbs energy, under the conditions stated above, decreases with time until it reaches the minimum possible value G_{min}, corresponding to the thermodynamic equilibrium state.

Assume again that $T = \text{const}$ and $p = \text{const}$. Then, from (2.18),

$$dG \leq \delta W_{\text{other}} \tag{2.21}$$

If the process is reversible and no usable energy is lost into heat, then all work done on the system goes into usable work stored in the system. In this case, the inequality of (2.21) becomes equality

$$dG = \delta W_{\text{other}} \tag{2.22}$$

By integrating (2.21) over a finite period of time, it is possible to find the work that must be performed on the system to evolve it from the initial state, having Gibbs energy G_{in}, to the final state, having Gibbs energy G_f

$$W_{\text{other}} \geq G_f - G_{\text{in}} \tag{2.23}$$

For given initial and final states, the required work is the minimum if the process is reversible. In such case,

$$W_{\text{other}} = G_f - G_{\text{in}} \tag{2.24}$$

This equation will be useful in the calculations of probabilities of thermodynamic fluctuations in the following section.

2.3
Fluctuation Probabilities

The Little's phase slip can be analyzed quantitatively by treating it as a nucleation process, in which a new phase (normal) nucleates in the parent phase, which is the superconducting phase. Generally speaking, the nucleation phenomenon, through which droplets of a new thermodynamic phase form within a parent phase, is well studied. It is applicable to such general situations as a gas condensation, liquid evaporation, or a crystal growth. The classical nucleation theory was outlined in the pioneering research of Volmer and Weber [73]. The theory was further developed by Farkas [74], Becker and Döring [75], and Volmer [76]. In this and the following sections, we outline the basic ideas of the nucleation theory in a simplified form, applicable to the nucleation of normal spots on a superconducting wire. We start the discussion with a more familiar ideal gas example and then focus on nanowires in subsequent sections.

The usable energy stored in the gas tends to dissipate in time even if the gas does no work. For example, if the gas is rotating, then it can perform work. Thus, a rotating gas contains some amount of usable energy. Over time, the rotation slows down and the amount of usable energy diminishes and approaches zero. The dissipation is the conversion of the usable energy into heat. The evolution stops only when the free energy reaches its lowest possible value, which is defined by the external conditions imposed on the system.

Thermal fluctuations have the opposite effect, namely, they convert heat into an equal amount of usable work, which propels the system into a nonequilibrium state. This happens temporarily and thus-obtained elevated thermodynamic potential cannot be used to perform any long-term net work on outside objects.

The fluctuations can be important in a different sense: Suppose the system is trapped in a metastable state separate by a potential barrier from its stable state. A fluctuation can give the system sufficient energy and thus can, with some probability, allow the system to overcome the barrier and to escape from the metastable state. Thus, thermal fluctuations provide an insurance that no metastable condition can last forever. Sooner or later, a sufficiently large fluctuation happens and allows the system to escape from the metastable nonequilibrium state and to evolve closer to its equilibrium state. In what follows, we describe a simple quantitative analysis of fluctuations and their probabilities.

A fluctuation is a stochastic event, usually a short-lasting one, which contradicts the second law of thermodynamics; that is, it leads to a temporary decrease of entropy of the system and its environment taken as a whole. Assume $p = $ const and $T = $ const. Then, the second law leads to the conclusion that G of the system is minimized over time, assuming no external work is done on the system (i.e., $\delta W_{\text{other}} = 0$). The minimization of Gibbs energy is the manifestation of the second law, for processes such that $p = $ const and $T = $ const.

On average, the total entropy of the system under investigation and the environment increases with time, until full equilibrium is reached. This principle of the entropy increase constitutes the second law of thermodynamics. The total entropy of the system and the environment decreases for a short time when a thermodynamic fluctuation occurs. The essence of a thermal fluctuation is that some thermal energy from the environment gets converted into usable work, which propels the system into a nonequilibrium (i.e., excited) state, having Gibbs energy G_{ex}, which is always larger than the equilibrium Gibbs energy G_{eq}. The entropy of any nonequilibrium state is lower than the entropy of the equilibrium state.

The free energy is at its minimum in equilibrium. Here we consider fluctuations out of equilibrium. A fluctuation occurs when, by chance, the particles of the environment act collectively to do positive work on the system, so that its free energy increases according to (2.24). The free energy increase, $\Delta G = G_{ex} - G_{eq}$, does not have to be infinitesimally small; it can be finite and rather large.

In some sense, a thermal fluctuation can be compared with a perpetual motion machine of the second kind. Such a hypothetical machine takes thermal energy from the environment and converts it into work, which then increases the usable energy of the system. It should be added, of course, that the fluctuation does not last long. After a short operation time, the fluctuation-based "perpetual motion machine of the second kind" halts its action and the free energy of the system drops to its equilibrium minimum again. One cannot extract a long-term work from such equilibrium thermal fluctuations. Yet, they play a very important role in various transitions since the fluctuations can allow the system to jump over a potential energy barrier and thus transit to a different state. Without fluctuations, thermal or quantum, the system would not be able to cross any barrier and the metastable states would be absolutely stable. Such logic is applicable only if the free energy of the system has many (at least two) minima. Then, fluctuations allow the system to "jump" from one minimum to the other. Note that in practice, while thermal fluctuations can be made arbitrary small by cooling the system, its quantum fluctuations cannot be eliminated. In the analysis, on the other hand, we sometimes postulate that quantum fluctuations do not exist for the purpose of better understanding the thermal fluctuations.

During a thermal fluctuation, some heat ΔQ is taken from the environment. It is then converted into work, δW_{th} (the subscript "th" indicates that this work originates from thermal fluctuations), which is applied to the system, resulting in an increase of its Gibbs energy according to (2.24).[2] As was discussed in the previous section, the change of the system's usable energy equals the work reversibly performed on the system. For the purpose of computing the fluctuation probabilities, we first treat thermal fluctuations as reversible events. Thus, (2.24) is applicable. Therefore, the work of thermal fluctuations equals the change of Gibbs energy, that is, $W_{th} = \Delta G$.

According to the energy conservation principle, if the work W_{th} is done by environment, its energy has to decrease by the same amount. The environment can only give away heat since it is in equilibrium. The thermal fluctuation converts this heat into work. Thus, the heat taken from the environment is $\delta Q = W_{th}$.

---

2) Note that to use Gibbs energy as a measure of the usable energy stored in the system, we have to limit the discussion to systems having a fixed number of microscopic particles, maintained under a constant temperature and pressure.

Taking into account the equality $W_{th} = \Delta G$, we obtain the change in entropy of the environment $\Delta S_{env} = -\delta Q/T = -\Delta W_{th}/T = -\Delta G/T$. Here, $T = $ const is the equilibrium temperature of the environment. For now, let us assume that the entropy of the system itself does not change since the considered thermal fluctuation does a reversible work on the system. Then, the total entropy change equals the entropy change of the environment. Again, using (2.24), we get $\Delta S_{tot} = \Delta S_{env} = -\delta Q/T = -W_{th}/T = -\Delta G/T$. Thus, the change in entropy is simply $\Delta S_{tot} = -\Delta G/T$. This expression only involves the temperature and the increase of Gibbs energy, defined by the given initial and final states connected the fluctuation.

In the above analysis we have assumed that the entropy of the system does not change. Let us make the consideration more general now, that is, with the goal of demonstrating that the result would not change. Assume that in addition to a reversible work on the system, W_{th}, a heat Q_s flows to the system during the fluctuation. Then, according to the energy conservation, $\delta Q = \delta Q_s + W_{th}$. The entropy change is, in such a case, $\Delta S_{tot} = -\delta Q/T + \delta Q_s/T = -\delta Q/T + (\delta Q - W_{th})/T = -W_{th}/T$. Since, by definition, the work on the system equals the change of the free energy of the system, we have $\Delta G = W_{th}$ and therefore it follows again that

$$\Delta S_{tot} = -\Delta G/T \tag{2.25}$$

This is identical to the result obtained under the assumption that no heat flows from the environment to the system. Thus, that assumption was not necessary and the result is generally applicable.

The equilibrium state is the state in which the total entropy of the system plus the environment equals the maximum possible value, $S_{tot} = S + S_{env} = S_{max}$. The entropy of the system is defined as $S = k_B \ln(N)$, where N is the number of microstates corresponding to a particular macrostate of the system. The state in which the energy is shared equally (on average) between various degrees of freedom is more probable and has the largest entropy since it can be represented by the largest number of microstates. That is why, in equilibrium, all parts of the system must have the same temperature, that is, the average kinetic energies of all particles in the system are the same. Note that in thermodynamics, a microstate is a certain state of the system in which the states of all particles (i.e., all electrons and all ions) in the system are specified up to the maximum possible precision. And a macrostate is a state defined by macroscopic parameters only, such as, volume, total energy, temperature, pressure, total supercurrent, or the density of the condensed electrons. For example, for a fixed amount of a gas, a macrostate can be defined by specifying only two parameters, namely, the volume and the temperature. All other macroscopic parameters, for example, the pressure, the total energy, or the entropy, can be computed by known gas equations.

The expression for the total entropy, that is, the system plus the environment, is $S_{tot} = k_B \ln(N_{tot})$. In thermodynamic equilibrium, this expression reads $S_{tot,eq} = k_B \ln(N_{tot,eq})$. The corresponding total equilibrium number of the microstates is $N_{tot,eq} = \exp(S_{tot,eq}/k_B)$. If a fluctuation occurs, as was discussed previously, the total entropy changes by the amount $\Delta S_{tot} = -\Delta G/T$. As expected, the change of the entropy is negative since the largest entropy corresponds to the equilibrium and a lower entropy corresponds to any nonequilibrium state. The number of microstates corresponding to the considered nonequilibrium state is

$$N_{tot,ex} = \exp\left(\frac{S_{tot,eq} + \Delta S_{tot}}{k_B}\right) = N_{tot,eq} \exp\left(-\frac{\Delta G}{k_B T}\right) \qquad (2.26)$$

Now, we come close to deriving the probabilities of fluctuations. The important step is to use the usual thermodynamic assumption that all microstates are equally probable. The probability is an additive quantity, proportional to the number of positive outcomes. Therefore, the probability of the nonequilibrium macrostate is proportional to the number of microstates representing the macrostate. Thus, the probability of the nonequilibrium state is

$$P_{ex} = \frac{N_{tot,ex}}{N_{tot,eq}} = \exp\left(-\frac{\Delta G}{k_B T}\right) = \exp\left(-\frac{G_{ex} - G_{eq}}{k_B T}\right) \qquad (2.27)$$

where ΔG is the change of Gibbs energy of the system associated with the transition of the system from equilibrium to the nonequilibrium state. The value ΔG also equals the minimum work required to excite the system. In the derivation, we have divided $N_{tot,ex}$ by $N_{tot,eq}$ because if the thermal equilibrium macrostate includes $N_{tot,eq}$ microstates, then the probability of each microstate is $1/N_{tot,eq}$ (all microstates are always assumed equally probable). The nonequilibrium macrostate includes a subset of $N_{tot,ex}$ microstates, out of the total $N_{tot,eq}$. Therefore, the probability of the nonequilibrium macrostate is $N_{tot,ex}(1/N_{tot,eq})$. In this discussion, the nonequilibrium macrostate is assumed to be a subset of the equilibrium macrostate. Such an assumption is reasonable since we are considering thermal fluctuations, which are an integral part of the equilibrium thermodynamic macrostate.

The assumption that all heat ΔQ, taken from the system by fluctuations, gets converted into work and increases Gibbs energy as $\Delta G = \Delta Q$ is not, in fact, necessary for (2.27) to be valid. To see this, assume that $\Delta G + \Delta Q_s = \Delta Q$, where ΔQ is the thermal energy (heat) *taken* from the environment in the course of a fluctuation, ΔQ_s is the heat released in the system, and ΔG is the change of Gibbs free energy of the system. In this case, the total entropy of the system and the environment changes as $\Delta S_{tot} = -\Delta Q/T_{env} + \Delta Q_s/T$, where T_{env} is the temperature of the environment, and T is the temperature of the system. The processes described by Gibbs free energy are such that the temperatures of the system and the environment are fixed and equal, that is, $T = T_{env}$. Therefore, $\Delta S_{tot} = (-\Delta Q + \Delta Q_s)/T = -\Delta G/T$. Thus, even in this more general case, we arrive at the conclusion that the total entropy change is given by (2.25). Therefore, the reduction of the number of microstates produced by the fluctuation is given by (2.26) and the probability of the fluctuation is given by (2.27).

Thus far, we have discussed the systems fluctuations under the following conditions: $T = T_{env} = $ const and $p = p_{env} = $ const, where T and p are the temperature and the pressure of the system, and T_{env} and p_{env} are the temperature and the pressure of the environment. Suppose now that the conditions are such that $T = T_{\text{env}} = \text{const}$ and $V_s = \text{const}$. In other words, the process is isothermal and the volume of the system is conserved, meaning that no work associated with volume change can be done on the system. In this case, thermal fluctuations also occur. Again, the thermal energy of the environment is converted into work, which leads to the increase of the appropriate thermodynamic potential, which is now the Helmholtz free energy F. As with the previous case, thermal fluctuation push the system into a nonequilibrium states, such that $\Delta F = W_{\text{th}}$ and $\Delta S = -\Delta F/T$. This conclusion is based on the knowledge that under the conditions considered, the change of the free energy for any change of the system equals the minimum work required to achieve the change. The probability of a fluctuation to a nonequilibrium state is given as

$$P_{\text{ex}} = \exp\left(-\frac{\Delta F}{k_B T}\right) \qquad (2.28)$$

where $\Delta F = F_{\text{ex}} - F_{\text{eq}}$ is the difference between the free energy of the nonequilibrium state and the free energy of the equilibrium state. Note that if $p = 0$ (i.e., the system, for example a solid object, is placed in vacuum), then $F = G$. Note also that $V_s = \text{const}$ means that no work can be done on the object through the change of its volume. However, thermal fluctuations can still do other kinds of work. For example, they can create a macroscopic flow of atoms or electrons in a gas or in a solid. In the case of a nanowire, thermal fluctuations can cause, for example, the occurrence of a normal spot on the nominally superconducting wire.

2.4
Work Performed by a Current Source on the Condensate in a Thin Wire

Here, we consider the external work on the condensate arising from a voltage V between the ends of the wire. The voltage accelerates the condensate, and the corresponding work is

$$dW = V I_s dt \qquad (2.29)$$

where I_s is the supercurrent in the wire and V is the voltage difference between the ends of the wire. Notice that the amount of work is given by the same formula as the one usually used for Joule heat in normal conductors, except that here we put the supercurrent into the formula, not the normal current, as in the case of Joule heating. The difference is that in the case of a superconductor, the work is directed at increasing the kinetic energy of the condensate rather than simply generating heat, as is the case in normal wires.

To justify (2.29), we first assume that the axis are chosen such that the electric field inside the wire is positive, $E > 0$, and the voltage is also positive, $V > 0$. The electric field averaged along the wire is $E = V/L$. The supercurrent, on the other hand, can be either positive or negative, depending on the phase gradient, according to (1.3). Note also that the supercurrent is related to the supercurrent density as

$I_s = A_{cs} j_s$, where A_{cs} is the wire cross-section and j_s is the supercurrent density, which only depends on the position along the wire since the wire is assumed to be 1D. In other words, the component of the supercurrent perpendicular to the wire axis is zero in 1D wires.

Assume that the system under consideration includes a thin superconducting wire and two bulk superconducting electrodes electrically well connected to the ends of the wire (see Figure 1.1). We will assume an idealized situation where each electrode is sufficiently thick so that the electric potential and the phase are both constant within each electrode. The electric field is consequently zero within each electrode, although it is not necessarily zero within the wire. Let n_e^+ be the number of electrons passing through the wire each second in the positive direction, and n_e^- be the number of electrons passing in the negative direction each second. The work done on one electron is the force eE acting on the electron times the distance traveled by the electron. If the electron passes through the wire of length L in the positive direction, then the distance is L (i.e., the total length of the wire). If the electron passes in the negative direction, then the displacement of each electron is $-L$. Thus, for each positive-moving electron passing through the entire wire, the work is $P^+ = eE * L = eV$ (which is negative since $e = -|e|$) and for each negative-moving electron, the work is $P^- = eE * (-L) = -eV$ (this work is positive). Thus, the total work per second P on all electrons is $P = P^+ * n_e^+ + P^- * n_e^- = eV(n_e^+ - n_e^-)$.

The electric current equals the total charge passing through the wire in the positive direction. The total charge per second is the charge of one electron times the difference of the number of electrons passing each second in the positive and in the negative direction. Thus, $I = e(n_e^+ - n_e^-)$. Therefore, $(n_e^+ - n_e^-) = I/e$, and the total work performed during each second is $P = eV * I/e = VI$. Thus, the work during the time interval dt is $dW = VI dt$.

It is much easier and frequently quite sufficient to limit the analysis to only the superconducting electrons. The work done on the condensate can be derived in the same way as above, but the total current must be replaced by the supercurrent. Thus, the work done on the condensate is $dW = VI_s dt$, confirming (2.29). If the normal current is significant, it can be modeled approximately as a current in a fictitious parallel normal resistor, which acts as a parallel shunt to the condensate of the wire. Here, we do not analyze this more complicated case.

The next step is to eliminate the explicit time dependence from (2.29). This is easily done using the Gor'kov phase evolution equation (1.5). Namely, $V = (\hbar/2e)(d\phi/dt)$, and thus $dW = (\hbar/2e)(d\phi/dt) I_s dt$. Since $(d\phi/dt)dt = d\phi$, one gets

$$dW = \left(\frac{\hbar}{2e}\right) I_s d\phi \qquad (2.30)$$

for the work done on the wire. This equation can be easily integrated if the supercurrent is maintained constant:

$$W = \left(\frac{\hbar}{2e}\right) I_s \phi \qquad (2.31)$$

Note that for $I_s > 0$, the work of the current source on the wire is positive if the phase change is negative because the condensate is negatively charged. Of course, the total work of the source also includes the work on the normal electrons, which equals $dW_n = I_n V dt$ where I_n is the current of the normal electrons, that is, the current of bogoliubons. This is a dissipative term which can influence the dynamics of the system but does not influence the energy barrier values for the fluctuations of the order parameter or the expressions for the free energy. In this respect, the BCS condensate is analogous to a simple harmonic oscillator immersed into a dissipative medium. In the latter case also, the potential (or usable) energy of the oscillator does not depend on the viscosity coefficient of the medium, although the equation of motion of the oscillator does depend on the viscosity.

2.5
Helmholtz Energy of Superconducting Wires

Now, let us turn our attention to superconducting nanowires again and analyze the usable energy stored in them. Here, we only discuss 1D or quasi 1D wires. We will assume that the volume of the wire is constant, that is, $V_s = \text{const}$ and $dV_s = 0$, because the wire is a solid object, usually measured in vacuum or under a very low pressure of the heat exchange gas.[3] Therefore, the work caused by the volume change is zero; $p\,dV_s = 0$ because $dV_s = 0$.

Taking into account the work performed on the condensate by the applied voltage (2.30), the first law of thermodynamics can be written as

$$dU = \delta Q + \delta W = \delta Q + \left(\frac{\hbar}{2e}\right) I_s d\phi \tag{2.32}$$

Therefore, Helmholtz energy changes as

$$dF = d(U - TS) = \delta Q - d(TS) + \left(\frac{\hbar}{2e}\right) I_s d\phi \tag{2.33}$$

From this equation, we can consider various limiting cases. Let us assume first that $T = \text{const}$ and $V = 0$, meaning that $\phi = \text{const}$ according to (1.5). Therefore, $dF = \delta Q - TdS$. From the second law we know that $\delta Q - TdS \leq 0$. Therefore,

$$dF \leq 0 \tag{2.34}$$

---

3) The term "heat exchange gas" is related to the helium gas which sometimes is injected into the Faraday cage in which the sample is located. The heat exchange gas serves the function of equilibrating the temperatures of different parts of the setup. In particular, it allows the sample to exchange heat with cold parts of the refrigerator and thus allows the sample to cool down to the base temperature. The pressure of the exchange gas is usually much lower than the atmospheric pressure. Its pressure can be made low without compromising its efficiency as a heat exchange agent because the thermal conductivity of a gas does not depend on its pressure, unless the pressure is so low that the mean free path of the gas molecules is comparable to the size of the volume in which the gas is contained.

Since we have agreed to consider positive time increments $dt > 0$, the inequality above can also be formulated as

$$\frac{dF}{dt} \leq 0 \qquad (2.35)$$

Thus, if no work is done on the condensate and its temperature is maintained constant, then its free energy decreases in time until it reaches thermodynamic equilibrium. After that, all macroscopic quantities remain constant (we neglect thermodynamic fluctuations in this section), including Helmholtz energy which obeys $dF = 0$ in equilibrium. That is why, when deriving the equations governing the BCS condensate, we must minimize the free energy. Such is the case, for example, when the Ginzburg–Landau equations are derived.[4]

An example of a nonequilibrium state would be a wire having a inhomogeneous distribution of the order parameter. As time increases, the order parameter equilibrates and approaches a homogeneous distribution. The usable energy stored in the gradients of the order parameter dissipates into heat. This is the essence of the equilibration process in a system that performs zero work.

Of course, when thermodynamic relations such as (2.32) are derived, the fluctuations are neglected. When the fluctuations are included into the analysis, we find that dF/dt can be positive over short periods of time because the thermal bath (the environment) can do work on the system and increase its usable energy temporarily. It is also clear that such increases cannot be used to extract usable energy from the environment since the fluctuations that increase the free energy and the fluctuations that decrease the free energy have equal probabilities. If that would not be the case, the free energy would evolve in time. However, we assume here that the system is in equilibrium, and thus it does not evolve in any particular direction.

Consider now a different situation. Suppose all stages of the processes are reversible. In this case, $\delta Q = TdS$, where, as always, S is the entropy of the wire and δQ is the heat entering the wire. One important requirement for the *total* entropy not to increase is the condition that no heat flows between subsystems having different temperatures. If a heat flow occurs between two parts which have the same temperature, the total entropy does not increase and the process can be reversible. Such is the case in the well-known cyclic process proposed by Nicolas Carnot. The Carnot cycle has the maximum possible efficiency of converting heat into work and it is reversible. In our case, if the free energy is not dissipated, then (2.33) becomes

$$dF = d(U - TS) = \left(\frac{\hbar}{2e}\right) I_s d\varphi - S dT \qquad (2.36)$$

Now, the free energy appears to only depend on two free variables, T and φ. The volume of the wire is assumed constant as always. Since dF is a full differential of F, Helmholtz energy must be a function of two variables, namely, $F = F(T, \varphi)$.

---

4) Of course there are always some thermodynamic fluctuations which can change the free energy slightly in both positive and negative directions. The rule of the free energy minimization is related to the mean value of the free energy averaged over rapid thermodynamic fluctuations. We will address the role of the fluctuations later.

Generally speaking, the full differential is $dF = (\partial F/\partial T)dT + (\partial F/\partial\varphi)d\varphi$. Comparing this with (2.36), we arrive at the important formula for the supercurrent in the wire, expressed through the partial derivative of its Helmholtz free energy

$$I_s = \frac{2e}{\hbar}\frac{\partial F}{\partial \varphi} \tag{2.37}$$

If the temperature is constant, then the $F(T, \varphi)$ becomes $F(\varphi)$. Then, the supercurrent can also be written as a full derivative as

$$I_s = \frac{2e}{\hbar}\frac{dF}{d\varphi} \tag{2.38}$$

This free energy function can change, of course, as the temperatures changes. For example, at $T = T_c$, the superconducting condensate free energy is zero, $F(\varphi) = 0$, because the superfluid density becomes zero at the critical temperature, and because it was assumed that $f_n = 0$.

2.6
Gibbs Energy of Superconducting Wires

In the first two paragraphs, we repeat some properties of the free energy defined in the standard way through the pressure of the gas p and the gas volume V_s. Then, we will make new definitions of the free energy useful for superconducting thin wires. For this, we will use the analogy in which the bias current plays the role of the pressure and the phase difference plays the role of the volume. The actual volume of the wire will always be assumed constant.

As we have previously argued, Helmholtz energy $F = U - TS$ is the appropriate thermodynamic potential if the system (e.g., some volume of gas) is allowed to exchange heat with the environment, but is not allowed to exchange work through a change in its volume. All processes must be slow enough so they are isothermal. The condition of no work through the volume change means that the volume is fixed. The term "appropriate" with respect to F means that F represents the usable energy stored in the system and that it is minimized in equilibrium.

If the volume of the gas V_s is allowed to change but the pressure and the temperature are kept constant, then the appropriate potential describing the gas is Gibbs energy G, which is defined as $G = U - TS + pV_s$. Again, this means that G represents the usable energy stored in the gas and that G is minimized in equilibrium. Under the same restrictions, if the value of Gibbs energy in the thermodynamic equilibrium is G_{eq} and G is Gibbs energy in a nonequilibrium state, then $G - G_{eq}$ equals the maximum work that can be performed by the system on external objects. Moreover, to propel the system from its equilibrium to the nonequilibrium state characterized by Gibbs energy $G > G_{eq}$, the minimum amount of work that must be performed is $G - G_{eq}$. In what follows, we redefine G in order to apply these principles of general thermodynamics to describe fluctuations in current-biased superconducting nanowires.

The first law of thermodynamics (2.9) can be written, in the case of a superconducting wire, as

$$dU = \delta Q + \left(\frac{\hbar}{2e}\right) I_s d\phi + \delta W_{\text{other}} \qquad (2.39)$$

where ϕ is the phase difference between the ends of the wire, $(\hbar/2e) I_s d\phi$ is the work done on the condensate by the current source (see (2.30)), and δW_{other} includes all other types of work performed during a time interval dt. Here, δQ is the heat entering the wire during the time interval dt. Our goal is to define Gibbs energy in such a way that it is the appropriate thermodynamic potential for the case of a fixed bias current, $I_s = \text{const}$. Thus, our assumption in what follows is $I_s = \text{const}$, while the voltage can change. This type of restriction is chosen since I_s is expected to be analogous to p. Such an analogy is suggested by a comparison of the expressions for the work on a gas, $-p d V_s$, and the work on a superconducting wire, $(\hbar/2e) I_s d\phi$. What is most important for us here is that if $I_s = \text{const}$, then $I_s d\phi = d(I_s \phi)$.

The second assumption of equal importance is that the temperature of the wire is also fixed, $T = \text{const}$, that is, the analysis here is restricted to isothermal processes. In other words, the wire is assumed to be in equilibrium with some large thermal reservoir of temperature T. Under the conditions specified above, the first law becomes

$$d\left(U - \hbar I_s \frac{\phi}{2e}\right) - \delta W_{\text{other}} = \delta Q \qquad (2.40)$$

On the other hand, the second law is $\delta Q \leq T d S$. Combined with the first law, it leads to $d(U - \hbar I_s \phi/2e) - T d S - \delta W_{\text{other}} \leq 0$. Taking into account that $T = \text{const}$ and, consequently, $T d S = d(T S)$, one gets

$$d\left(U - TS - \hbar I_s \frac{\phi}{2e}\right) - \delta W_{\text{other}} \leq 0 \qquad (2.41)$$

Now, we define Gibbs energy as

$$G \equiv U - TS - \hbar I_s \frac{\phi}{2e} = F - \hbar I_s \frac{\phi}{2e} \qquad (2.42)$$

Then, we get

$$dG - \delta W_{\text{other}} \leq 0 \qquad (2.43)$$

This inequality can be analyzed further under the following two types of assumptions: (1) no other work is done on the system while irreversible processes are allowed to happen, and (2) all processes are reversible, but the work on the wire can be done not only by the current source, but also by other agencies. Note that since the bias current is not zero and the voltage on the system can vary, the work exchange with the source is never zero, but it is taken into account automatically by virtue of the definition of G.

2.6 Gibbs Energy of Superconducting Wires

Consider the first class of processes in which the work on the wire is zero ($\delta W_{\text{other}} = 0$), with the exception of the current source work, which is allowed to accept any value. Remember that, as always when discussing thermodynamic equations or inequalities, the effect of thermal fluctuations is neglected. Then, one concludes that

$$dG \leq 0 \tag{2.44}$$

or, since we always assume $dt > 0$, it leads to the equivalent condition

$$\frac{dG}{dt} \leq 0 \tag{2.45}$$

Since, by definition, all thermodynamic quantities are constant in thermodynamic equilibrium, the equality condition $dG/dt = 0$ corresponds to equilibrium. The system can be driven out of equilibrium by allowing $\delta W_{\text{other}} \neq 0$ for a limited period of time. As soon as the condition $\delta W_{\text{other}} = 0$ is restored, G begins to decrease again, and it does so until it reaches the minimum. In such processes of equilibration, the usable energy typically dissipates into heat, which then flows to the environment.

In some cases, the system can be trapped in a metastable nonequilibrium state so that the equilibration process might be extremely slow and $dG/dt \approx 0$. An example is a superconducting solenoid with a nonzero persistent supercurrent. However, by allowing enough time, one can still achieve the condition of the G potential to decrease in time if the initial state is not the equilibrium state. For the solenoid example mentioned above, it is the Little's phase slips which are responsible for the decay of the metastable states having nonzero supercurrents and the evolution to the equilibrium state having zero supercurrent.

Now, consider the second important example: All processes are reversible and the system is allowed to exchange work with the environment. Then, the second law (2.11) takes the form of the equality

$$\delta Q = T dS \tag{2.46}$$

If we remember that $T = \text{const}$, and thus $T dS = d(TS)$ and combine (2.46) and (2.40), we get

$$d\left(U - TS - \hbar I_s \frac{\phi}{2e}\right) = dG = \delta W_{\text{other}} \tag{2.47}$$

Therefore, we arrive at an important conclusion that in the case of reversible processes, the change of G equals the work done by all agencies acting on the wire, except the current source, according to the definition of δW_{other}. Thus, if no work is done on the wire, except the work done by the current source, then $dG = 0$ (if all processes are reversible also). Suppose now that all processes are reversible and no work is done on the wire except that by the current source. Then, $G = \text{const}$. This result of Gibbs energy being constant holds of course because we have assumed that $T = $ const and $I_s = $ const. In experiments, one can change the external conditions, including the temperature and the bias current. According to the definition of (2.42), the equilibrium Gibbs energy varies as

$$dG = dU - SdT - TdS - \left(\frac{\hbar}{2e}\right) I_s d\phi - \left(\frac{\hbar}{2e}\right) \phi dI_s \qquad (2.48)$$

If the system is in equilibrium before and after the change, then $TdS = \delta Q$ and, from the first law (which is the energy conservation principle), $dU = \delta Q + \delta W = TdS + \delta W$. In this first law equation, we have to include the total work on the system, which is $\delta W = (\hbar/2e) I_s d\phi + \delta W_{\text{other}}$. By combining (2.48), the first law, and the second law one gets

$$dG = \delta W_{\text{other}} - SdT - \left(\frac{\hbar}{2e}\right) \phi dI_s \qquad (2.49)$$

This equation determines how the Gibbs energy evolves if the system remains in equilibrium, while external conditions change.

Assume again that no other work is done on the wire ($\delta W_{\text{other}} = 0$). Then, the equation above signifies that the equilibrium Gibbs energy is the function of two independent variables, T and I_s, and it can be written, in general, as $G = G(T, I_s)$. Therefore, its most general form the differential is $dG = (dG/dT)T + (dG/dI_s)dI_s$. In comparison to (2.49), one gets the following useful equations $S = dG/dT$ and $\phi = (2e/\hbar)(dG/dI_s)$, which are valid for reversible equilibrium processes, such that $\delta W_{\text{other}} = 0$.

To better understand the meaning of δW_{other}, consider an example in which some mechanical device brings a piece of normal metal into a direct contact with the superconducting wire. Then, due to the proximity effect [65–68], the normal electrons will diffuse into the superconducting wire, thus reducing its order parameter and increasing its free energy. Also, as the contact is established, the superconducting electrons diffuse into the normal metal, thus making it somewhat superconducting [65–68]. This process increases the free energy of the normal metal since in equilibrium it is normal, not superconducting. Thus the mechanical devices making the connection does a positive work, $\delta W_{\text{other}} > 0$. This also means that the proximity effect causes the piece of normal metal to be repelled from the superconductor. As the normal metal is suddenly removed, the distribution of the order parameter in the wire will be a nonequilibrium one. With time, the free energy will be converted into heat unless it is somehow used to do work on outside objects, which is possible, in principle.

An attentive reader might notice that in all equations defining the Gibbs potential, we have used the supercurrent. The normal current I_n was neglected. It is because the normal current only defines the loss (or, in other words, the dissipation) of the usable energy. The Gibbs potential on the other hand defines the maximum amount of work or the maximum amount of usable energy stored in the system. The maximum amount can only be extracted if the loss is zero, meaning that the normal current is zero, which means that the bias current equals the supercurrent. That is why the normal current is neglected in the definitions of the potential.

The situation is somewhat similar to the classical mechanics description of a particle experiencing a viscose force or friction. The viscosity is known to cause damping or loss of mechanical usable energy, for example, gravitational energy of an object, into heat. Such damping is analogous to the dissipative normal currents in a superconducting device which transform the usable energy of the condensate into heat due to the Joule heating which is always associated with normal currents. The dynamic equation describing the motion of a classical particle (i.e., the Newton equation) does include the viscose term if any damping is present in the system. Whereas the expression for the potential energy, which is the usable energy possessed by the particle, does not include the damping coefficient (even if the damping coefficient is not zero). Similarly, the usable energy stored in the wire depends on the order parameter distribution, and, if present, on the charge density distribution, but it does not depend on the normal current which only defines the rate of loss of the usable energy into heat. The rate of loss of the usable energy equals to the Joule power, namely, to $I_n V = V^2/R_n = (\hbar/2e)^2(d\phi/dt)^2/R_n$. Obviously, the loss can be made negligible whenever it is possible to make the rate of change of the phase sufficiently small (here, R_n is some effective normal resistance).

2.7
Relationship between Gibbs and Helmholtz Energy Densities

To find the relationship between Gibbs and Helmholtz energy *densities*, denoted g and f, we start with their definitions expressed as $F = \int_{V_s} f(r) d^3r$ and $G = \int_{V_s} g(r) d^3r$. Assume also for simplicity that the magnetic field is zero or negligible. That is why the free energy integrals are restricted to the volume of the wire itself. According to (2.42), we can write $G - F = -\hbar I_s \phi / 2e$. The phase difference can be expressed as the integral over the length of the wire $\phi = \int_0^L \mathbf{e}_x \nabla \phi \, dx$. We assume that the wire is straight, parallel to the x-axis, and its ends are located at $x = 0$ and $x = L$. The phase gradient $\nabla \phi$ is directed along the wire axis because the wire is assumed thin and therefore one-dimensional, meaning that all of its characteristics can only change along the wire and remain constant within perpendicular cross-sections. In the case that the wire is long and the effect of the electrodes connected to the ends of the wire can be neglected, the phase gradient is also constant along the wire, $|\nabla \phi| = $ const. (assuming the wire is in equilibrium). This homogeneous distribution of the phase gradient (as well as the order parameter amplitude) corresponds to the minimum of the free energy.[5] Then, the phase difference can be expressed as $\phi = L|\nabla \phi|$. Similarly, the total current can be expressed through the current density $\mathbf{j}_s$ as $I_s = A_{cs} j_s = \int_{A_{cs}} \mathbf{e}_x \mathbf{j}_s \, dy \, dz$, where A_{cs} is the cross-section area of the wire and $j_s = |\mathbf{j}_s|$. Since for thin wires the order parameter and therefore the current density are constant within the cross-section of the wire, it must be directed along the axis of the wire, coinciding with the x-axis.

---

[5] This is similar to the ideal gas example, the thermodynamic equilibrium of which is characterized by constant pressure and density within the entire gas volume.

Using the integral presentations, we can write

$$G - F = \int (g - f) d^3 r = -\left(\frac{\hbar}{2e}\right) I_s \phi$$

$$= -\left(\frac{\hbar}{2e}\right) \int_{A_{cs}} e_x j_s dy dz \int_0^L e_x \nabla \phi dx$$

$$= -\left(\frac{\hbar}{2e}\right) \int_{V_s} j_s |\nabla \phi| d^3 r \quad (2.50)$$

Under the assumption that all considered physical quantities, namely, g, f, $\nabla \phi$, j_s, are constant along the wire (because the wire is assumed long and the effects of electrodes, if any, are negligible), the expressions under the integrals must be equal because the integrals are equal. Thus,

$$g = f - \left(\frac{\hbar}{2e}\right) j_s |\nabla \phi| \quad (2.51)$$

The equation $G = F - \hbar I_s \phi / 2e$ can also be transformed using (1.2) and assuming that the superfluid velocity is constant along the wire, leading to

$$G = F - \left(\frac{mL}{e}\right) I_s v_s \quad (2.52)$$

where L is the wire length. Taking into account the equations for the homogeneous case $j_s = I_s/A_{cs}$, $g = G/A_{cs}L$ and $f = F/A_{cs}L$, one arrives at the relationship

$$g = f - \left(\frac{m}{e}\right) j_s v_s \quad (2.53)$$

which is the form used in the Tinkham book [1].

2.8
Relationship between Thermal Fluctuations and Usable Energy

To compute the fluctuation probabilities, assume that $T = \text{const}$ and $I_s = \text{const}$. Then, (2.47), which reads $dG = \delta W_{\text{other}}$, is applicable. As was discussed above, a thermal fluctuation contradicts the second law. It takes a heat δQ from the environment and converts its reversibly into work $\delta W_{\text{other}} = \delta Q$ performed on the wire. In this case, the entropy of the environment decreases by the amount $dS = \delta Q/T$, where T is the temperature of the environment (which is assumed equal to the temperature of the wire and is constant). If any heat flows from the environment to the wire, this fact along does not change the total entropy, provided that their temperatures are equal. Thus, the total entropy changes as $\delta S_{\text{tot}} = \delta W_{\text{other}}/T$. On the other hand, the work performed on the wire leads to a change of its Gibbs energy from the equilibrium level G_{eq} to a nonequilibrium (excited) state having Gibbs energy

G_{ex}. According to (2.47), which can be integrated, the increase of the Gibbs energy equals the work done by external forces (such as thermal fluctuations), with the exclusion of the work done by the current source. Thus, the change of Gibbs energy is $\Delta G = G_{ex} - G_{eq} = \delta W_{other}/T$. Therefore, if we choose to consider a certain type of fluctuation for which it is know how to compute the change of Gibbs energy, we can predict by how much entropy reduction such fluctuation requires and therefore compute the probability of such fluctuations according to (2.27), which is simply $P = \exp(-\Delta G/k_B T)$. Here, ΔG is the increase of Gibbs energy of the system caused by the considered fluctuation and counted from the minimum (i.e., the equilibrium) level.

As was first elucidated by McCumber [64], the experiments (and theoretical models) on nanowires can be classified into two main categories, namely, those in which the current through the wire is maintained constant and those in which the phase difference between the ends of the wire is kept constant. The expression above for the probability of fluctuations is derived under the assumption that the current is constant. It is also useful to consider a different situation, namely, such in which the voltage on the wire is maintained at zero ($V = 0$) while the current I_s is not fixed and allowed to change, for example, due to thermal fluctuations. Such a situation would be realized if the wire is connected to a very large capacitor, the voltage on which is zero. In the limit of infinite capacitance, any current through the wire would change the charge Q on the shunt capacitor, but the change of the voltage would be zero since $V = Q/C$ and we assume here $C \to \infty$. So, for such large capacitor, if the voltage is set to zero, it will remain zero for a long time, even if the wire connected to it has a nonzero supercurrent. In addition, it can be assumed that the capacitor connected to the nanowire is made of the same superconducting material as the wire itself. This would lead to a conclusion that since $V = 0$, then ϕ = const, that is, the phase difference between the plates of the capacitor remains constant (because $\dot{\phi} \sim V$).

Under such conditions, the free energy which is minimized as the wire reaches equilibrium is F, that is, Helmholtz energy. Also, if any external object or device does a work δW_x on the wire (note that in the example considered the capacitor itself does not do any work since the voltage on it is zero) and if the work is done reversibly, then Helmholtz energy changes as $\Delta F = W_x$. If now thermal fluctuations convert some heat ΔQ taken from the environment into work, then $\Delta F = \Delta Q$. On the other hand, as in the arguments above, the total entropy change is $\Delta S_{tot} = -\Delta Q/T$ in this case because the temperature of the environment T (which is equal to the temperature of the wire) remains constant since the environment is always assumed infinitely large, and thus any heat taken from it does not change its temperature. Since the heat taken from the environment in the course of a thermal fluctuation is converted reversibly into work, we can assume that the temperature of the wire does not change, only its free energy increases as $\Delta F = \Delta Q$. Thus, the change of the total entropy is $\Delta S_{tot} = -\Delta F/T$. As previously argued, the number of the microstates corresponding to the thus created nonequilibrium state is $N_{ex} = N_{tot} \exp(\Delta S_{tot}/k_B)$. Thus, the probability of such fluctuation is $P = N_{ex}/N_{tot} = \exp(-\Delta F/k_B T)$. The processes in which a large amount of heat

2.9
Calculus of Variations

The free energy of a superconductor (2.7) is a functional, not just a function. The meaning of this statement is that to each permissible choice of functions $\psi(r)$ and $h(r)$, a single number, denoted F, is put into correspondence. Functionals are different from functions is the following sense: If one consider a function, for example, $y(x)$, than to each permissible choice of the number x, another number, y, is put into correspondence. However, in functionals, numbers are put into correspondence to functions. One can say that each function "maps" a space of numbers into another space of numbers, while a functional maps a space of functions into a space of numbers. The task of minimization of a functional is somewhat more complicated than minimization of a function. For example, to minimize the free energy F of a superconductor, we need to consider all possible shapes of the functions $\psi(r)$ and $h(r)$, and out of this infinite variety of functions, find those functions which correspond to the smallest value of F.

The minimization of the free energy can be done using methods of the calculus of variations, [69] which is a generalization of the ordinary calculus, and is applicable to functionals. Remember that in the case of a function, say $y(x)$, the extrema (i.e., the maxima and the minima of the function) are found by changing the free variable x by a small amount dx, that is, by making a substitution $x \to x + dx$ and by requiring that the corresponding variation dy of y is zero in the first approximation, that is $dy \equiv y(x + dx) - y(x) = 0$ in the limit of dx approaching zero. Another representation of the same method of finding minima (and maxima) is to write $dy = (dy/dx)dx$ and to require that the derivative $dy/dx = 0$. Each point x where such a condition is satisfied is the point of a minimum or a maximum (or a flat region of the function $y(x)$).

By analogy, one can understand the minimization of functionals. Consider some functional $F = F[z(x)]$ which puts into correspondence a real number F to each choice of the function $z(x)$. In this situation, the free "variable" is the function $z(x)$ itself. Thus, we have to subject to small arbitrary variations the entire function $z(x)$, not just x, as we did in the previous section. Indeed, to find $z(x)$, which minimizes the functional, we have to allow a small "virtual" change to the function $z(x)$. This change, called arbitrary infinitesimal variation, is denoted as $\delta z(x)$. It is an infinitesimally small, but not zero, function of x. Next, one needs to make a replacement of the type $z \to z + \delta z$ in the functional $F[z]$. The variation of the functional is $\delta F \equiv F[z(x) + \delta z(x)] - F[z(x)]$. Then, the minima and the maxima of the functional F correspond to functions $z(x)$ for which $\delta F = 0$. In other words, the maxima and minima are the stationary points, characterized by the condition that any infinitesimally small (but distinct from zero) variation of the argument function z does not change the functional.

Consider an example $F = \int z^2 dx$, where $z = z(x)$ can be any smooth function. The variation of this functional can be found by differentiating the function under the integral and taking the small variation of z at each point x to be δz. Thus, at each point x, we have $\delta(z^2(x)) = 2z(x)\delta z(x)$, similar to the ordinary differentiation. Therefore, the variation of the functional is $\delta F = \delta(\int z^2 dx) = \int 2z\delta z dx$. One needs to remember that the variation symbolized by the δ operator is independent and can exchange order with the integration as well as with the ordinary differential operator d (as in dx). This independence is due to the fact that the choice of δz is in no way influenced by the value of z itself or by the dependence of z on x, or by the value of dx. In other words, although dz depends on dx (namely, as $dz = z'_x dx$), yet $\delta z(x)$ and dx are completely independent. Thus, for example, $\delta(dx) = 0$ since the variation δz has no effect on the value of the differential dx. To find the stationary solution, which brings F either to a minimum or to a maximum, we have to require $\delta F = 0$, which is equivalent to requiring that $\int 2z\delta z dx = 0$ for any choice of the infinitesimal variation $\delta z(x)$. The latter equation can always be true only if $z(x) = 0$. Thus, the solution of the considered simple minimization problem is $z(x) = 0$.

2.10
Ginzburg–Landau Equations

One important function of Helmholtz free energy (2.7) is that it allows one, by virtue of its minimization, to derive a second order differential equation that governs the spatial dependence of the complex order parameter $\psi(r)$ inside superconductor. As mentioned before, in the most general case, the order parameter can also depend on time. Such time dependence cannot be derived from the minimization of Helmholtz free energy. The minimization of the free energy expression (2.7) only allows one to derive solutions for stationary states, that is, states which do not change in time. The reason that the states corresponding to local minima of the free energy do not evolve in time is that if they would evolve in time the free energy would increase, which would contradict the general thermodynamic equation (2.35). Equation (2.35) tells us that the free energy either decreases as the system progresses to its equilibrium state or it is not changing with time, which is the case when the system has reached its equilibrium. This is of course true only if no external work is done on the system since by performing work it is always possible to increase the free energy of the system above its equilibrium minimum. Note that if external work can be performed without limitations, then one can manipulate the system and its free energy in any way desired, and thus the minimization principle would not apply.

If the conditions are such that a nonzero work can be performed on the sample by external devices, then a different thermodynamic potential can in many cases be constructed which would obey the principle of minimization in the equilibrium. For example, if the wire is biased with a constant current, then Gibbs free energy is minimized in the equilibrium. The importance of different thermodynamic potentials is that they govern fluctuations of the order parameter. Until specified explicitly, we will assume that no work is done on the superconducting sample and that the temperatures of the sample and the surrounding thermal bath are equal and constant in time. With these conditions, Helmholtz free energy is minimized in the equilibrium.

The knowledge of stationary states is sufficient to understand a great deal about the properties of a superconducting system. For example, the rate of Little's phase slips in thin wires is governed by the energy separation of two stationary states, namely, the state having a homogeneous order parameter and the state in which at one spot on the wire the order parameter is zero. The latter state (i.e., the state having a normal spot on the wire) is also stationary in the sense that it corresponds to an extremum (namely a saddle point) of the free energy functional. The energy difference between these two states of the wire governs the probability that a thermal fluctuation is powerful enough to create Little's phase slip on the wire.

To describe the superconductor completely, it is not enough to know the distribution of the order parameter; one also needs to know the value of the supercurrent and the magnetic field. The distribution of the supercurrent inside the superconductor can also be derived from Helmholtz free energy, as will be shown below.

To apply the calculus of variations to the free energy minimization, we notice that the free energy functional (2.7) explicitly depends on the following functions: ψ, h, and A. The expression for Helmholtz free energy (in Gaussian units) can be written explicitly as

$$F = \int \left(f_n + \alpha |\psi|^2 + \frac{\beta}{2} |\psi|^4 + \frac{1}{4m} \left| \left(\frac{\hbar}{i} \nabla - \frac{2e}{c} A \right) \psi \right|^2 + \frac{h^2}{8\pi} \right) d^3 x \tag{2.54}$$

This total Helmholtz free energy has to be minimized in order to find the equilibrium state of the superconductor. The integral is taken over the entire space. Outside the superconductor, the order parameter is zero. Thus, only the magnetic field contributes to the free energy integral outside the volume of the superconducting sample. Here and everywhere, $h = |h| = \sqrt{h \cdot h}$. Note that the two vector fields involved, namely, A and h, are not independent because $h = \nabla \times A$. We will choose the variation $\delta A = \delta A(r)$ of the vector potential as the independent function which is assumed to be arbitrary and infinitesimal everywhere and zero at infinity. The variation of the vector potential being zero at infinity is due to the fact that the superconducting sample is assumed to be finite in size so that any variation of its state has zero effect at infinity. The magnetic field is assumed to be created by a large (compared to the superconducting sample) but finite solenoid, so that the magnetic field is zero at infinity. The variation of the magnetic field is $\delta h = \delta(\nabla \times A) = \nabla \times \delta A$. Let us once more be aware that the goal of the variation procedure is to find such $\psi(r)$ and $A(r)$ that for any choice of the variation functions $\delta \psi(r)$ and $\delta A(r)$, the corresponding variation of the free energy, δF, equals zero ($\delta F = 0$). Such a condition signifies that the free energy is at its minimum (or the maximum). The other parameters, such as the volume of the sample and the temperature, are assumed fixed, and thus they are not subject to the variation procedure.

As a start, let us consider the last term in the free energy functional, (2.54), which is the free energy of the magnetic field $F_m = \int (h^2/8\pi)d^3x$. Its variation is

$$\delta F_m = \delta \int \frac{h^2}{8\pi} d^3r = \int \frac{h\delta h}{4\pi} d^3r \qquad (2.55)$$

By introducing the vector potential, we get

$$\delta F_m = \frac{1}{4\pi} \int h(\nabla \times \delta A) d^3r \qquad (2.56)$$

Here the integration must be taken over the entire space since the magnetic field associated with the supercurrent exists inside the wire as well as outside the wire. To transform this integral, we will use the divergence theorem and the theorem of the rotational known from the vector analysis [72]. They can be written for any smooth vector field $P(r)$ as

$$\int_{V_s} \nabla P d^3r = \int_{A_s} P da \qquad (2.57)$$

and as

$$\int_{V_s} (\nabla \times P) d^3r = \int_{A_s} da \times P \qquad (2.58)$$

correspondingly, where the first integral is taken over some volume V_s and the second integral is taken over the surface A_s that defines the volume considered. The vector da represents an infinitesimal surface element of the sample surface. This vector is perpendicular to the surface and it is directed outward. The magnitude of the vector da equals the area of the surface element represented by this vector. The volume integral in the above two theorems can be taken over the entire space – in this case, the surface A_s is the surface at infinity.

We will also use a known expression involving vector and scalar products of any two smooth vectors h and A, namely, [72]

$$h(\nabla \times A) = \nabla(A \times h) + A(\nabla \times h) \qquad (2.59)$$

Therefore, the variation of the magnetic field energy (2.56) becomes[6]

$$\delta F_m = \frac{1}{4\pi} \int [\delta A(\nabla \times h) + \nabla(\delta A \times h)] d^3r$$

$$= \frac{1}{4\pi} \int \delta A(\nabla \times h) d^3r + \frac{1}{4\pi} \int (\delta A \times h) da$$

6) Note that, in general, the energy of a static magnetic field equals its free energy since the entropy of a static magnetic field is zero. All of its energy can be used to perform work.

Here, the volume integrals are taken over the entire space and the last integral is the surface integral which is taken over an infinite surface, all elements of which are at infinity. Thus, it is zero since the variation δA is zero at infinity, as was discussed above. Therefore, we arrive at $\delta F_m = \frac{1}{4\pi} \int \delta A (\nabla \times h) d^3 r$. Finally, after one of the Maxwell's equations is used (here, we use Gaussian units), namely,

$$\nabla \times h = \left(\frac{4\pi}{c}\right) j + \left(\frac{1}{c}\right) \partial_t E$$

the variation of the magnetic field free energy becomes (assuming the static case $\partial_t E = 0$)

$$\delta F_m = \frac{1}{c} \int j \delta A d^3 r \qquad (2.60)$$

where $\delta A(r)$ is an arbitrary, infinitesimally small, and a smooth function.

At this point, let us consider a simple example: assume that $\psi = 0$, that is, the considered metal is not superconducting and $f_n = $ const. In this case, the total free energy $F = F_m + F_n$ is the sum of the magnetic free energy F_m and the free energy of the normal electrons $F_n = \int f_n d^3 r$, where the integral is taken over the entire space, as in (2.54). Here and everywhere, we assume that F_n is field-independent (which is not really correct and we will see the consequences soon). The variation of the free energy is therefore $\delta F = \delta F_m = (1/c) \int j \delta A d^3 r$. However, if no work is done on the system, then F is the minimum in the equilibrium, which means that $\delta F/\delta A = 0$. This condition is possible to satisfy universally for any choice of $\delta A(r)$ only if $j = 0$. Such a conclusion seems quite obvious since we have assumed from the beginning that the sample is not superconducting, so there is no possible supercurrent in it. Yet, a curious reader might question the current in the solenoid and the apparent absence of its contribution to the presumably total current density j. The answer to such question is that we explicitly remove the normal currents from the consideration since we assume $f_n = $ const. On the other hand, if one desires to include the normal electrons and the currents created by them into the free energy (2.54), one needs to introduce a term similar to the fourth term in (2.54) for each normal electron. To simplify our discussion, we do not make such inclusions. Thus, the minimization of the considered free energy in which f_n is assumed, somewhat artificially, to be independent of A, only gives the expression for the supercurrent and not for the normal currents. In other words, the free energy of normal currents will not be minimized and not considered anywhere. Thus under j we understand here the supercurrent.

The variation of the GL functional, including all terms, magnetic and nonmagnetic, with respect to the vector potential is

$$\delta F = \int \left(\delta F_{kin} + \frac{1}{c} j \delta A\right) d^3 r \qquad (2.61)$$

where we have defined the term

$$F_{kin} = \frac{1}{4m} \left|\left(\frac{\hbar}{i} \nabla - \frac{2e}{c} A\right) \psi\right|^2 \qquad (2.62)$$

2.10 GinzburgLandau Equations

corresponding to the kinetic contribution into the free energy. Note that other terms are not included since they are not dependent on A, which is the only function that is varied for now (variations of the order parameter will be analyzed separately, which is a legitimate approach since the variations of the order parameter and the vector potential are independent).

With these assumptions, we get

$$4m\delta F_{kin} = \delta \left|\left(\frac{\hbar}{i}\nabla - \frac{2e}{c}A\right)\psi\right|^2$$

$$= \left[\left(\frac{\hbar}{i}\nabla - \frac{2e}{c}A\right)\psi\right]^* \delta\left[\left(\frac{\hbar}{i}\nabla - \frac{2e}{c}A\right)\psi\right] + \text{c.c.} \quad (2.63)$$

where "c.c." stands for a term which is the complex conjugate of the term before it. This term arises because $|Z|^2 \equiv ZZ^*$ and $\delta|Z|^2 = Z^*\delta Z + Z\delta Z^* = Z^*\delta Z + \text{c.c.}$ for any complex function Z. Remembering that A is the only function that is varied by our assumption here and therefore $\delta\psi = \delta\psi^* = 0$ and also $\delta\nabla = 0$ since the coordinates are not varied. Thus we get

$$\delta F_{kin} = \left(-\frac{2e\psi}{4mc}\right)\delta A\left[\left(\frac{\hbar}{i}\nabla - \frac{2e}{c}A\right)\psi\right]^*$$

$$+ \left(-\frac{2e\psi^*}{4mc}\right)\delta A\left[\left(\frac{\hbar}{i}\nabla - \frac{2e}{c}A\right)\psi\right] \quad (2.64)$$

Further algebra leads to

$$\delta F_{kin} = \left(-\frac{e}{2mc}\right)\delta A\left[i\hbar(\psi\nabla\psi^* - \psi^*\nabla\psi) - \left(\frac{4en_s}{c}\right)A\right] \quad (2.65)$$

where $n_s = \psi\psi^*$ is the density of electronic pairs in the condensate which is also called superfluid density or superpair density. Now, we can combine all the terms sensitive to the variation of the vector potential and obtain the following expression $\delta F = \delta F_m + \delta F_{kin}$ which translates into

$$\delta F = \int \left(-\frac{e}{2mc}\right)\delta A\left[i\hbar(\psi\nabla\psi^* - \psi^*\nabla\psi) - \frac{4en_s}{c}A + \frac{4m}{c}\left(-\frac{c}{2e}\right)j\right]d^3r \quad (2.66)$$

Finally, since the free energy must be at its minimum when the superconductor is in the equilibrium state, the variation must be zero ($\delta F = 0$) for any choice of the variation function δA. This is only possible if the expression in the square brackets in the equation above equals zero. Thus, one obtains the expression for the current:

$$j = 2e\left(\frac{1}{4m}\right)i\hbar(\psi\nabla\psi^* - \psi^*\nabla\psi) - (2e)\frac{en_s}{mc}A \quad (2.67)$$

It still remains to be discussed what components contribute to the total current j. In general, the current involves the supercurrent density j_s and the normal current density (also called quasiparticle current or the current of bogoliubons) j_n. The normal current experiences dissipation according to Ohm's law. Thus, even if it is not zero for any reason in some initial nonequilibrium state of the superconductor, it would drop to zero due to the dissipation very quickly as the system approaches the equilibrium. Since here we are deriving the equations for the equilibrium state, the normal current must be set to zero. Thus, the total current equals the supercurrent. Therefore we set $j = j_s$ and get the final formula for the equilibrium supercurrent, that is,

$$j_s = \frac{ie\hbar}{2m}(\psi \nabla \psi^* - \psi^* \nabla \psi) - \frac{2e^2|\psi|^2}{mc}A \qquad (2.68)$$

Note that it is exactly the same as the expression for the probability density current for a single quantum particle of mass $2m$. The probability current is multiplied by the charge of a superpair, $2e$, to convert the probability current into the electric current.

To derive the main GL equation, we now assume that $\delta A = 0$. The order parameter is a complex function so we can write in the general case $\psi = \psi_1 + i\psi_2$ where ψ_1 is the real part of the order parameter and ψ_2 is the imaginary part. The variation procedure can be carried with respect to two independent components, namely, the real and the imaginary components of the order parameter, $\delta \psi_1$ and $\delta \psi_2$. The variation of the order parameter itself is $\delta \psi = \delta \psi_1 + i \delta \psi_2$. For the complex conjugate of the order parameter $\psi^* = \psi_1 - i\psi_2$, the variation is $\delta \psi^* = \delta \psi_1 - i \delta \psi_2$. It will prove more convenient to choose $\delta \psi$ and $\delta \psi^*$ as independent variations. The variations of the real and the imaginary components are dependent and can be expressed as $\delta \psi_1 = (\delta \psi + \delta \psi^*)/2$ and $\delta \psi_2 = (\delta \psi - \delta \psi^*)/2i$. Let us illustrates that these two choices are equivalent. Consider, for example, the condensate density $n_s = |\psi|^2 = \psi \psi^* = (\psi_1 + i\psi_2)(\psi_1 - i\psi_2) = \psi_1^2 + \psi_2^2$. The variation is $\delta n_s = 2\psi_1 \delta \psi_1 + 2\psi_2 \delta \psi_2$. On the other hand, if the order parameter itself and its complex conjugate are chosen as independently varied functions, then the variation of the same quantity is $\delta n_s = \delta(\psi \psi^*) = \psi \delta \psi^* + \psi^* \delta \psi$. One can verify by direct substitution that these results are identical since $2\psi_1 \delta \psi_1 + 2\psi_2 \delta \psi_2 = \psi \delta \psi^* + \psi^* \delta \psi$. The free energy (2.54) is

$$F = \int \left(f_n + \alpha|\psi|^2 + \frac{\beta}{2}|\psi|^4 + \frac{1}{4m}\left|\left(\frac{\hbar}{i}\nabla - \frac{2e}{c}A\right)\psi\right|^2 + \frac{h^2}{8\pi}\right)d^3x$$

The terms containing the order parameter must be integrated over the volume of the sample since the order parameter is not defined outside the sample. Therefore we get

$$\delta F = \int \left(\alpha \delta n_s + \beta n_s \delta n_s + \left(\frac{\hbar^2}{4m}\right)\delta |(-i\nabla - \gamma_0 A)\psi|^2\right)d^3x \qquad (2.69)$$

The first two terms are obtained by using $n_s = |\psi|^2$. The newly introduced constant is $\gamma_0 = 2e/c\hbar$. Let us consider the third term under the integral. It can be

2.10 Ginzburg Landau Equations

transformed using the following algebra

$$\delta \, |(-i\nabla - \gamma_0 A)\,\psi|^2$$
$$= (i\nabla\psi^* - \gamma_0 A\psi^*)\delta(-i\nabla\psi - \gamma_0 A\psi)$$
$$+ (-i\nabla\psi - \gamma_0 A\psi)\delta(i\nabla\psi^* - \gamma_0 A\psi^*)$$
$$= \nabla\psi^*\nabla\delta\psi - i\gamma_0 A\nabla\psi^*\delta\psi + i\gamma_0 A\psi^*\nabla\delta\psi + \gamma_0^2 A^2\psi^*\delta\psi$$
$$+ \nabla\psi\nabla\delta\psi^* + i\gamma_0 A\nabla\psi\delta\psi^* - i\gamma_0 A\psi\nabla\delta\psi^* + \gamma_0^2 A^2\psi\delta\psi^*$$
$$= \nabla\psi\nabla\delta\psi^* + i\gamma_0 A\nabla\psi\delta\psi^* - i\gamma_0 A\psi\nabla\delta\psi^* + \gamma_0^2 A^2\psi\delta\psi^* + \text{c.c.}$$
$$\tag{2.70}$$

In the latter expression, we only explicitly keep the last four terms, namely, those which involve $\delta\psi^*$. The notation "c.c." denotes the terms which are complex conjugate all the terms preceding the symbol c.c. in the equation. All the terms collectively denoted as c.c. involve the infinitesimal variation function $\delta\psi$, which we treat independently from the variation $\delta\psi^*$. One can explicitly check that the results are equivalent to the case if ψ_1 and ψ_2 are treated as independent variation functions.

Our goal now is to collect all terms multiplied by $\delta\psi^*$ and zero them. The first term in (2.70), when placed into the integral of (2.69), can be transformed as $\int(\nabla\psi\nabla\delta\psi^*)d^3x = \int(\nabla(\delta\psi^*\nabla\psi))d^3x - \int(\delta\psi^*\Delta\psi)d^3x$ where $\Delta = \nabla\cdot\nabla$. The first term in the right side of the latter equality reduces to the surface integral of the sort $\int(\nabla(\delta\psi^*\nabla\psi))d^3x = \int(\delta\psi^*\nabla\psi)da$. It can be shown that all the integrals taken over the surface of the sample sum up to zero if one assumes that the supercurrent crossing the sample suface is zero. Thus in what follows we keep only the volume integrals and neglect all surface integrals. In a similar fashion, using the condition that $\delta\psi^* = 0$ at the surface of the volume over which the free energy density is integrated, one can transform the integral of the third term in (2.70). Namely, $\int(A\psi\nabla\delta\psi^*)d^3x = -\int(\delta\psi^*\nabla(A\psi))d^3x + \int(\delta\psi^*A\psi)da$.

Taking into account the calculation details discussed above, we can write the variation of the free energy as

$$\delta F$$
$$= \int \delta\psi^* \left(\alpha\psi + \beta n_s \psi + \frac{\hbar^2}{4m}(-\Delta\psi + i\gamma_0(A\nabla\psi + \nabla(A\psi))) + \gamma_0^2 A^2 \psi \right) d^3x$$
$$+ \int \delta\psi(\text{c.c.})d^3x \tag{2.71}$$

To simplify the expression above, the following identity is useful, that is,

$$(-i\nabla - \gamma_0 A)^2 \psi = (-i\nabla - \gamma_0 A)(-i\nabla\psi - \gamma_0 A\psi)$$
$$= -\Delta\psi + i\gamma_0 \nabla(A\psi) + i\gamma_0 A\nabla\psi + \gamma_0^2 A^2 \psi \tag{2.72}$$

Thus, the variation of the free energy of the superconductor is

$$\delta F = \int \delta \psi^* \left[\alpha \psi + \beta n_s \psi + \left(\frac{\hbar^2}{4m} \right) (-i\nabla - \gamma_0 A)^2 \psi \right] d^3x$$
$$+ \int \delta \psi (c.c) d^3x \quad (2.73)$$

As was discussed in the previous sections, Helmholtz energy reaches minimum in the thermodynamic equilibrium, provided that the external conditions are such that no work is done on the system. Typically, this means that there is no macroscopic electric field. This requirement is identical to a requirement that the phase gradients of the wavefunction do not change in time and the magnetic field is fixed also at all points of the sample. This is because if the magnetic field is changing, then an electric field is generated (according to the Maxwell equations), which can accelerate the condensate and thus can increase the energy above the minimum. Thus, we expect that in the equilibrium, the condition $\delta F = 0$ must be true for any arbitrary variation of $\delta \psi^*$. Therefore, the expression in the parentheses, which are multiplied by $\delta \psi^*$, must be zero. Thus, we arrive at the GL equation

$$\left[\alpha + \beta n_s + \left(\frac{\hbar^2}{4m} \right) (-i\nabla - \gamma_0 A)^2 \right] \psi = 0 \quad (2.74)$$

Note that if one requires that the terms multiplied by $\delta \psi$ also give zero when taken together (these terms are marked as c.c. in (2.73)), one would get a complex conjugate of the GL equation. In other words, one would get no new information, and this is because the variations $\delta \psi$ and $\delta \psi^*$ are not, in fact, independent.

The equation has a number of useful limiting solutions. If, for example, the gradients and magnetic fields are both zero, then $(\alpha + \beta n_s) \psi = 0$, which means that the superfluid density assumes its bulk equilibrium value $n_{s0} = -\alpha/\beta$.

2.11
Little–Parks Effect

The GL equation can be used to analyze the magnetic field effect on samples having multiply connected geometries. If the temperature is very near T_c, then $n_s \to 0$ and therefore the second term in the GL equation can be neglected. Then, the equation becomes linear in ψ, namely, as

$$\left[\alpha + \left(\frac{\hbar^2}{4m} \right) (-i\nabla - \gamma_0 A)^2 \right] \psi = 0 \quad (2.75)$$

Such a linearized equation can be used to find the temperature at which the order parameter starts to deviate from zero. Such temperature is defined by the condition that the expression in the parentheses is zero (then $\psi \neq 0$ is allowed).

Through this approach, one can understand the Little–Parks effect. Consider a thin-wire circular loop having its center at the origin of coordinates. The loop is resting in the x–y plane and the magnetic field is applied perpendicular to the loop, that is, along the z-axis. The solution can be found assuming that $\psi = \sqrt{n_s}\exp(i\phi)$ and assuming that n_s is constant along the loop. Then, the gradient reduces to the gradient of phase as $-i\nabla\psi = \psi\nabla\phi$. This vector is directed along the loop and its magnitude is $\sqrt{n_s}\Delta\phi/2\pi R = \sqrt{n_s}n/R$, where R is the radius of the loop and $\Delta\psi = 2\pi n$ is the phase accumulation along the loop, where n is an integer[7]. The vector potential can be taken as $\mathbf{A} = [\mathbf{B}\times \mathbf{r}]/2$, where $\mathbf{B} = \langle \mathbf{h}\rangle$ is the macroscopic magnetic field perpendicular to the loop, while $\mathbf{h}$ is the microscopic magnetic field, and $\langle \ldots \rangle$ denotes averaging over distances larger than the distance between the atoms, though smaller than the scale over which the order parameter typically changes. The vector $\mathbf{A}$ is directed along the loop also (as the gradient of the phase), and its magnitude is $A = BR/2$, where B is the magnitude of the magnetic field, which is assumed constant in space. Thus, the second power of the vector in the GL equation gives the square of it magnitude, and we get

$$\left[\alpha + \left(\frac{\hbar^2}{4mR^2}\right)\left(n - \gamma_0 B\frac{R^2}{2}\right)^2\right]\sqrt{n_s} = 0 \tag{2.76}$$

Remember that the magnetic flux through the loop is $\Phi = B\pi R^2$ and the magnetic flux quantum, in cgs units, is $\Phi_0 = hc/2e = 2\pi/\gamma_0$ or $\gamma_0 = 2\pi/\Phi_0$.[8] Also, we introduce the length scale for the order parameter variations, the so-called coherence length $\xi(T)$ defined through $\xi^2(T) = -\hbar^2/4m\alpha$, where $\alpha = \alpha(T) = (T-T_c)a_0 = a_0 T_c(T/T_c - 1)$ and a_0 is a material-specific constant. The coherence length can also be written as $\xi(T) = \xi(0)/\sqrt{1 - T/T_c}$, where $\xi(0) = \sqrt{\hbar^2/ma_0 T_c}$ is called extrapolated zero-temperature coherence length. With this new notation, the GL equation becomes

$$\alpha\left\{1 - \left[\frac{\xi^2(T)}{R^2}\right]\left[n - \left(\frac{\Phi}{\Phi_0}\right)\right]^2\right\}\sqrt{n_s} = 0 \tag{2.77}$$

As the temperature is decreased, the superfluid density starts to deviate from zero at the temperature at which the expression in parenthesis is zero. Such temperature is called the field-dependent critical temperature $T_{LP}(B)$ and is defined by the condition

$$\frac{R^2}{(n - (\Phi/\Phi_0))^2} = \xi^2(T) = \frac{\xi^2(0)}{1 - T_{LP}(H)/T_c} \tag{2.78}$$

Finally, we can write the expression for the Little–Parks field-dependent critical temperature as

$$T_{LP}(H) = T_c\left\{1 - \left[\frac{\xi^2(0)}{R^2}\right]\left[n - \left(\frac{\Phi}{\Phi_0}\right)\right]^2\right\} \tag{2.79}$$

7) Note that the increase of the phase around the loop must be equal to zero or 2π multiplied by some integer in order to make sure that the complex order parameter is single-valued
8) The magnetic flux quantum in SI units is $\Phi_0 = h/2e = 2.07 \times 10^{-15}$ Wb

Here, as everywhere, the T_c is the critical temperature of the material of the considered thin wire loop, which gives the critical temperature of the loop for zero magnetic field. The integer n must be chosen such that the Little–Parks critical temperature accepts the highest possible value, although in any field, it is lower or equal to T_c. The magnetic flux through the loop is strictly defined by the external field since the supercurrent in the loop is infinitesimally small at the critical temperature $T_{LP}(B)$. Note that the LP critical temperature oscillates with the flux because the integer number n is chosen to maximize $T_{LP}(B)$ and thus its value depends on the applied field. For example, if $\Phi = \Phi_0$, then the maximum is achieved by taking $n = 1$ and this maximum is such that $T_{LP}(\Phi_0) = T_c$. The same is true for any $\Phi = n\Phi_0$. Thus, the period of the oscillation of the LP critical temperature is the magnetic flux quantum. The minima of this periodic function correspond to the most "frustrated states," namely, to $\Phi/\Phi_0 = n_1 + 1/2$, where n_1 is an integer. At those points, the critical temperature is $T_{LP}(\Phi_0/2) + n_1\Phi_0 = T_c\{1-(n-n_1-0.5)^2[\xi^2(0)/R^2]\} = T_c[1-\xi^2(0)/4R^2]$. The maximum of the critical temperature is achieved by taking $n = n_1$ That means that the superconducting state that first occurs in the ring is such that the loop contains n_1 vortices. Note that $n = n_1 + 1$ is also a valid solution. Under a state having a vortex, we understand a state of the loop in which the phase increases by 2π on a path circling the loop. If the phase increase around the loop is 4π, then we will say that there are two (coreless) vortices trapped in the loop and so on. The estimates above show that the LP oscillation of the critical temperature has an amplitude which diminishes quadratically with the loop radius, namely, as $\propto \xi^2(0)/R^2$. The oscillations are strong if the radius is of the order of magnitude of the coherence length. The oscillations are weak if the radius of the loop is much larger than the coherence length at zero temperature.

In the discussion above, we encounter the usual complication related to the fact that $e < 0$, which lead to $\Phi_0 < 0$, as defined above. However, this is not a problem since n can also be chosen negative so that equalities like $\Phi = n\Phi_0$ remain valid even if $\Phi > 0$. An alternative way to deal with such complications is to assume $e = |e| > 0$. Such a method will be used in the next chapter to avoid citing negative critical current values.

In systems with nanowires, the LP effect was observed also. The geometry studied involves two thin film electrodes connected by two parallel wires fabricated using DNA templates. It is illustrated in Figure 2.1. It is clear from the measured $V(I)$ curves (Figure 2.1b) that the switching current shows LP oscillations while the retrapping current (which is on the negative side of the plot) does not. This is in agreement with the general view that after the switching the wire enters into the normal state sustained by the Joule heating. And, of course, the LP effect does not exist in normal-metal loops.

The most unusual characteristic of the observed oscillation is its period. For the usual LP effect, the period is defined by the area of the closed superconducting loop. If the length of the wires is L and the distance between the two wires is Y, then the expected period in magnetic field is $\Delta B = \Phi_0/LY$. This is not the case for the device under consideration.

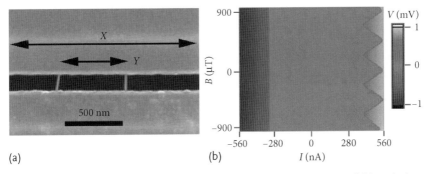

Figure 2.1 (a) An SEM micrograph of a sample with two parallel wires [105, 110, 111] made using DNA molecules as templates. The same device is illustrated on the cover of the book. The width of the electrodes is X and the distance between the wires is Y. (b) Color-coded plot of the voltage as a function of the bias current and the magnetic field, applied perpendicular to the thin film electrodes. The Little–Parks oscillation, induced by the magnetic field, is visible [105, 110, 111] as periodically occuring gray triangles which correspond to the normal state. Please find a color version of this figure in the color plates.

The experiments show that the actual period is much smaller than the estimation above. The explanation is in the occurrence of the Meissner currents in the thin film electrodes which impose strong phase gradients at the points where the wires are connected to the electrodes [105, 111, 146]. The actual period is $\Delta B \approx 0.916 \Phi_0/XY$, where X is the width of the electrodes. As unexpected as this might be, the period does not depend on the length of the wires, while it does depend on the total width, X, of the electrodes approaching the wires [105, 110, 111]. The result is valid if $X \gg Y$ and the perpendicular magnetic field penetration depth is much larger than the width of the electrodes $X \ll \lambda_\perp$. It is not difficult to satisfy these conditions. For example, in the experiments of Hopkins and coworkers [105, 110, 111], the parameters were $Y \sim 1\,\mu m$, $X \sim 15\,\mu m$, and $\lambda_\perp \sim 70\,\mu m$. See the "Problems and Solutions" section at the end of the book (Appendix C) for a more detailed discussion of the double-wire systems.

Another interesting recent development is the observation of Little–Parks oscillations in the kinetic inductance at temperatures much lower than the critical temperature [148]. The connection of the wires was similar to Figure 2.1a. The wires were included into a coplanar waveguide resonator, which is a thin film analogue of the Fabry–Pérot interferometer operating at a frequency of a few GHz. The device is so sensitive that it allows one to detect single phase slips entering the loop formed by the two wires and the two electrodes, to which the wires are connected (Figure 2.1a). In these experiments, the periodic structure of the transmission dependence on the magnetic field has the same origin as the LP critical temperature periodic oscillations. They are due to the oscillation of the supercurrent magnitude and, correspondingly, to the free energy of the superconducting loop. Such free energy oscillations cause the oscillations of the resistance at high temperatures (standard LP effect) as well as the oscillations of the kinetic inductance (low-temperature LP effect) [148].

2.12
Kinetic Inductance and the CPR of a Thin Wire

As will be shown in Section 3.1, the kinetic inductance of superconducting samples can be defined as

$$L_k = \left(\frac{\hbar}{2e}\right)\left(\frac{d I_s}{d\phi}\right)^{-1}$$

The meaning of the kinetic inductance is also explained in that section. Here, we provide expressions applicable to nanowires and argue that the kinetic inductance of a superconducting wire is proportional to its length.

To find the inductance, we need to know the current–phase relationship (CPR) of a wire. According to Tinkham [1], the supercurrent density is (Appendix C)

$$j_s = 2 e n_s v_s \left(\frac{1 - m v_s^2}{|\alpha|}\right)$$

where $n_s = |\alpha|/\beta$ and $v_s = (\hbar/2m)(\phi/L)$ according to (1.2). Here, as always, L is the wire length and ϕ is the phase difference between the ends of the wire. The critical current density is always defined as the maximum possible supercurrent. Thus, it is defined by the equation $d j_s/d v_s = 0$, or, equivalently, $d j_s/d\phi = 0$. By solving this, one gets

$$j_c = \left(\frac{4}{3\sqrt{3}}\right) e n_s \sqrt{\frac{|\alpha|}{m}} \quad \text{or} \quad e n_s = \left(\frac{3\sqrt{3}}{4}\right) j_c \sqrt{\frac{m}{|\alpha|}}$$

One more expression we require is the formula for the coherence length

$$\xi = \frac{\hbar}{2\sqrt{m|\alpha|}} \quad \text{or} \quad \sqrt{|\alpha|} = \frac{\hbar}{2\xi\sqrt{m}}$$

All of the above formulas are based on the GL theory and most of them are temperature-dependent. For example, $\alpha = \alpha_o(T - T_c)$ and $\beta = $ const.

Thus, we can exclude α and get the supercurrent density in a more convenient form, that is,

$$j_s = \frac{3\sqrt{3}}{2} j_c \frac{\xi\phi}{L}\left[1 - \left(\frac{\xi\phi}{L}\right)^2\right]$$

By multiplying the left and the right sides by the wire cross-section A_{cs}, we get the CPR of the nanowire, which is also called the Likharev CPR, as

$$I_s = \frac{3\sqrt{3}}{2} I_c \left[\left(\frac{\xi\phi}{L}\right) - \left(\frac{\xi\phi}{L}\right)^3\right]$$

To find the inductance at weak currents, we first find the derivative of the CPR for the case when $\xi\phi/L \ll 1$. In such a case,

$$\frac{d I_s}{d\phi} = \frac{3\sqrt{3}}{2} I_c \left(\frac{\xi}{L}\right)$$

Finally, the kinetic inductance of the nanowire is

$$L_k = \frac{2}{3\sqrt{3}} \frac{L}{\xi} \frac{\hbar}{2e I_c} \tag{2.80}$$

It scales linearly with the wire length.

2.13
Drude Formula and the Density of States

Here, we review some well-known formulas related to normal conductivity of metals. Let us begin with the well-known Drude formula[9] for the normal-state resistivity of the wire, namely, $\rho_n = m v_F / n e^2 l_e$, where n is the density of electrons, v_F is the Fermi velocity, and l_e is the electronic elastic mean free path. It is also known that the diffusion constant for electrons able to move in three dimensions (this is the case for nanowire which are a few nanometers in diameter) is $D = l_e v_F / 3$. Therefore, $l_e = 3D/v_F$, and so

$$\rho_n = \frac{m v_F^2}{3 n e^2 D} = \frac{2 E_F}{3 n D e^2} \tag{2.81}$$

The electronic density n is equal to the total number of states in the Fermi sphere of radius $p_F = m v_F$. Quite generally, the number of states is $n = 2 V V_F / (2\pi\hbar)^3$, where V is the volume in real space and $V_F = 4\pi p_F^3 / 3$ is the volume in the momentum space. Here, we make an implicit assumption that the number of states equals the number of electrons since the sphere in the momentum space is chose of radius p_F and all states are filled within such a sphere according to the Fermi distribution at zero temperature. The factor 2 appears because there are two possible spin orientations, namely, up and down. Since n, by definition, is the number of electrons in a unit volume, we set $V = 1$. Thus, $n = p_F^3 / 3\pi^2 \hbar^3$. The Fermi momentum p_F is related to the Fermi energy E_F as $p_F = (2 m E_F)^{1/2}$. Thus, $n = (2 m E_F)^{3/2} / 3\pi^2 \hbar^3$. The total density of states is, by definition, $dn/d E_F$. We will define N_0 as the density of states per a single spin direction which is twice smaller than the total density of states. Therefore, $N_0 = (1/2)(dn/d E_F) = m(2m E_F)^{1/2} / 2\pi^2 \hbar^3 = (3/4)(2 m E_F)^{3/2} / 3\pi^2 \hbar^3 E_F$. This can also be written as

$$N_0 = \frac{3n}{4 E_F} \tag{2.82}$$

which is the most useful presentation for us here. Another form for the density of states of electrons of a given spin orientation is $N_0 = m p_F / 2\pi^2 \hbar^3 = m^2 v_F / 2\pi^2 \hbar^3$.

Combining (2.81) and (2.82), one gets $\rho_n = 1/2 D N_0 e^2$. We also take into account that $R_n = \rho_n L / A_{cs}$. Thus, we get $R_n = L / 2 A_{cs} D N_0 e^2$. The quantum resistance

---

9) The Drude formula was proposed in 1900 and has proved to be remarkably successful. An interesting historic fact is that Paul Drude was responsible for introducing, in 1894, the symbol c for the speed of light in vacuum.

$R_q = h/4e^2 = \pi\hbar/2e^2$. Therefore,

$$D N_0 A_{cs} = \frac{L}{2 R_n e^2} = \left(\frac{R_q}{R_n}\right)\left(\frac{L}{\pi\hbar}\right) \qquad (2.83)$$

In theoretical units, $\hbar = 1$, and thus the formula becomes

$$D N_0 A_{cs} = \left(\frac{L}{\pi}\right)\left(\frac{R_q}{R_n}\right) \qquad (2.84)$$

3
Stewart–McCumber Model

The Stewart–McCumber (SM) [95, 96] model was developed to describe Josephson junctions, though it has wide applicability, including superconducting nanowires. In particular, it provides a good approximation to the potential energy of superconducting wires which are sufficiently short, so that the charging of the capacitance between the electrodes connected to the wire ends contributes significantly to the total free energy of the device. The essence of the SM model is to split the total bias current into three components, namely, the supercurrent, the normal current, and the displacement current, that is, the current charging the capacitance formed between the electrodes of the device. These three components are equated according to the current conservation law, to the total bias current I entering the junction (which could be a superconducting current also, provided that the electrodes are made of a superconducting material). The model is relatively simple, allows analytic solutions, and is very efficient in describing a great multitude of phenomena occurring in superconducting devices. All equations of this chapter are valid in a SI system of units.

To understand the model, we first introduce Josephson junctions for the description of which the SM model was developed in the first place. A Josephson junction (JJ) is a devices made of two superconducting electrodes separated by a thin layer of an insulator [15]. Such superconductor-insulator-superconductor (SIS) "sandwiches" are also called SIS junctions. They are similar to Giaever junctions (GJ) [77]. The difference is the following: In the GJ, the insulating layer separating two superconductors is rather thick (a few nanometers typically), and thus only single electrons, but not the pairs, have a noticeable rate of tunneling. Thus, in GJs, the current is mainly carried by electrons which are not included into the BCS condensate, that is, by quasiparticles or bogoliubons. Therefore, at low temperatures, a noticeable current through a GJ occurs only at voltages larger than the superconducting energy gap expressed in electron volts [1]. Note that according to the BCS theory, the gap in the spectrum of a superconductor is related to its critical temperature as $\Delta = 1.76 k_B T_c$.

The most popular type of SIS JJs is the aluminum-aluminum oxide-aluminum junction (Al-Al$_2$O$_3$-Al), pioneered by Giaever [77]. Such junctions are used in many applications, for example, superconducting quantum interference devices (SQUID), photon detectors, and superconducting qubits. To fabricate small scale

Superconductivity in Nanowires, First Edition. Alexey Bezryadin.
© 2013 WILEY-VCH Verlag GmbH & Co. KGaA. Published 2013 by WILEY-VCH Verlag GmbH & Co. KGaA.

devices having SIS junctions, the conventional Niemeyer–Dolan technique [78, 79] is typically used. This approach involves such fabrication steps as electron beam lithography, angular evaporation of pure Al, oxidation in situ, under a low pressure of oxygen (unlike in the original experiments by Giaever where the oxidation was done in air), and lift-off. Any detailed discussion of such standard fabrication techniques would go beyond to scope of this book. The Giaever junctions allow one to measure the superconducting energy gap since they exhibit no current flow (or a very low current) for voltages less than $2\Delta(T)/e$ and a greatly increased current for $V > 2\Delta/e$. On the other hand, properties of Josephson junctions are qualitatively different from GJs since JJs exhibit a measurable supercurrent, that is, a current flowing at zero voltage. This is because the oxide in JJs is thinner, typically about 1 nm or less, and thus the probability of simultaneous tunneling of two electrons is significant. Since the number of electrons in the BCS condensate is even, and taking out just one electron from the condensate costs much energy (i.e., the energy gap $\Delta(T)$), the possibility of moving two electrons at once is crucial for the supercurrent to flow. As is usual for the BCS condensate, if it flows, it flows without friction because the condensate is the ground state of the system, and there is no dissipation when the system is in the ground state. Therefore, the current flows through JJs without any voltage applied, which means the resistance is zero. Note that a nonzero resistance can occur if phase slips are present since they break the coherence of the condensate. If the phase of the wavefunction of the condensate is constant within each of the electrodes, implying that magnetic field is zero, then the supercurrent I_s through the JJ is defined only by the phase difference ϕ (also called the phase of the junction) between the superconducting electrodes and is expressed as $I_s = I_c \sin \phi$ [15]. Such an expression is called the current–phase relationship (CPR) of the superconducting device. The critical current of the SIS junction is expressed through its normal-state (i.e., measured above T_c) resistance $R_{n,JJ}$ of the tunnel junction as $I_c = (\pi \Delta / 2 e R_{n,JJ}) \tanh(\Delta / 2 k_B T)$ (Ambegaokar–Baratoff formula). If a voltage is applied to JJ, the phase evolves according to Gor'kov's phase evolution formula (1.5).

At this point, we have to resolve some technical difficulties. According to (1.5), if $V > 0$, then $\dot{\phi} < 0$ since the electrons are negatively charged and so $e < 0$. Suppose at $t = 0$, we have $\phi = 0$ and so $I_s = 0$. If we allow a small time increment, then ϕ becomes negative because $\dot{\phi} < 0$. That implies $\sin \phi < 0$ and $I_s < 0$, assuming $I_c > 0$. This is not right since a small positive voltage should accelerate the condensate slightly and produce a small positive current. To fix the paradox, one could chose $I_c < 0$, taking into account the fact that $e < 0$. Technically, this is the correct resolution of the difficulty. However, this makes formulas unnecessarily complicated. Thus, we follow the tradition and assume $I_c > 0$. Additionally, to avoid contradictions, we will assume in the SM model and in related topics that $e > 0$. In this case, if $V > 0$, then the phase increases according to $\dot{\phi} = 2eV/\hbar$ and therefore, according to $I_s = I_c \sin \phi$, the supercurrent remains positive (as long as $\phi < \pi$).

One might have an impression that the paradox reoccurs as $\phi > \pi$ since the current becomes negative, although the voltage remains fixed and positive by our initial assumption. This is a separate question. To resolve it, remember that the phase is defined modulo 2π. It is thus correct to say that when it reaches the point $\phi = \pi$, the order parameter in the middle of the junction goes to zero and the phase difference shifts by 2π and becomes negative, that is, $-\pi$. Thus, the current shifts to negative values. So far there is no contradiction. The usual convention is different though. It is assumed that the phase keeps increasing above π (assuming, as before, $V = $ const and $V > 0$), and this convention creates the paradoxical effect of negative currents at positive phase difference values.

To summarize, in the discussion of the SM model of JJs, the following formulas will be used: $e = |e|$, $I_c > 0$, $I_s = I_c \sin \phi$, and $\dot{\phi} = 2eV/\hbar$. Such a set of formulas provides correct predictions for the JJ behavior. The purpose of accepting the positive electronic charge is to ensure a positive critical current and also a positive magnetic flux quantum, $\Phi_0 = h/2e > 0$ (in SI units) or $\Phi_0 = hc/2e > 0$ (in Gaussian units).

If $V = $ const and $V \neq 0$, then the supercurrent is not constant but, rather, it oscillates periodically in time, as $I_s(t) = I_c \sin(2eVt/\hbar)$. The time average of such dependence is zero, $\overline{I_s(t)} = 0$. Therefore, it appears as though the junction is insulating since for a fixed nonzero voltage, one expects zero net current. This fact ensures that the dissipation or Joule power P_J is zero since $\overline{P_J} = \overline{VI_s} = V\overline{I_s} = 0$.

A variety of interesting and important phenomena occur in superconducting circuits incorporating JJs. The list includes such exotic phenomena as macroscopic quantum tunneling [33, 80, 81], the Schmid–Bulgadaev transition, which is a superconductor to insulator quantum phase transition [30, 82, 83, 92, 93] (the last reference is the experimental confirmation), and various qubit effects [36].

To understand these and other phenomena, one needs a realistic model describing SIS junctions (Figure 3.1a). Such a universal model was developed by Stewart and McCumber (SM) [95, 96]. The SM model takes into account the shunting of the supercurrent by the normal and the displacement currents.[1] Here, we present a basic level discussion of the SM model, illustrated in Figure 3.1b. It is based on the current conservation principle and assumes that the total current can be presented as a sum of the displacement current $\dot{Q}$ charging the electric capacitance C formed between the electrodes of the junction, the supercurrent I_s and the total dissipative (also called "normal," meaning nonsuperconducting) current I_n. Thus, the current-conservation equation for the considered device is $I = I_s + I_n + dQ/dt$, where I is a constant bias current and Q is the charge on the capacitor formed between the superconducting electrodes of the junction. Another important assumption of the SM model is that the phase is constant within each electrode. Thus, the condensate state can be described by just one number, namely, the phase difference ϕ between the plates of the junction. It is also assumed that the voltage within each of the plates (electrodes) is the same. Then, the supercurrent is $I_s = I_c \sin \phi$ and the normal current is $I_n = V/R_n$. Here, R_n is the normal resistance, that is, the resistance which includes all nonsuperconducting channels, such as the current

---

1) For a more technical in-depth theoretical analysis of the SM model, the reader is referred to the paper by Joyez et al. [13].

of bogoliubons through the tunnel barrier forming the junction and the current of normal electrons through the shunt if the device has some shunt (the normal shunt is shown schematically as a red band in Figure 3.1a). The voltage on the junction is related to the phase as $V = \hbar\dot\phi/2e$ and it also relates to the charge as $CV = Q$. By combining these formulas with the current conservation law, we get the equation of motion of the phase difference (the Stewart–McCumber or SM equation)

$$\frac{\hbar C \ddot\phi}{2e} + \frac{\hbar\dot\phi}{2e R_n} + I_c \sin\phi - I = 0 \tag{3.1}$$

For $I \leq I_c$, the SM equation has a time-independent (stationary) solution for ϕ defined by $I_c \sin\phi = I$, which means that in equilibrium, all bias current flows as the supercurrent through the junction and the voltage is zero. If $I > I_c$, then all solutions are time-dependent, and the voltage is not zero on the dives, meaning that a part of the bias current flows as a dissipative current.

In many practically interesting cases, the device is operated at low temperatures, where the concentration of bogoliubons in the electrodes is negligible, and thus the current I_{nb} can be neglected. If also there is no normal metal shunt resistor in the device, then the junction is not damped. In this case, the equation of motion is

$$\frac{\hbar C \ddot\phi}{2e} + I_c \sin\phi - I = 0 \tag{3.2}$$

It is now possible to analyze small oscillations near the equilibrium solution, which is ϕ_0 defined by $I_c \sin\phi_0 = I$, where I is some constant bias current supplied to the device such that $I < I_c$. Now, we denote the deviation from the equilibrium as θ, such that $\phi(t) = \phi_0 + \theta(t)$. If the deviation is small, then the Taylor expansion gives $\sin(\phi) \approx \sin(\phi_0) + (d\sin\phi/d\phi)(\phi - \phi_0)$. Here, the derivative in the last term is supposed to be computed for the point of equilibrium, that is, $\phi = \phi_0$. Therefore, we get $I_c \sin\phi \approx I + I_c \theta \cos\phi_0$. In this linear approximation, assuming that $\theta(t)$ is infinitesimally small, we can write the SM equation as

$$\frac{\hbar C \ddot\theta}{2e} = -I_c \theta \cos\phi_0 \tag{3.3}$$

The general solution is (one can verify that this is the solution by substituting it into the equation)

$$\theta(t) = \theta_m \cos(\omega_p t + c_1) \tag{3.4}$$

where

$$\omega_p = \sqrt{\frac{2e I_c \cos\phi_0}{\hbar C}} \tag{3.5}$$

is called the plasma frequency because it corresponds to small plasma oscillation on a Josephson junction. Basically, the JJ acts as a harmonic oscillator having the

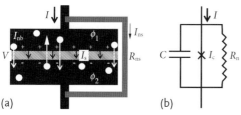

Figure 3.1 Stewart–McCumber model for Josephson junctions. (a) A schematic cross-section of a superconductor-insulator-superconductor (SIS) sandwich junction. It includes two superconducting electrodes (black) separated by a very thin insulating oxide layer (dashed region). The bias current, I, is shown by the large black arrow. The current through the junction contains the supercurrent I_s (small black arrows) and the bogoliubon current I_{nb} (white arrows). The bogoliubons are shown as white circles. The optional Ohmic normal shunt, shown in red, has a resistance R_{ns} and can carry a current of normal electrons I_{ns}. Moreover, the electrodes separated by the oxide form a capacitor which becomes charged if the currents through the junction and the shunt do not altogether amount to the bias current. (b) An equivalent electrical circuit for a JJ, called the Stewart–McCumber (SM) model [95, 96] or the resistively and capacitively shunted junction (RCSJ) model. The model is based on the current conservation law $I = I_n + I_s + \dot{Q}$. Here, $I_s = I_c \sin\phi$ is the supercurrent and I_n is the total dissipative current, which includes the current of bogoliubons through the junction and the current of normal electrons in the shunt, that is, $I_n = I_{nb} + I_{ns}$ (see part (a) for the notations). The displacement current $\dot{Q}$ is the current that does not flow through the junctions, but produces extra charge accumulated on the plates of the capacitor, that is, on the electrodes forming the junction. One of the electrodes gets charge Q and the other always gets charge $-Q$. The total normal current resistance R_n is defined as a parallel connection of the shunt R_{ns} and the quasiparticle tunnel resistance R_{nb} as $R_n^{-1} = R_{nb}^{-1} + R_{ns}^{-1}$. The normal resistance can also be defined such that the total normal current is $I_n = V/R_n$ and the Joule heating in the entire system is $P_J = V^2/R_n = I_n^2 R_n$. Here, the voltage V is, as usual, the difference between the electric potentials of the electrodes forming the junction. The potential as well as the superconducting phases ϕ_1 and ϕ_2 are assumed constant within each of the electrodes. The phase difference is simply $\phi = \phi_1 - \phi_2$, where ϕ_1 is the phase of the order parameter in the top electrode and ϕ_2 is the phase in the bottom electrode. Please find a color version of this figure in the color plates.

resonance frequency that can be tuned with the bias current.[2] The amplitude θ_m and the phase shift c_1 of the oscillation are determined by the initial conditions. The term "plasma" is used to emphasize the fact that during these oscillations, the electrodes of the junction get partially ionized since the electrons (in the form of the BCS condensate) flow periodically from one electrode to the other, while positively charged ions stay in place, of course.

The equilibrium value of the phase difference is defined by the bias current which is assumed fixed. The equilibrium phase is governed by the following simple equations $\cos\phi_0 = \sqrt{1-\sin^2\phi_0} = \sqrt{1-I^2/I_c^2}$. Thus, the small-amplitude

---

2) At large amplitudes, the frequency of these plasma oscillations changes. Such oscillations are used as the basis for phase qubits.

oscillation frequency can be written as

$$\omega_p(I) = \left(\frac{2eI_c}{\hbar C}\right)^{1/2} \left(1 - \frac{I^2}{I_c^2}\right)^{1/4} \tag{3.6}$$

The zero-bias plasma frequency is therefore $\omega_p(0) = \sqrt{(2eI_c/\hbar C)}$.
For the solution of (3.4), we can find the voltage and the charge as

$$Q = CV = \frac{C\hbar\dot{\phi}}{2e} = \frac{C\hbar\dot{\theta}}{2e} = -\theta_m \left(\frac{C\hbar}{2e}\right) \omega_p \sin(\omega_p t + c_1) \tag{3.7}$$

Thus, if the amplitude of the phase oscillations is θ_m, then the amplitude of the charge oscillations is

$$Q_m = \theta_m \left(\frac{C\hbar}{2e}\right) \omega_p = \theta_m \left(\frac{\hbar C I_c}{2e}\right)^{1/2} \left(1 - \frac{I^2}{I_c^2}\right)^{1/4} \tag{3.8}$$

(The amplitude is just the maximum deviation from the equilibrium or the factor by which the sin or cos function is multiplied.)

Consider an example. Suppose the critical current is 1 μA and the capacitance is 10 fF. Then, one can quickly compute that the plasma frequency at zero-bias is 88 GHz (remember that to get the frequency in Hz, one has to divide the angular frequency ω_p by 2π). Suppose one needs to reduce it by factor 10. What current has to be chosen? From (3.6), one gets $I = 0.99995 I_c$. Approaching the critical current with such precision is usually not possible in practice, and thus the bias current cannot be used to control the plasma frequency in such a wide range. What could be practically possible is to set the bias current somewhere within a small percentage below the critical current. Assume, for example, $I = 0.99 I_c$. Then, $\omega_p(I)/\omega_p(0) = 0.38$. Thus, it is quite realistic to reduce the plasma frequency roughly by a factor of two or three using the bias current method, but not by much more than that.

Now, let us understand the relationship of the phase and the energy. As always, it is up to us to choose which state corresponds to zero energy. Here, we choose the zero energy reference state as the state which has $Q = 0$ and the phase defined by the bias current as $\phi = \phi_0 = \arcsin(I/I_c)$ so that the bias current equals the supercurrent. Suppose now that at $t = 0$, the system is prepared in a nonequilibrium state such that $\phi = \phi_0$ and the charge on the capacitor is $Q = Q_m > 0$. This state is considered a nonequilibrium state since the equilibrium state is the state having $\phi = \phi_0$ and $V = Q/C = 0$. The usable energy of such a nonequilibrium state is $G = Q_m^2/2C$, which is simply the electrostatic energy stored in the capacitor. We use Gibbs energy since the system is current-biased. Since the state having $Q = Q_m > 0$ is a nonequilibrium state, an oscillation of the charge and phase occurs at $t > 0$, according to the SM equation, such that $Q(t) \leq Q_m$ and $\theta(t) \leq \theta_m$, where θ_m is the amplitude of the phase oscillations. In the course of such oscillations, as usual for oscillatory systems, the electrostatic (or the "kinetic") energy is transformed into the energy of the condensate (to be called the "potential energy") and vice versa. Since, here, we assume zero dissipation in the system (i.e.,

$R_n \to \infty$), the energy is conserved. If $Q(t_1) = 0$ for some moment in time t_1, then $\theta(t_1) = \theta_m$, meaning that all energy of the system is stored at this particular moment of time in the potential energy of the condensate and no energy is stored as the Coulomb energy of the charged capacitor. Since $G(t) = \text{const}$ for this device, because there is no dissipation, we can write $G(0) = G(t_1)$ or $G_C(Q_m) = G_s(\theta_m)$, in which case we denote G_C as the Coulomb energy of the junction and G_s is the superconducting condensate energy of the junction. Our goal now is to find the function G_s. Using (3.8), it is possible to express the Coulomb energy through the phase difference as

$$G_C(Q_m) = \frac{Q_m^2}{2C} = \theta_m^2 \left(\frac{\hbar I_c}{4e}\right)\left(1 - \frac{I^2}{I_c^2}\right)^{1/2} \tag{3.9}$$

In some cases, such as the computations of the energy barrier for phase slips, one needs to know the minimum amount of work needed to change the phase from the equilibrium value to the value corresponding to the maximum of the free Gibbs energy under fixed bias currents. The expression above gives the answer to such questions if the Coulomb energy at the arrival to the destination condition is zero. Thus, Gibbs energy, for the cases when the charging (Coulomb) energy is zero, is

$$G_s(\theta) = \theta^2 \left(\frac{\hbar I_c}{4e}\right)\left(1 - \frac{I^2}{I_c^2}\right)^{1/2} \tag{3.10}$$

The total usable energy, including the Coulomb energy, is

$$G(Q, \theta) = \theta^2 \left(\frac{\hbar I_c}{4e}\right)\left(1 - \frac{I^2}{I_c^2}\right)^{1/2} + \frac{Q^2}{2C} \tag{3.11}$$

The first term will be referred to as the potential energy and the second term as the kinetic energy since it is proportional to the square of the time derivative of the phase, that is, proportional to $\dot{\theta}^2 = \dot{\phi}^2$ (remember $Q \sim V\dot{\phi}$).

Now, just consider the potential energy, which, generally speaking, is used to compute barriers. The corresponding expression is the one in which the charging energy is set to zero, that is, (3.10). Traditionally, only this term is referred to as Gibbs energy of the Josephson junction and usually denoted by G, not G_s. We will follow this tradition and in what follows, write G instead of G_s, while the complete Gibbs energy, which includes the charging energy, will not be used at all, unless explicitly introduced into the discussion. Gibbs energy can be written therefore as

$$G(\phi) = (\phi - \phi_0)^2 \left(\frac{\hbar I_c}{4e}\right)\left(1 - \frac{I^2}{I_c^2}\right)^{1/2} \tag{3.12}$$

On the other hand, for arbitrarily large values of the phase difference, it is known that

$$G = \left(\frac{\hbar}{2e}\right)[I_c(1 - \cos\phi) - I\phi] \tag{3.13}$$

This function is usually called the "washboard potential" because it is similar to a tilted washboard, the tilt being proportional to the bias current. Let us verify that this general expression is in agreement with (3.12), which is true only for small deviations from the equilibrium. The equilibrium point corresponds to the minimum of G, meaning $dG/d\phi = 0$ or $I_c(\sin \phi) - I = 0$. The solution, as we have seen above, is defined by $\sin \phi_0 = I/I_c$. Very near this point, according to Taylor, we can write $\cos \phi = \cos \phi_0 - \sin \phi_0 (\phi - \phi_0) = \cos \phi_0 - (I/I_c)\theta$. Thus, for small deviations θ, $G = (\hbar/2e)((I_c - I_c \cos \phi_0 + I\theta) - I\theta - I\phi_0) = (\hbar/2e)(I_c - I_c \cos \phi_0 - I\phi_0) = G(\phi_0)$, which is a constant since I is assumed constant. So, in the first (linear) approximation, Gibbs energy is constant. This is what one expects since the Taylor series is taken near the minimum of Gibbs energy. This result does not contradict (3.12) because that expression deviates from zero only in the second (quadratic) order and also because the minimum value itself was set to zero (the energy was counted from the bottom of the minimum).

Therefore, let us try to take into account the second correction as $\cos \phi = \cos \phi_0 - (I/I_c)(\phi - \phi_0) - (1/2)(\phi - \phi_0)^2 \cos \phi_0$. Within this (quadratic) approximation, Gibbs energy is $G = (\hbar/2e)(I_c - I_c \cos \phi_0 - I\phi_0 + (1/2)I_c(\phi - \phi_0)^2 \cos \phi_0) = G(\phi_0) + (\hbar I_c/4e)(\phi - \phi_0)^2 \cos \phi_0$. This is exactly what we got in (3.12), except the constant offset $G(\phi_0)$, which can be chosen arbitrarily in fact. This agreement justifies, to some extent, the general expression (3.13). In particular, it justifies the choice of the second term in (3.13) as $I\phi$ and not as $I_s\phi = I_c\phi \sin \phi$. Note that in the case of a long superconducting wire, the Gibbs energy has a term $I_s\phi$ (see (2.42), (2.52), and (2.53)) [1]. Thus, there is a difference between two types of considered superconducting systems, namely, in Gibbs energy of JJs, the linear term involves the total current while the linear term of Gibbs energy of superconducting wires involves the supercurrent only. This difference is due to the fact that in JJs, the current that does not flow as the supercurrent through the junction charges the junction and thus increases its Coulomb energy, which is a part of its usable free energy. This fact might explain why the barrier for phase slips in a JJ scales as $(1 - I/I_c)^{3/2}$ while it is predicted to scale as $(1 - I/I_c)^{5/4}$ in infinitely long nanowires, in which case the Coulomb energy was neglected.

3.1
Kinetic Inductance and the Amplitude of Small Oscillations

Generally speaking, the meaning of inductance is the inertia of an electric circuit to the acceleration of the current in it. For example, if one considers a solenoid, the current in it is known to accelerate in time if voltage is applied according to the equation (in SI units) $L_s \dot{I} = V$ or $dI/dt = V/L_s$, where L_s is the inductance of the solenoid. Thus, the larger the inductance is, the slower the current acceleration is, for a given voltage. Note that in the normal solenoid, the kinetic energy of the electrons is negligible, while the energy of the magnetic field associated with the current is large. Thus, the current acceleration is a process in which the source does work on the system (solenoid) and the work goes into the creation of the magnetic field in the solenoid.

In nanoscale superconducting devices, the currents are usually small and the associated magnetic fields are negligible. Thus, the work of the source goes into the acceleration of the condensate, that is, the work is equal to the increase of the kinetic energy of the condensate. By analogy with the normal solenoid example above, the kinetic inductance of a superconducting device is defined as $L_k = V/\dot{I}_s$. Let us assume that the CPR is, in general, some smooth function of the phase difference, $I_s = I_s(\phi)$ and, therefore, $dI_s/dt = (dI_s/d\phi)(d\phi/dt)$. As always, the phase difference rate of change is defined by the voltage as $V = \hbar\dot\phi/2e$. Then, the kinetic inductance (the self-inductance to be more precise) is

$$L_k = \frac{V}{\dot{I}_s} = \frac{\hbar\dot\phi}{2e(dI_s/d\phi)\dot\phi} = \frac{\hbar}{2e(dI_s/d\phi)} \quad (3.14)$$

Note that a system of a superconducting weak link connected to two electrodes having mutual electric capacitance C is usually characterized by so-called plasma frequency ω_p. The plasma frequency is defined such that $\omega_p/2\pi$ is the frequency of small current and charge oscillations which occur in the superconducting device if, for example, the initial condition is such that the voltage on the electrodes is not zero. If the plasma frequency is known, then the kinetic inductance can be defined as usual as

$$L_k \equiv \frac{1}{C\omega_p^2} \quad (3.15)$$

Since dissipation or friction change the resonance frequency of an oscillator, the equation above is only exact if there is no dissipation. This means that there is no normal shunt in parallel with the weak link. The plasma frequency of a superconducting device is a very important characteristic since it defines the attempt frequency for phase slips and, also, defines the energy scale of zero-point current and charge fluctuations in the device.

For a Josephson junction, the CPR is sinusoidal, $I_s(\phi) = I_c \sin\phi$, so that $dI_s/d\phi = I_c \cos(\phi)$. Thus, the kinetic inductance of a JJ is

$$L_k = \frac{\hbar}{2eI_c \cos\phi} \quad (3.16)$$

In the case when the current is weak, the kinetic inductance is

$$L_k = \frac{\hbar}{2eI_c} \quad (3.17)$$

because $\cos\phi \approx 1$ if $\phi \approx 0$, corresponding to $I_s \approx 0$.

So the plasma frequency for the JJ for zero bias is

$$\omega_p = \frac{1}{\sqrt{CL_k}} = \sqrt{\frac{2eI_c}{C\hbar}} \quad (3.18)$$

For example, if the critical current is 1 µA, the inductance in the limit of zero current is 0.3 nH. If the capacitance is 10^{-13} F, then the plasma frequency is $\omega_p/2\pi = 28\,\text{GHz}$.

Knowing the inductance, one can compute the plasma frequency for any current. It is known that for an LC-circuit, the angular frequency is generally $1/\sqrt{LC}$. For a JJ, the inductance is mostly contained in the kinetic inductance of the device. Thus, the plasma frequency is $\omega_p = 1/\sqrt{L_k C} = \sqrt{2eI_c \cos\phi/\hbar C}$, which is the same as was derived above from the current-conservation equation (see (3.5)). Note that here, that is, in the formulas involving the kinetic inductance, the phase difference ϕ represents the equilibrium value defined by the bias current.

3.2
Mechanical Analogy for the Stewart–McCumber Model

The mechanical analogy is useful for the purpose of developing physical intuition about JJs. The mechanical analogy, that is, a one-to-one correspondence with the so-called "phase particle" will help us to apply the laws of quantum mechanics to to the description of JJs. The analogy follows from the SM dynamic equation (3.1), which can be written as

$$\left[\frac{\hbar^2 C}{(2e)^2}\right]\ddot{\phi} + \left[\frac{\hbar^2}{R_n(2e)^2}\right]\dot{\phi} + (I_c \sin\phi - I)\left(\frac{\hbar}{2e}\right) = 0 \tag{3.19}$$

or

$$m_p \ddot{\phi} = -\eta \dot{\phi} - \left(\frac{I_c \hbar}{2e}\right)\sin\phi + I\frac{\hbar}{2e}$$

where the effective "phase particle" mass is defined as $m_p = \hbar^2 C/(2e)^2$, so that the left term appears as mass times acceleration. Note that the effective mass is proportional to the parallel capacitance of the junction C. We also have introduced the damping coefficient

$$\eta = \frac{\hbar^2}{(2e)^2}\frac{1}{R_n} = G_n \frac{\hbar^2}{(2e)^2} = \frac{\hbar}{2\pi}\frac{R_q}{R_n}$$

The definition shows that the damping is proportional to the normal conductance of the shunt, $G_n = 1/R_n$ and inversely proportional to the normal resistance R_n. This means that plasma oscillations become more and more damped if more and more normal electrons couple to the oscillating condensate. The reason is that as the condensate flows, a charge builds up on the electrodes, causing some voltage and the electric field to occur. This voltage accelerates normal electrons, which then dissipate their kinetic energy into heat. Thus, the energy of the plasma oscillations converts into heat with a rate proportional to G_n.

Now, to stress the analogy with classical mechanics, we introduce three effective forces, namely, the damping (friction) force $F_d = -\eta\dot{\phi}$, the phase difference force

$F_\phi = -(I_c\hbar/2e)\sin\phi$, and the bias current force $F_b = I\hbar/2e$. Also, let us look at ϕ as though it is the position of the effective phase particle. We will denote this position as x_p, and define it simply as $x_p = \phi$. In that case, $F_d = -\eta \dot{x}_p$ and $F_\phi = -(I_c\hbar/2e)\sin x_p$. Then, the dynamics equation of the JJ is the Newton equation

$$m_p \ddot{x}_p = F_d + F_\phi + F_b = -\eta \dot{x}_p - \left(\frac{I_c\hbar}{2e}\right)\sin x_p + \frac{I\hbar}{2e} \quad (3.20)$$

The fact that a classical particle and the JJ are both described by the same differential equation implies that the known classical-particle solution and methods of solving the equation apply to JJs.

As is usual in classical physics, the conservative forces, that is, F_ϕ and F_b, can be represented by corresponding potentials. The potential corresponding to the force F_ϕ is $U_\phi = E_J(1 - \cos x_p)$, so that, as usual, $F_\phi = -dU_\phi/dx_p$. Here, $E_J = \hbar I_c/2e$ is called the Josephson energy. The bias potential energy is $U_b = -I\hbar x_p/2e$ so that $F_b = -dU_b/dx_p = I\hbar/2e$. The total potential energy, which is, in fact, the Gibbs free energy of the current-biased junction, is

$$U_{wb} = U_\phi + U_b = E_J[(1 - \cos x_p) - i_b x_p] \quad (3.21)$$

where $i_b = I/I_c$ is the normalized bias current which defines the tilt. The potential energy U_{wb} is frequently called the "tilted washboard potential" (Figure 3.2) because it resembles a hand-washing board.

Using these notations, the equation of motion of the SM model becomes

$$m_p \ddot{x}_p = -\frac{dU_{wb}}{dx_p} - \eta \dot{x}_p \quad (3.22)$$

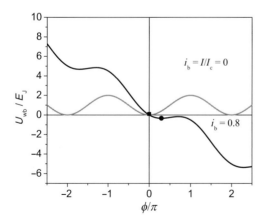

Figure 3.2 The washboard potential of the Stewart–McCumber model. The normalized potential energy of the phase particle is plotted versus the normalized phase difference $\phi/\pi = x_p/\pi$. The untilted curve corresponds to the zero-bias current. The tilted curve corresponds to the bias current $I = 0.8I_c$. The equilibrium position of the phase particle, which is denoted by x_{eq}/π in this normalized graph is shown by the black square, for $I = 0$, and by the black circle, for $I = 0.8I_c$.

Equation (3.22) is the same as the Newton equation for a single particle moving in a tilted washboard potential. The analogy between the MS model and the Newtonian particle is contained in the fact that the same differential equation describes both systems. The equation tells us that the mass of the particle times its acceleration equals the force. The damping force, which is proportional to the velocity, slows down the particle and dissipates its energy. In the rest of this section, we will assume that there is no dissipation in order to simplify the analysis.

At the critical current, $i_b = 1$, local minima in the potential U_{wb} disappear. In other words, the equation for the minima $dU_{wb}/dx_p = E_J(\sin x_p - i_b) = 0$ does not have solutions for $i_b > 1$. Without metastable states, the particle cannot be stationary. Since the voltage on the junction is $V = (\hbar/2e)(dx_p/dt)$, the absence of time-independent solutions means that the voltage on the junction cannot be zero. That means that superconductivity in the simple sense of $V = 0$ at $I > 0$ cannot exist if $I > I_c$.

If $i_b < 1$, then the stationary solution, $x_p = x_{eq}$, can be easily found. It is given by $dU_{wb}/dx_p = E_J(\sin x_{eq} - i_b) = 0$, which means $\sin x_{eq} - i_b = 0$. We can immediately find the frequency of small oscillations near the point of equilibrium. Indeed, according to the analogy with the physics of a single point mass particle, the frequency is $\omega = \sqrt{k_p/m_p}$, where $k_p = d^2 U_{wb}/dx_p^2|_{x_p=x_{eq}}$ is the effective "spring constant" corresponding to the equilibrium position of the particle, x_{eq}. Note that the potential energy has a cosine term and a linear term. The linear term disappears when the second derivative is taken. Thus, we simply have $k_p = E_J \cos x_{eq}$. Since $\sin x_{eq} = i_b$, therefore $\cos x_{eq} = \sqrt{1 - i_b^2}$. Thus, the frequency is

$$\omega = \sqrt{\frac{k_p}{m_p}} = \sqrt{\frac{E_J \cos x_{eq}}{m_p}} = \sqrt{\frac{E_J}{m_p}} \left(1 - i_b^2\right)^{1/4} \tag{3.23}$$

By reversing the analogy and returning to the usual variable, we can derive the final expression for the small oscillations of the phase difference near the equilibrium as

$$\omega_p(I) = \frac{2e}{\hbar}\sqrt{\frac{E_J}{C}}\left(1 - \frac{I^2}{I_c^2}\right)^{1/4} = \sqrt{\frac{2eI_c}{\hbar C}}\left(1 - \frac{I^2}{I_c^2}\right)^{1/4} \tag{3.24}$$

This is in agreement with (3.6), as expected. The expression is valid at any $I < I_c$. The oscillations correspond to a periodic flow of the condensate from one electrode to the other one and back, like a pendulum. Since the charge displacement is involved, such oscillations are called plasma oscillations and the frequency is the plasma frequency. The expression for the frequency can be also written as

$$\omega_p(I) = \omega_p(0)\left(1 - \frac{I^2}{I_c^2}\right)^{1/4} \tag{3.25}$$

where the plasma frequency at zero-bias is defined as $\omega_p(0) = \sqrt{\frac{2eI_c}{\hbar C}}$.

3.2.1
Defining the Supercurrent Through Helmholtz Free Energy

The forces introduced above, for example, F_ϕ and F_b, should be regarded as generalized forces. Their meaning is that the work on the system is the force times displacement in the space of the junction's phase variable ϕ. For example, the infinitesimal work of the phase difference force is $F_\phi d\phi$. Our definition of the phase difference force is $F_\phi \equiv -dU_\phi/d\phi = -E_J \sin\phi$ since $U_\phi = E_J(1-\cos\phi)$. Thus, $\sin\phi = (1/E_J)(dU_\phi/dx_p)$. Even if the phase is not at its equilibrium value, it is still true that the supercurrent in the junction is given by the Josephson equation $I_s = I_c \sin\phi$. This expression simply defines the rate of the condensate flow through the tunnel barrier of the junction under a given value of the phase difference. This is a property of the junction itself. A simple combination of the equations above leads to

$$I_s = I_s(\phi) = \frac{I_c}{E_J}\frac{dU_\phi}{d\phi} = \frac{2e}{\hbar}\frac{dU_\phi}{d\phi} \qquad (3.26)$$

The function $I_s(\phi)$ is called the current–phase relationship (CPR) of the device. The formula written above provides a link between the CPR and the phase-difference potential energy of the device. Note that the bias current potential energy is not included in the definition of the CPR.

Such a result is equivalent to the general result that the supercurrent equals the derivative of Helmholtz free energy with respect to the phase difference, multiplied by $2e/\hbar$ (see (2.38)). Such a coincidence is not an accident. In fact, U_ϕ is the usable energy of the junction under the condition that the phase difference is the controlled variable. Thus, U_ϕ is the Helmholtz energy of the junction. On the other hand, if the external bias current I is controlled, then the usable energy is given by $U_{wb} = U_\phi - \hbar I\phi/2e$. Thus, U_{wb} is Gibbs energy of the system.

Note parenthetically that I can also be viewed as a supercurrent if the leads connected to the junctions are all superconducting, which is typically the case in experiments. Thus, a simple interpretation of I is that it is the total supercurrent supplied to the device. Some part of the total current tunnels through the junction. Here, we denote this part as I_s. The rest of it, assuming no dissipation and thus zero normal current, charges the capacitor. Such interpretation allows one to make a more direct comparison of the results obtained here and the results obtained for a nanowire in the Sections 2.5 and 2.6, in which "I_s" stands for the total current supplied to the nanowire device (again assuming that the dissipation is zero and so the normal current is zero). Note that in the case of a Little's phase slip the order parameter amplitude and the phase evolve in such a way that $I_s = $ const.

Equation (3.26) provides a simple relation between the effective spring constant, k_p, of the washboard potential, U_{wb}, and the CPR. Indeed, the spring constant, defined in the section above, is

$$k_p = \left.\frac{d^2 U_{wb}}{dx_p^2}\right|_{x_p=x_{eq}} = \left(\frac{\hbar}{2e}\right)\left.\left(\frac{dI_s}{d\phi}\right)\right|_{\phi=x_{eq}}$$

This equation is general and can be applied to the nanowire device of the type shown schematically in Figure 1.1. The interesting development now is that if the wire is long, its CPR is linear at all currents except very near I_c. Thus, the spring constant is expected to be approximately independent of the amplitude of oscillations. The effective mass m_p, which is proportional to the capacitance between the electrodes, is expected to be very small because the electrodes are far away from each other (the wire length sets the distance between the electrodes). Thus, the plasma frequency is not easy to predict for nanowires. One can also use this analysis to measure the CPR. Indeed, the capacitance is expected to be fixed and so $m_p = \text{const}$. Therefore, $(\hbar/2e)(dI_s/d\phi)|_{\phi=x_{eq}} = \omega_p m_p$. Thus, one can obtain information about the CPR (its derivative, to be more precise) by measuring the resonance frequency of the device of Figure 1.1. Here, the free variable is x_{eq}. It can be set at will by applying more or less a bias current, as $I = I_c \sin x_{eq}$.

3.2.2
Cubic Potential

An interesting property of U_{wb} is that its inflection point, defined by $d^2 U_{wb}/dx_p^2 = 0$, does not move with i_b. Indeed, according to the definition of the potential, (3.21), the position of the infection point x_{in} is given through $d^2 U_{wb}/dx_p^2 \propto \cos x_p = 0$. Therefore, the inflection position is $x_p = x_{in} = \pi/2$ for any $i_b < 1$. Notice that $\arcsin(1) = \pi/2$. Therefore, since the condition of equilibrium of the particle is $x_p = \arcsin i_b$, it then follows that if $i_b \to 1$, then $x_{eq} \to \pi/2 = x_{in}$. Thus, we arrive at the conclusion that the fixed inflection point x_{in} and the current-dependent equilibrium point are near each other at high currents. Therefore, it is reasonable to apply the Taylor expansion near the inflection point and expect that it will be applicable, approximately, at the equilibrium point also. Such expansion starts with the cos of the washboard potential, which develops as

$$\cos x_p = \cos x_{in} - \sin(x_{in})s - \frac{1}{2}\cos(x_{in})s^2 + \frac{1}{6}\sin(x_{in})s^3$$

where $s \equiv x_p - x_{in} = x_p - \pi/2 = \phi - \pi/2$ is the distance to the inflection point. Since the inflection point of the potential is always $x_{in} = \pi/2$, the expression above becomes very simple, $\cos x_p = -s + s^3/6$.

The Taylor series for the potential energy, expressed through $s = \phi - \pi/2$, is

$$\frac{U_{wb}}{E_J} = 1 - \frac{\pi i_b}{2} + (1-i_b)s - \frac{s^3}{6} = 1 - \frac{\pi i_b}{2} + s\left(1 - i_b - \frac{s^2}{6}\right) \quad (3.27)$$

The points of maxima and minima are defined by $dU_{wb}/ds = 0$. This is equivalent to $s^2 = 2(1-i_b)$. The minimum, that is, the metastable point, of the potential energy occurs at $s_{min} = -(2-2i_b)^{1/2}$. The potential energy at the minimum is

$$\frac{U_1}{E_J} = 1 - \frac{\pi i_b}{2} - \frac{2\sqrt{2}}{3}(1-i_b)^{3/2}$$

The maximum, that is, the top of the barrier, is $s_{max} = (2 - 2i_b)^{1/2}$. Note that the maximum and the minimum are positioned symmetrically with respect to the to the inflection point. The potential energy at the maximum is

$$\frac{U_2}{E_J} = 1 - \frac{\pi i_b}{2} + \frac{2\sqrt{2}}{3}(1 - i_b)^{3/2}$$

Thus, the energy barrier, that is, the work needed to move the particle from the minimum to the maximum is

$$\Delta U = U_2 - U_1 = \frac{4\sqrt{2}E_J}{3}(1 - i_b)^{3/2} = \frac{2\sqrt{2}\hbar I_c}{3e}\left(1 - \frac{I}{I_c}\right)^{3/2} \quad (3.28)$$

Taking into account that the barrier at zero current is $\Delta U(0) = 2E_J = \hbar I_c/e$, the barrier at high currents can be presented as

$$\Delta U = \frac{2\sqrt{2}}{3}\Delta U(0)\left(1 - \frac{I}{I_c}\right)^{3/2} \quad (3.29)$$

The frequency of small oscillations near the equilibrium point, $\omega_p(I)$, which is called the plasma frequency, was already derived in previous sections, see, for example, (3.24).

The analysis above is also valid, approximately, for short superconducting nanowires, connected to electrodes forming a capacitor, if the CPR is parabolic near its maximum. The CPR of a nanowire is sinusoidal at high temperatures, while at low temperatures, it deviates somewhat from the sinus function of the phase difference. Thus, the model presented above could only be used to get a estimate for short wires.

Note also that the analysis of the potential energy barrier does not require the knowledge of the dissipative term defined by the normal resistance. In the phenomena in which the dissipation is crucial, the analogy between the SM model and a thin wire is not easy to build because there is no external shunt in the wire. On the contrary, the damping is caused by the normal electrons or bogoliubons which occur in the wire itself when the normal core of a phase slip crosses the wire.

3.2.3
Thermal Escape from the Cubic Potential

The rate of thermal escape is given by the Kramers theory [21]. It depends on the damping in the system. Yet, in any case, it can be presented in the following general form

$$\Gamma = \Omega \exp\left(\frac{-\Delta U}{k_B T}\right) \quad (3.30)$$

According to the Garg summary [181], in the case of moderate damping, $\Delta U = (2\sqrt{2}/3)\Delta U(0)(1 - I/I_c)^{3/2}$. In this case, the attempt frequency is $\Omega = \omega_p(0)(1 -$

$I^2/I_c^2)^{1/4}$. In the cases of weak damping as well as strong damping the exponential factor $\exp(-\Delta U/k_B T)$ of the escape rate remains the same. It does not depend on the damping coefficient η appearing in (3.20). On the contrary, the attempt frequency depends on the damping significantly. Indeed, according to Garg, if the damping is weak, the attempt frequency is $\Omega = 18\eta \Delta U/5\pi k_B T$. Interestingly enough, it does not depend on the plasma frequency in the weak damping limit. In the case of strong damping, the attempt frequency is $\Omega = \omega_p^2/2\pi\eta$. The derivations of these various attempt frequencies are usually based on the Fokker–Planck equation which is beyond the scope of this introductory monograph.

Following [181], we can express the escape rate in different limits using a single general formula as

$$\Gamma = \Omega_0 \left(\frac{1-I}{I_c}\right)^{a_2} \exp\left[-a_1 \frac{\hbar I_c}{2ek_B T}\left(\frac{1-I}{I_c}\right)^{a_1}\right] \quad (3.31)$$

The exponential or Arrhenius activation factor is classified according the power-law-exponent a_1 that determines how the barrier approaches zero as the current approaches I_c. For JJs, the power is $a_1 = 3/2$, independent on damping, or, in other words, independent on the shunt value in the SM model. The barrier is always proportional to the critical current I_c and its exact value is also determined by a system specific constant a_1. For JJs, the constant a_1, according to (3.28), is $a_1 = 4\sqrt{2}/3 \approx 1.88$. Such values for a_1 and α_1 would be found for any system which has a CPR defined by the Josephson equation $I_s = I_c \sin\phi$. The attempt frequency is dependent on the damping level as explained above. The value of the exponent for the prefactor is $a_2 = 3/2$ or $1/4$ or $1/8$ for weak, intermediate and strong damping correspondingly. The constant prefactor is $\Omega_0 = 18\eta \Delta U(0)/5\pi k_B T$ or $\Omega_0 = \omega_p(0)$ or $\Omega_0 = \omega_p(0)^2/2\pi\eta$ for weak, intermediate and strong damping, correspondingly.

3.3
Macroscopic Quantum Phenomena in the Stewart–McCumber Model

The SM model of a Josephson junction is very convenient, especially if one wants to introduce macroscopic quantum effects, such as tunneling, into the analysis. Yu. M. Ivanchenko, a Ukrainian scientist, played a key role in the development of both the classical and the quantum dynamics models of the JJ. First, he anticipated the SM model in his publication of 1966, which presents equations analogous to the SM model equations [159]. The quantum mechanical description of the macroscopic phase variable in JJs was also first advanced by Ivanchenko [160], as early as 1967.[3] Later, a number of authors have perfected the quantum description of the superconducting circuits [162–164].

The analysis is easy to carry out using the language of the mechanical analogy which is summarized in (3.22). Let us consider the case when the dissipation is

---

[3] Curiously enough, in the same year, William Little proposed the idea of a phase slip [17], Dmitrenko and collaborators published the report on electromagnetic radiation from a Josephson junction [161], and the present author was born.

3.3 Macroscopic Quantum Phenomena in the Stewart–McCumber Model

negligible, $\eta = 0$. Then, the Newton equation is $m_p \ddot{x}_p + dU_{wb}/dx_p = 0$. The corresponding time-dependent Schrödinger equation is

$$i\hbar \frac{\partial \Psi_p}{\partial x_p} = \hat{H}\Psi_p \tag{3.32}$$

It is written by assuming that the variable $x_p \equiv \phi$ is the position of the quantum particle and $\Psi(x_p)$ is the wavefunction of the particle.[4] The probability density to find the particle at point x_p is $|\Psi(x_p)|^2$. Thus, the probability density that the considered JJ has a phase difference ϕ is $|\Psi(\phi)|^2$, which can be found easily if the Schrödinger equation is solved and $\Psi(x_p)$ is found. Below, we will give some useful simple solutions to the equation.

As usual, the Hamiltonian operator is the sum of the kinetic energy operator and the potential energy

$$\hat{H} = -\frac{\hbar^2}{2m_p} \frac{\partial^2}{\partial x_p^2} + U_{wb} \tag{3.33}$$

The Hamiltonian is written assuming that $m_p \equiv C\hbar^2/(2e)^2$ is the particle mass and U_{wb} is the potential energy "landscape" into which the particle is placed.

If the wavefunction of the particle (and the corresponding JJ) is stationary, it means that it is an eigenstate of the Hamiltonian. Then the corresponding energy E can be found from the time-independent Schrödinger equation

$$\hat{H}\psi_p = E\psi_p \tag{3.34}$$

The time-independent part of the wavefunction ψ_p is related to the complete wavefunction as

$$\Psi_p = \psi_p \exp\left(-\frac{iEt}{\hbar}\right) \tag{3.35}$$

One can check by direct substitution that this expression is a solution of the time-dependent Schrödinger equation.

Let us take a simple example of zero-bias current. Then, the potential energy, (3.21), is $U_{wb} = E_J[(1 - \cos x_p)] \approx (E_J/2)x_p^2$. Thus, the Schrödinger equation for the phase particle is

$$-\frac{\hbar^2}{2m_p} \frac{\partial^2 \psi_p}{\partial x_p^2} + \frac{E_J}{2} x_p^2 \psi_p = E\psi_p \tag{3.36}$$

As for any harmonic oscillator, the ground-state wavefunction can be expected in the form

$$\psi_p = C_1 \exp\left(-\frac{x_p^2}{4\Delta x^2}\right) \quad \text{and} \quad \psi_p^2 = C_1^2 \exp\left(-\frac{x_p^2}{2\Delta x^2}\right) \tag{3.37}$$

[4] The terms "wave function" or "wavefunction" refer to the function describing the time and spatial dependence of the de Broglie wave, associated with the particle. It could have been named the de Broglie function as well.

where C_1 and Δx are some unknown constants to be found. We need to choose C_1 such that the integral of the probability density over the entire x_p-axis is unity. We can remember that $|\psi_p|^2$ gives the probability density and compare it to the properly normalized Gaussian probability density, (B.3), concluding that $C_1^{-2} = \Delta x \sqrt{2\pi}$.

Below, we illustrate that the chosen ψ_p is indeed a solution of the Schrödinger equation. Also ψ_p is the ground-state solution because the Gaussian does not have zeros. The first derivative of ψ_p is

$$\frac{\partial \psi_p}{\partial x_p} = -\left(\frac{C_1}{2\Delta x^2}\right) x_p \exp\left(-\frac{x_p^2}{4\Delta x^2}\right)$$

The second derivative is

$$\frac{\partial^2 \psi_p}{\partial x_p^2} = -\left(\frac{C_1}{2\Delta x^2}\right)\left[1 - \left(\frac{x_p^2}{2\Delta x^2}\right)\right] \exp\left(\frac{-x_p^2}{4\Delta x^2}\right)$$

If we now put the second derivative into the Schrödinger equation above and divide it by ψ_p, we get

$$\frac{\hbar^2}{2m_p}\frac{1}{2\Delta x^2} - \frac{\hbar^2}{2m_p}\frac{x_p^2}{4\Delta x^4} + \frac{E_J}{2}x_p^2 = E \tag{3.38}$$

Since x_p is a free variable, the equation is valid only if the factor multiplying x_p^2 is zero. Thus, we arrive at the solution for the standard deviation (i.e., the Heisenberg uncertainty) of the position of the phase particle

$$-\frac{1}{4\Delta x^4}\frac{\hbar^2}{2m_p} + \frac{E_J}{2} = 0 \tag{3.39}$$

Thus, we find the constant Δx which defines the effective width of the ground state of the wavefunction

$$\Delta x = \left(\frac{\hbar^2}{4m_p E_J}\right)^{1/4} \tag{3.40}$$

This concludes the derivation of the ground state of the wavefunction. The excited states can also be found using the analogy with a harmonic oscillator. The energy difference between the ground state and the first excited is approximately $\hbar\omega_p = \hbar\omega_p(0)(1 - I/I_c)^{1/4}$. Approximately, because the potential is not exactly a parabola but, rather, it is a washboard potential. In order to make the system jump to the excited state, one needs to apply a microwave radiation having the frequency $\omega_p/2\pi$ because ω_p is the angular frequency, not just frequency.

The wavefunction of the ground state allows us to find the standard deviation of the position of the phase particle. Noticing that the wavefunction of the ground state has the same shape as the Gaussian normal distribution, one concludes that the standard deviation of the position equals Δx. Remember also that x_p represents the phase difference, $m_p = \hbar^2 C/4e^2$, and $E_J = \hbar I_c/2e$. Therefore, the standard deviation of the phase difference dues to quantum fluctuations is

$$\Delta\phi = \left(\frac{2e^3}{\hbar C I_c}\right)^{1/4} = \left[\pi \frac{e}{I_c} \frac{1}{R_q C}\right]^{1/4} \qquad (3.41)$$

For a nanowire device, having typical parameters of $I_c = 1\,\mu\text{A}$ and $C = 10\,\text{fF}$, one gets the standard deviation of the phase difference $\Delta\phi \approx 0.3$, which might be a significant fluctuation for short wires for which a phase difference of $\phi = \pi = 3.14$ is enough to suppress the order parameter to zero, according to the sinusoidal CPR, $I_s = I_c \sin\phi$, approximately applicable to short wires [18].

Presence of phase fluctuations can reduce the average supercurrent for a given phase difference (which is also called a superconducting response) [84]. In other words, fluctuations can increase the kinetic inductance of the system since the inductance is inversely proportional to I_c. To see this, consider sinusoidal CPR and average over fluctuations, $\delta\phi$, as $\overline{I_s} = \overline{I_c \sin(\phi + \delta\phi)}$. If fluctuations are not too large, the Taylor theorem applies and $\sin(\phi + \delta\phi) = \sin\phi + \delta\phi\cos\phi - (1/2)(\delta\phi)^2 \sin\phi$. When doing averaging, assume that thermal and quantum fluctuations go both ways with equal probabilities and equal zero on average, and thus $\overline{\delta\phi} = 0$. Also, the dispersion is $\Delta\phi$ by definition, so that $\overline{(\delta\phi)^2} \equiv \Delta\phi^2$. Thus, we conclude that $\overline{\sin(\phi + \delta\phi)} = \sin\overline{\phi} - (1/2)\Delta\phi^2 \sin\phi$. Therefore, the CPR becomes $\overline{I_s} = I_c[1 - (1/2)\Delta\phi^2]\sin\overline{\phi} = I_c^* \sin\overline{\phi}$. In other words, the new critical current, reduced due to fluctuations of the phase, is $I_c^* = I_c(1 - \Delta\phi^2/2)$.

One can use such an analysis to approximately estimate the critical current I_c^{**} at which superconductivity becomes washed out by quantum fluctuations and the superconducting response becomes zero. From the above, the effective critical current becomes zero if $\Delta\phi^2 = 2$. Using the result for the dispersion, (3.41), it is possible to estimate that $I_c^{**} = \pi e/4 R_q C$. Taking the typical capacitance, $C \approx 10\,\text{fF}$, for a device having a nanowire, Figure 1.1, one gets an estimate for the value of the critical current at which quantum fluctuations are expected to wash out superconductivity, $I_c^{**} \sim 2\,\text{nA}$. Such an estimate has not been checked quantitatively, but, on the qualitative level, it is indeed true that superconductivity disappears if wires are made sufficiently thin.

Note that in the quantum picture the charge on the JJ is an operator which is $\hat{q} = -2ei(\partial/\partial x_p) = -2ei\partial/\partial\phi$. The commutation relation [85] is $[\hat{\phi}, \hat{q}] = 2ei$, where $\hat{\phi} = \phi$. The Anderson–Heisenberg uncertainty then is $\Delta\phi\Delta q \approx e$.

3.3.1
MQT in a Cubic Potential at High Bias Currents

If the bias current I is large, the barrier for the phase particle to escape from its metastable equilibrium state becomes smaller according to (3.28). As the barrier is reduced, the quantum tunneling becomes the dominant process for the escape, provided that the temperature is sufficiently low. Here, we discuss the regime of strong currents, namely, $I_c - I \ll I_c$. In such a regime, the cubic potential approximation for the barrier shape is applicable.

The tunneling rate can be found using the single-particle analogy combined with the WKB expression, valid for negligible damping, in which case the tunneling rate is

$$\Gamma = \Omega \exp\left(-\frac{S}{\hbar}\right) \tag{3.42}$$

According to WKB, the quantum action is

$$S = 2 \int_{s_{\min}}^{s_{\max}} \sqrt{2m_p[E - U_{wb}(s)]}\, ds$$

where $s = \phi - \pi/2$, as defined in Section 3.2.2, and E is the total energy of the particle. Assuming that the barrier is much higher than the zero-point energy of the particle, one can write $E \approx U_{wb}(0) = E_J(1 - \pi i_b/2)$ so that $E - U_{wb}(s) = E_J(1 - i_b)s - E_J s^3/6$. Under such assumptions, the action is

$$S = 2\sqrt{2m_p E_J} \int_{s_{\min}}^{s_{\max}} [s(1 - i_b) - s^3/6]^{1/2}\, ds \tag{3.43}$$

The expression for the attempt frequency in the WKB approximation is

$$\Omega = C_b \omega_p \sqrt{S/\hbar}$$

where C_b is a constant dependent on the exact shape of the barrier.

Solving this problem in detail is rather complicated. Various aspects of this problem have been discussed in [159, 160, 162–164]. A useful summary of the results is given in [34], which, by the way, provides a solid experimental evidence that the quantum, rather than classical, description of JJs is the correct one. Below, we list relevant formulas.

The quantum action for a phase slip in the case of zero damping, S_{zd}, is

$$\frac{S}{\hbar} = \frac{S_{zd}}{\hbar} = \frac{36}{5} \frac{\Delta U}{\hbar \omega_p} \tag{3.44}$$

This equation is valid for negligible damping, that is, if the macroscopic degree of freedom, namely, the phase variable, does not interact with any microscopic degrees of freedom.

To find the action, we have to combine (3.29) for the barrier at high currents and (3.25) for the frequency of small oscillations which defines the strength of zero-point fluctuations. The result for the action is then

$$\frac{S_{zd}}{\hbar} = \frac{24\sqrt{2}}{5} \frac{\Delta U(0)}{\hbar \omega_p(0)} \frac{(1 - I/I_c)^{3/2}}{(1 - I^2/I_c^2)^{1/4}} \tag{3.45}$$

The expression above is valid near the critical current only. In that case, $1 - I^2/I_c^2 = (1 + I/I_c)(1 - I/I_c) \approx 2(1 - I/I_c)$ because $(1 + I/I_c) \approx 2$. Therefore, $(1 - I/I_c) \approx$

$(1 - I^2/I_c^2)/2$. Using this the action becomes

$$\frac{S_{zd}}{\hbar} = \frac{12}{5} \frac{\Delta U(0)}{\hbar \omega_p(0)} \left(1 - \frac{I^2}{I_c^2}\right)^{5/4} = \frac{6}{5} \frac{\sqrt{2\hbar C I_c}}{e\sqrt{e}} \left(1 - \frac{I^2}{I_c^2}\right)^{5/4} \qquad (3.46)$$

Note that it scales with the normalized current introduced to the power 5/4.

The attempt frequency in the case of zero damping, Ω_{zd}, equals the plasma frequency $\omega_p/2\pi$ multiplied by the square root of the normalized action and multiplied by a constant. Namely,

$$\Omega_{zd} = \sqrt{120\pi} \frac{\omega_p}{2\pi} \sqrt{\frac{36}{5} \frac{\Delta U}{\hbar \omega_p}} \qquad (3.47)$$

If damping is not negligible but weak, then the action increases according to Caldeira–Leggett theory [30]. The weak-damping regime action, S_{wd}, is

$$\frac{S}{\hbar} = 7.2 \frac{\Delta U}{\hbar \omega_p} \left(1 + \frac{0.87}{Q}\right) \qquad (3.48)$$

Here, the quality factor is defined, as usual, as $Q = \omega_p RC$. As always, ΔU and ω_p are current dependent functions according to the cubic potential high-current limit. Namely, they are defined by (3.29) and (3.25). The attempt frequency in the case of the weak damping has the same shape as for the case of zero damping, that is, (3.47).

In the case of strong damping, the attempt frequency is

$$\Omega_{sd} = 8\sqrt{6} \omega_p^{-3} (2R_n C)^{-7/2} \sqrt{\frac{\Delta U}{\hbar}}$$

and the MQT action is

$$\frac{S_{sd}}{\hbar} = \frac{R_q}{R_n} (\Delta\phi)^2/9$$

where $\Delta\phi$ is the distance under the barrier, or, in other words, the size of the classically forbidden region in the space of the phase difference ϕ. For zero-bias currents, $\Delta\phi = 2\pi$, which is the distance from one minimum to the next in the washboard potential having zero tilt. In the limit $I_c - I \ll I_c$, the SM model gives the distance under the barrier as $\Delta\phi = \pi(1 - I/I_c)^{1/2}$. The barrier height, as in the case of no damping, equals $\Delta U = (2\sqrt{2}\hbar I_c/3e)(1 - I/I_c)^{3/2}$ since the potential energy does not depend on damping or friction.

Note that the total action is the sum of the action due to the barrier and the action due to the damping. Thus, the barrier action can be neglected and the strong damping limit formula can be used only if the damping action is dominant, that is, if $S_{sd} \gg S_{zd}$.

3.4
Schmid–Bulgadaev Quantum Phase Transition in Shunted Junctions

The Schmid–Bulgadaev (SB) transition was predicted by Albert Schmid [83, 84] and constitutes a transition of a JJ from superconducting to insulating behavior. This phase diagram for this quantum phase transition was extended by Bulgadaev [90, 91]. The SB transition is a quantum phase transition. Generally speaking, quantum phase transitions take place at zero temperature. They are driven by quantum fluctuations, the magnitude of which changes in response to a variation of some control parameter, which could be, for example, external magnetic filed, the diameter of a nanowire, or a shunting resistor. The SB transition was studied by multiple authors [86–89] and was confirmed experimentally [93, 94]. A detailed numerical study of this quantum transition was undertaken by Werner and Troyer [92].

The essence of the SB transition is the phenomenon of localization of a particle in periodic potential occurring as the damping coefficient exceeds some critical value. The theoretical bases to the SB quantum transition is given by Cakravarty [82], and Caldeira and Leggett [30, 31]. A rigorous derivation of the transition critical point requires some advanced methods of quantum mechanics, which are beyond the scope of this book. Below, we present an heuristic derivation based on semiquantitative arguments.

Consider a particle in a periodic potential, as is illustrated in Figure 3.3. Assume that a viscous drag (friction) force is acting on the particle, $F_{vd} = -\eta v$, where $v = dx/dt$ is the velocity. The x-axis is horizontal and is pointed to the right in Figure 3.3. To move from one well to another in the periodic potential, the particle needs to overcome a certain potential barrier ΔU. Since quantum phase transitions occur at $T = 0$, thermal fluctuations are not involved. Thus, the barrier can only be crossed by quantum tunneling. The tunneling process occurs because the particle can borrow the necessary energy from quantum fluctuations and thus travel over the barrier.

Let us calculate how much energy the particle loses due to the friction force generated by the environment as it moves from one minimum to the next in the periodic potential. The amount of dissipated energy equals the work of the damping force, which is $W_{vd} = F_{vd} X_0 = \eta v X_0 = \eta X_0^2/t_t$. Here, X_0 is the distance from one minimum to the next one, and t_t is the time the particle spends on average as it propagates from one minimum to the next. To be able to climb up the potential energy barrier, the particle needs to "borrow" some energy from the quantum fluctuations. The process of borrowing is understood in the sense that the law of the energy conservation can be violated for a short time so the particle can move over the barrier. This energy is defined by the time over which the borrowing takes place, through the Heisenberg uncertainty principle, as $E_t = \hbar/t_t$. Another way to get an estimate of the energy related to the tunneling event is to use Planck's formula, which relates frequency and the energy. If the tunneling time is t_t, then the effective tunneling frequency is $f_t = 1/t_t$. Then, the energy scale related to the tunneling from one minimum to another is, according to Planck, $E_t = h f_t = h/t_t$.

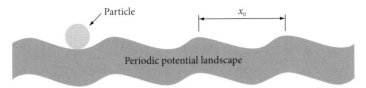

Figure 3.3 A macroscopic but still quantum particle (gray circle) is placed into a periodic potential of period X_0. The Schmid–Bulgadaev (SB) transition occurs in such a system if the viscosity η of the medium in which the particle moves reaches a critical value. The dissipation or viscosity results from the interactions of the particle with various oscillators present in the environment. The particle, as it moves, can excite some of the oscillators in the environment and thus lose some or all of its kinetic energy. This process is commonly called viscous motion. The quantum behavior of the particle is strongly influenced by the viscosity of the environment according to the Caldeira and Leggett theory [30]. At zero damping ($\eta = 0$), the eigenstates of the particle are the so-called Bloch waves, which are basically modulated plane waves. Thus, the particle is delocalized in the potential, that is, it is present everywhere. If $\eta > 0$ then the probability of tunneling from one potential well to the next one is reduced. The particle becomes completely localized in one of the minima if $\eta > h/X_0^2$. Such transition from the delocalized to a localized state is the SB quantum transition [83, 84, 90, 91].

We will use the Planck's relationship between the energy and the frequency for our purpose. Thus, the tunneling time is $t_t = h/E_t$. Note that the Heisenberg's principle would give a similar result, namely, $t_t = \hbar/E_t$. Using the Planck's formula, we can update the expression for the energy loss due to friction as $W_{vd} = \eta X_0^2 E_t/h$. The last step in our heuristic derivation is to realize that the particle cannot lose more energy than the energy E_t that it receives from quantum fluctuations. Thus, the tunneling process can only be possible if $E_t > W_{vd}$ or $E_t > \eta X_0^2 E_t/h$. By dividing both sides by E_t, we conclude that quantum tunneling in the periodic potential is possible only if

$$\eta < \frac{h}{X_0^2}$$

In the case $\eta > h/X_0^2$, the particle is localized in one of the wells (at $T = 0$) since tunneling is not allowed. These conclusions are in agreement with the exact theory of Schmid and Bulgadaev.

This result can be applied to resistively shunted superconducting junctions using the mechanical analogy, Section 3.2. Within this analogy, the friction or damping coefficient is expressed as $\eta = (\hbar/2e)^2(1/R_n)$. The period of the potential energy landscape is exactly 2π. Therefore, the condition for the phase particle to be delocalized is

$$\left(\frac{\hbar}{2e}\right)^2 \frac{1}{R_n} < \frac{h}{(2\pi)^2} \quad \text{or} \quad R_n > \frac{h}{(2e)^2} = R_q$$

Such delocalized state of the phase particle means that the phase and, correspondingly, the supercurrent, are not defined. This is similar to the position of the electron being not defined when it is descibed by a plane wave. Thus, the junction is resistive since it cannot carry a DC supercurrent. Thus, we arrive at the conclusion that the shunted Josephson junction is not superconducting if the shunt resistance is larger than the quantum resistance R_q.

On the contrary, if $R_n < R_q$, then the phase particle is localized, meaning the phase is a good quantum number. In such cases, the junction is a true superconductor, within our terminology, meaning that its resistance is exactly zero in the limit of zero temperature and zero bias current.

Thus, the SB transition occurs in shunted JJs and the critical point is $R_n = R_q$. As clarified by Bulgadaev, the critical point does not depend on the ratio E_c/E_J of the Coulomb ($E_c = e^2/2C$, where C is the capacitance) and the Josephson ($E_J = \hbar I_c/2e$) energies of the junction.

3.5
Stewart–McCumber Model with Normalized Variables

Here, we introduce a normalized time [97] and find the time-evolution equation for the phase difference $\phi(t)$ corresponding to the superconducting circuit of Figure 3.1. The usual equations apply $I_s = I_c \sin \phi$ and $2eV = \hbar \dot\phi$. The analysis begins, as usual, with the Kirchhoff equation of the current conservation

$$\frac{C\hbar}{2e}\frac{d^2\phi}{dt^2} + \frac{\hbar}{2eR_n}\frac{d\phi}{dt} + I_c \sin \phi = I + I_{no} \tag{3.49}$$

The new term here is the Gaussian white noise representing the Johnson–Nyquist thermal current noise I_{no} in the shunt resistor R_n,

$$\langle I_n(t) \rangle = 0$$

$$\langle I_n(t_1) I_n(t_2) \rangle = \frac{2k_B T}{R_n} \delta(t_1 - t_2)$$

where T is the temperature of the device. The meaning of the noise is simply that the current through the normal shunt resistor equals $V/R_n - I_{no}$, where the thermal fluctuations noise current I_{no} is assumed independent on V. This is modeled in the Kirchhoff equation above by making an equivalent assumption that the current through the resistor is V/R_n, but the bias current is $I + I_{no}$.

As discussed above, the plasma frequency of small charge oscillations of the device is $\omega_p = \sqrt{2eI_c/\hbar C}$ and the relaxation time is $\tau = R_n C$. The corresponding quality factor is defined as $Q = \tau \omega_p = R_n C/\sqrt{L_k C} = R_n \sqrt{C/L_k}$, which can be understood as the number of oscillations of the supercurrent within the relaxation time, multiplied by 2π. The quality factor also defines the dissipation rate, as $Q = 2\pi E_{stored}/E_{dissipated}$, where E_{stored} is the energy stored in the oscillator and $E_{dissipated}$ is the energy lost into heat within one cycle. The quality factor can also be written as $Q = \tau \omega_p = R_n C\sqrt{2eI_c/\hbar C}$. The McCumber parameter is defined as $\beta_c = Q^2 = 2eCR_n^2 I_c/\hbar$.

One way to introduce a normalized time is to define new, unitless time, as $t' = Q\omega_p t = \omega_p^2 \tau t = (2eI_c R_n/\hbar)t$. In such cases, the dynamics equation, (3.49), can be rewritten in terms of only dimensionless variables [97] as

$$Q^2 \frac{d^2\phi}{dt'^2} + \frac{d\phi}{dt'} + \sin\phi = i + i_{no} \tag{3.50}$$

where $i = I/I_c$ is the normalized bias current and $i_{no} = I_{no}/I_c$ is the normalized thermal noise current.

The equation of motion, (3.50), is the same as that of a damped pendulum driven by an external force proportional to $i + i_{no}$. The dissipative term $d\phi/dt'$ breaks the time-reversibility of the equation of motion. This equation is useful, among other things, for computer simulations. The Gaussian noise can be simulated using the method of [98]. The voltage on the junction, which is the same as the voltage on the shunt resistor, is $V = I_c R_n \langle d\phi/dt' \rangle_{t'}$.

At zero temperature, in the absence of thermal current noise I_n, the zero-voltage state, defined as $\dot{\phi} = 0$, is stable at any bias current less than the critical current, $|i| < 1$. Here, we assume that quantum fluctuations do not exist. The voltage state is stable at all bias currents greater than I_{r0}, which is the fluctuation-free retrapping current. For $Q < 0.838$, the damping is sufficient so that any running state (i.e., a state having $\langle \dot{\phi} \rangle \neq 0$) is not stable unless the potential decreases monotonically [97], meaning $I > I_c$. Thus, for $Q < 0.838$, one finds $I_{r0} = I_c$. For $Q_0 > 0.8382$, a running state is possible even for $I < I_c$, that is, when the effective potential has local minima. The presence of a stable running states means that $I_{r0} < I_c$ and that the V–I curve has a hysteresis. In the limit of large $Q \gg 1$, $i_{r0} = 4/\pi Q$ (for $Q > 3$) [95–97].

The time normalization can be done in a different way, namely, as $\tilde{t} = \omega_p t = \sqrt{2eI_c/\hbar C}\, t$, as in [1]. In such cases, a new dimensionless equation obtains, namely,

$$\frac{d^2\phi}{d\tilde{t}^2} + Q^{-1}\frac{d\phi}{d\tilde{t}} + \sin\phi = i + i_{no} \tag{3.51}$$

This form is convenient for strongly overdamped junctions, i.e., such that $Q \ll 1$. In this case, the term that contains the second time derivative, representing inertia, can be neglected using the analogy with a Newtonian particle moving in a tilted washboard potential under strong damping. For the particle, strong damping can be achieved if the particle is immersed in some very viscose oil, for example. Then, if we go back to the physical time t, the equation becomes

$$\sqrt{\frac{\hbar C}{2eI_c}}\frac{d\phi}{dt} = Q(i + i_{no} - \sin\phi) = (i + i_{no} - \sin\phi)\sqrt{\frac{2eCR_n^2 I_c}{\hbar}}$$

3 Stewart–McCumber Model

Consider a simple example of $T = 0$, which means $i_{no} = 0$. Then, the dynamics of the overdamped junction is governed by

$$\frac{d\phi}{dt} = \frac{2e R_n I_c}{\hbar}\left(\frac{I}{I_c} - \sin\phi\right)$$

This differential equation can be solved analytically. If V and I are averaged over time, then the resulting V–I dependence is $\overline{V} = R_n \sqrt{I^2 - I_c^2}$.

4
Fabrication of Nanowires Using Molecular Templates

To fabricate a homogeneous metallic wire of a diameter less than 10 nm, which is seamlessly connected to larger electrodes, it is usually not sufficient to use traditional fabrication techniques such as electron beam lithography [99]. Instead, one needs to use nontraditional methods or tricks. One such method is so-called "molecular templating" (MT) [100]. It has been used successfully to fabricate wires having diameters significantly less than 10 nm, having also a short length, in the range roughly between 30 nm and 1 μm. As will be explained in detail below, an advantage of molecular templating is that, as produced, the nanowires are already seamlessly connected to larger metallic electrodes which are composed of the same material as nanowires themselves. Such types of samples are especially convenient for carrying out various transport measurements since the contact resistance is essentially zero. Moreover, the wires can be produced suspended over a slit in a Si substrate, making the utilization of a transmission electron microscope beam for the purpose of nanometer-scale modifications of the wire geometry and morphology possible. Finally, nanowires can be fine-tuned by applying voltage pulses, with the possibility of precisely adjusting the critical current of the wire in situ [114]. This arsenal of techniques makes MT-fabricated wires a model system in studies of Little's phase slips.

To fabricate a nanowire using a single molecule as a template, one first needs to make a substrate having a narrow, nanometer scale, trench (see Figure 4.1) [101]. This is done on a Si wafer that has a 500 nm thick layer of oxide (SiO_2) grown on its surface (including a 100 nm thick film of "dry oxide" and a 400 nm of "wet oxide").[1] On top of the oxide film, a 60 nm film of SiN is deposited using a low pressure chemical vapor deposition (LPCVD) process. The LPCVD ensures that SiN has a low internal stress, so that when the SiN layer becomes free-standing, it does not curl up or gets distorted in any way. The trench, which is typically 100 nm wide and 5 mm long, is defined in the SiN film using electron beam (e-beam) lithography with PMMA resist, followed by reactive ion etching (RIE) in SF_6 plasma. In

---

1) The purpose of having this oxides is two-fold: to allow the formation of an undercut around the trench and to provide good electrical isolation of the metallic electrodes to be placed on top of the SiN layer from the underlying Si wafer, which can be slightly conducting (or even strongly conducting if it is dope on purpose, say with phosphorus, making it possible to use the wafer itself as a gate electrode).

Superconductivity in Nanowires, First Edition. Alexey Bezryadin.
© 2013 WILEY-VCH Verlag GmbH & Co. KGaA. Published 2013 by WILEY-VCH Verlag GmbH & Co. KGaA.

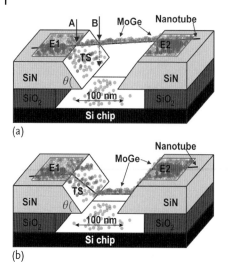

Figure 4.1 Schematic (not to scale) of molecular templating (MT) using a carbon nanotube [100]. (a) The templating nanotube is shown as the solid black line. The tube is coated with atoms of the desired metal (red disks). The atoms, deposited by DC magnetron sputtering, stick to the nanotube and form a continuous nanowire. The smallest achievable diameter of the wire is comparable to the diameter of the nanotube, which is ~ 2 nm. The wire forms naturally, perfectly connected to the electrodes which are superconducting films, marked "E1" and "E2," deposited simultaneously with the wire at the sides of the trench. The electrodes are shaped by photolithography. The segment of the nanowire located between the marks "A" and "B" is suspended over the tilted side (TS) of the trench. This is why the region AB appears brighter on scanning electron microscope (SEM) images. Such "white spots" (see Figure 4.3) at the ends of the wire are the indicators that the wire is straight and is well-connected to the electrodes. (b) A similar schematic, but having the molecule diving down into the trench. Such samples do not show "white spots" in SEM images. Such an arrangement is typical for DNA molecules which are flexible on the length scale of the trench width. The van der Waals force is sufficient to force the molecule to stick to the surface over the maximum possible length. Thus, the molecule crosses the trench at points where the gap is the narrowest, which, in this geometry, happens at the level of the bottom surface of the SiN film. Please find a color version of this figure in the color plates.

other words, the chip surface is spin-coated with an electron-beam sensitive resist (so-called PMMA resist, which is a long-chain polymer), a narrow strip is irradiated with an electron beam, the sample is immersed into a specific solution (the developer) so the resist in the irradiated region dissolves. After drying, the sample is placed into a chamber filled with ionized SF_6 gas for the purpose of etching the region of the SiN film that is not protected by the resist. Such dry etching does not provide vertical walls of the etched trench, but rather it spreads to the sides since etching of SiN in SF_6 is quite isotropic. This is why the sides of the trench are tilted, as shown in Figure 4.1, and the angle θ, defined on the figure, is less than $90°$.

The essential part of the MT fabrication approach is the formation of an undercut at both sides of the trench (see Figure 4.1). This is done by removing the underlying silicon oxide just under the SiN trench and to the sides from the trench. The SiO$_2$ is underetched by immersing the chip into HF acid of 49% concentration for about 10 s. (The acid should be used with great care, only in a fume hood. Any contact with skin or the inhalation of vapors must be strictly avoided.) The width of the undercut thus produced is about 300 nm. Its formation is due to the fact that HF etches the oxide much faster than it etches the nitride of silicon. After the HF etching step, the sample is rinsed in water, then HNO$_3$, then water again, and finally in isopropanol and then dried. When the electrodes are deposited over the trench, the undercut serves an important purpose of ensuring that the trench cuts the deposited metallic film efficiently, thus forming two electrically disconnected electrodes.

To form a metallic nanowire, a single elongated molecule needs to be placed over the trench and subsequently decorated evenly with a metal. Since manipulation of just one molecule is technically difficult, it is usually much easier to deposit many molecules, positioned relatively far from each other, and then destroy all but one using photolithography and etching. Molecules are deposited from a solution. For example, fluorinated carbon nanotubes [102, 103] are deposited from a solution in isopropanol. A drop of the solution is placed on the surface for a few seconds and then blown away using a flow of dry nitrogen gas. Such a procedure leaves a few molecules attached to the chip surface (by the van der Waals force), some crossing the trench. The concentration of the solution is tuned such that after the deposition, the distance between molecules crossing the trench is about $\sim$ 50 µm on average. If one uses regular carbon nanotubes rather than the fluorinated ones, then the tubes can be dispersed in dichloroethane by sonication and then deposited in the same way [100]. The properties of wires made on fluorinated and regular tubes appear to be the same [102, 104]. Another linear molecule that has been used as a template for nanowires is λ-DNA. The DNA can be deposited from a dilute solution in water followed by drying in a pumped desiccator and rinsing in pure water to remove salt deposits [105].

After rinsing and drying the chip, it is ready for a sputter-deposition of a desired metal or alloy. The wire and the metallic films connected to the wire are formed during the same sputtering run. This ensures the continuity of the sample and the absence of the contact resistance between the wire and the electrodes. The sputtering should be done at low base pressure. For example, for a Nb film to be superconducting, the base pressure should be as low as $\sim 10^{-9}$ Torr. To keep the deposited superconducting film free of impurities, the sputtering rate should be as high as possible. Most of the samples to be discussed here have been sputtered at a rate of $\sim$ 0.1 nm/s, which appears to be sufficiently fast for the MoGe film to be superconducting. These alloys can be sputtered using a DC magnetron sputtering setup equipped with a target made of compressed powder of Mo and Ge mixed in desired proportions. Typically, the composition is chosen as 21% of Ge and 79% of Mo by weight, to maximize the critical temperature. Yet, this composition is borderline since just a little less Ge percentage leads to a strong reduction in T_c of the alloy. Thus, a composition having one part of Ge per three parts of Mo might be more reliable. The components should be chosen as pure as possible, at least

99.95% pure for Mo and 99.99% pure for Ge. The sputtering system should be equipped with a cold trap which is typically cooled using liquid nitrogen. The trap is essential for reducing oxygen and organic impurities in the sputtered films. If the environment in the sputtering system is not clean enough, the sputtered MoGe film might have an expected normal resistance, but no superconductivity at low temperatures.

The sputter-deposition process decorates each suspended molecule with a thin metallic film, thus transforming each such molecule (e.g., a carbon nanotube crossing the trench) into a very thin wire (Figure 4.1). As the figure shows, the width of the trench (w_t) determines the length of the wire (L) by the formula $L = w_t/\sin\theta_{w_t}$, where θ_{w_t} is the angle between the tube and the trench. The diameter of the templating nanotube is roughly two or three nanometers. If the deposited metal would be accumulated on the top surface of the molecule (a carbon nanotube, which is typically used as a template for nanowires, has a form of a cylinder), then the width of the wire would be the same, that is, two or three nanometers. In reality, the sputtering process is rather isotropic, and thus some amount of metal sticks to the side of the nanotube. Thus, the wire becomes wider than the underlying molecule. If the nominal thickness of the sputtered film is, for example, 5 nm, the width of the resulting wire can be in the range roughly between 7 and 10 nm, which is still very thin. For comparison, note that electron beam lithography allows fabrication of wires wider than about 30 nm in most practical situations. The exact shape of the wire cross-section, made using MT, has not been investigated. It is likely that the shape is not an exact circle having the templating molecule in the center. Rather, the cross-section is probably widened at the top surface and the nanotube is located close to the bottom part of the wire (Figure 4.1). Imaging of wires under various angles, under a transmission electron microscope (TEM), qualitatively supports these speculations. The exact shape of the cross-section is not in fact important since, typically, the wire diameter is smaller than the coherence length so that the order parameter is homogeneous within the cross-section and thus the barrier for phase slips only depends on the cross-section area and not on its shape.

The sputter-deposition procedure converts all molecules crossing the trench into nanowires and there may be many of them. In order to select just one wire for transport experiments, one has to examine the Si chip under a scanning electron microscope (SEM) and select a wire having desired dimensions and shape. Typically, a homogeneous wire is desirable, that is, such that the apparent width of it is about constant. After a single defect-free wire is selected, its position is measured with respect to a periodic set of markers prefabricated along the trench (the markers can be, for example, numbers etched into the SiN in the same way as the trench itself). In order to keep just one wire on the sample and eliminate all other wires, the chip is spin-coated with a photoresist and standard photolithography is used to protect the desired wire and to pattern the electrodes connected to it. The photolithography mask alignment is guided by the markers which are large enough to be visible in an optical microscope. Although the resolution of the photolithography is typically no better than one micron, while the length of the wire is just one-tenth of a micron, the task of shaping of the electrodes attached to just one

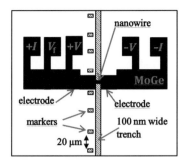

Figure 4.2 A typical layout of the electrodes such as the one used in [100]. The trench is the dashed area in the middle going through the entire Si chip. The electrodes and the nanowire are shown in black. The white background represents the SiN film which covers the entire chip, except in the trench where it is removed. The markers, placed in a row parallel to the trench and spaced by 20 μm, are shown as rectangles having a dashed pattern inside (same pattern as for the trench). In real samples, each marker has a number in front of it for identification purposes. These numbers are not show on this schematic drawing for simplicity. The picture is not to scale.

wire is quite doable because the concentration of nanotubes is chosen to be low so the average distance between the wires on the trench is large, typically between 20 and 100 μm. The markers are located periodically, having a step of 20 μm (Figure 4.2). The SEM microscope allows the examiner to see the nanowire. The goal is to find two markers between which there is just one single good wire. Let us say the numbers of such two markers turn out to be 14 and 15. To perform photolithography, the sample is spin-coated with a photoresist and baked. Next, the photomask is positioned such that the narrow bridge (typically 10 μm wide) which forms the electrodes connected to the wire (Figure 4.2) lands itself between the markers 14 and 15. This alignment needs to be done with a precision of ∼ 5 μm. In this case, the 10 μm bridge in the center of the wire will cover the selected nanowire. The photoresist under mask is not UV exposed, and thus it remains during the development steps, provided that the used photoresist has positive tone. Thus, the wire and the nearby regions remain protected with the photoresist while other regions on the sample become open during the development step (we presume that the reader is familiar with the basics of photolithography). Finally, one needs to find an agent that can etch the deposited metal. For MoGe, the most efficient agent is the hydrogen peroxide H_2O_2.[2] As the sample is immersed in concentrated H_2O_2, one observes that within just a few seconds the MoGe disappears from the chip surface, except in the regions which are protected with the photoresist. Thus, the electrodes are formed. The photoresist is then completely removed in acetone and the sample is ready to be measured.

When the sample is ready, it can be examined under the SEM one more time in order to make sure that the number of nanowires bridging the slit and connecting

---

2) Other metals can require other etchants. For example, Nb is well-etched in SF_6 plasma. Ar ion bombardment can also be used for various metals. The photoresist might provide more or less adequate protection depending on the choice of the metal and the etchant.

the electrodes is as needed. In the case when the number of wires is larger than what is wanted, one can still correct the problem by using the focused ion beam machine (FIB) to "cut" unwanted wires. This is a risky procedure though since the FIB has a tendency to shoot some number of Ga ions quite far from the point where the beam is focused. Thus, nanowires can be modified unintentionally. While using the molecular templating, in most cases, it is not necessary to use FIB since the number of wires connecting the electrodes can easily be chosen as desired by careful examination of the trench under the SEM before the photolithography is made.

A schematic of the typical layout of electrodes and contact pads relative to the nanowire is shown in Figure 4.2. The total size of the Si chip is approximately 5 mm × 5 mm. The ∼ 100 nm wide trench is shown as the dashed area in the middle of the chip, going through the entire chip. The undercut is not shown. The MoGe electrodes are black. Five contact pads are shown, three on the left and two on the right side. The electrical connection to these contact pads is typically done using gold wires (50 μm in diameter) and indium dots (200 μm in diameter). The nanowire is shown as the short black rectangle in the middle crossing the trench. The electrodes are about 10 μm wide, so they fit between the markers (although, if the electrodes overlap a marker, it is not a problem either).

The Si chip containing the sample is fixed using a carbon tape (intended for the SEM high vacuum environment) onto a plastic chip carrier having nonmagnetic metallic pins. As mentioned above, the connections between the contact pads of the sample and the pins of the chip carrier are made using thin gold wires which are glued to the pins of the chip carrier using a silver paste Electrodag-1415M, and attached to the contact pads using small indium spheres, which are soft and sticky. The procedure of mounting the sample requires certain precautions since static electricity can easily burn the nanowire. Thus, the person making the installation of the sample must be grounded during the sample installation.

The images captured using a high resolution Hitachi-4800 SEM have a resolution of roughly 5 nm. They indicate that the great majority of MoGe nanowires deposited on single carbon nanotubes are continuous and apparently quite homogeneous (Figure 4.3). On the other hand, other metals might give granular wires. Thus, the choice of the material used to decorate the molecule is important. For example, most of pure elemental metals result in granular nanowires if MT technique is used [106]. Nevertheless, some surface roughness is present on the wires even if MoGe is used because it is an amorphous alloy and thus the position of atoms is random in the wire. This roughness does not change the properties of the wire significantly if the scale of the roughness is smaller than the superconducting coherence length and if the fluctuations of the cross-section area are much smaller than the average area of the wire cross-section. Imaging in a transmission electron microscope (TEM) shows, with a high resolution, that the wires are quite homogeneous with a small variation of their cross-section (see Figure 4.4). The surface of the wire is oxidized due to its exposure to air during the fabrication and mounting. The thickness of this oxide is between 2 and 3 nm for $Mo_{79}Ge_{21}$ [102].

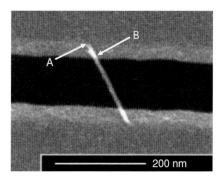

Figure 4.3 A scanning electron microscope (SEM) image of a MoGe nanowire (gray) crossing over the trench (black). The wire is smoothly connecting to the two MoGe electrodes (gray) on the top and on the bottom. The "white spots" are clearly present on each side of the wire. They indicate that the wire is straight and does not dive down the trench. The boundaries of one of the white spots are marked by arrows A and B, corresponding to the arrows A and B shown in Figure 4.1a. The image is obtained by A. Rogachev (University of Illinois at Urbana-Champaign).

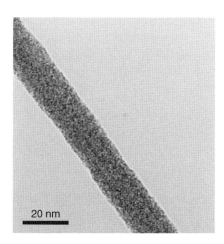

Figure 4.4 A transmission electron microscope (TEM) image of a nanowire deposited over a thin bundle of fluorinated carbon nanotubes. The wire appears amorphous and morphologically and geometrically homogeneous. No granularity is present on this wire. The image is obtained by A. Bollinger (University of Illinois at Urbana-Champaign). (For an illustration of how granular wires appear in TEM, see Figure 1 of [96]. Such granular systems might be used as models of weak link arrays. This book is devoted to homogeneous systems only.)

4.1
Choice of Templating Molecules

The molecule, used as a "substrate" in the MT method, must be straight and sufficiently rigid to withstand the metal deposition. The molecule must be sufficiently strong in order to be able to retain its shape in a suspended state, even in vacuum, which is needed for the metal deposition. The sputtering of the metal is a room-temperature process. It is mild enough so that even DNA molecules can withstand it and maintain their structural integrity during the deposition. It was found empirically that amorphous molybdenum-germanium (MoGe) alloy exhibits an excellent adhesion to DNA molecules as well as to carbon nanotubes [101, 105, 107]. To date, the MT technique has been successfully used with various types of molecules, including single-walled carbon nanotubes [100, 106, 108], fluorinated carbon nanotubes (sometimes abbreviated as "fluorotubes") [102, 103], multiwall carbon nanotubes [114, 115], and DNA molecules [105, 110, 111]. Nanorods have also been used as templates for the metal deposition [112, 113].

Fluorinated carbon nanotube is the easiest molecule to use and the most reliable substrate molecule. This molecule, unlike regular carbon nanotubes, is perfectly insulating because the π-electrons on its surfaces are bound by fluorine atoms [103]. Since the molecule is insulating, only the metallic coating contributes to the electrical conductivity of the device. Theoretical modeling of such devices is easier. The fluorotube can be dissolved in isopropanol, and thus getting single tubes is possible. On the contrary, there is no solvent for regular carbon nanotubes. They can only form a dispersed suspension in 1,2-dichloroethane, which is typically achieved by sonication. Thus, in most cases, if regular nanotubes are used, they come in bundles and not as single tubes. This fact contributes to the increased width of nanowires made with regular nanotubes.

4.2
DNA Molecules as Templates

Deoxyribonucleic acid (DNA) contains genetic instructions used in the development and functioning of all living organisms. It is robust because, as first discovered by James Watson and Francis Crick [117], the structure of DNA comprises two chemically linked helical chains spiraling around each other and therefore reinforcing each other. The diameter of this Crick–Watson double-helix is $\sim$ 2.5 nm, and thus it is suitable for templating nanowires. A new field, known as DNA nanotechnology, has recently emerged. It relies upon the unique molecular recognition properties of DNA molecules, which are able to create self-assembling DNA constructs having useful properties [118]. DNA is thus being used as a *structural template* rather than as a carrier of biological information. Such an approach has been used to create a great variety of two-dimensional periodic patterns and networks, as well as three-dimensional constructs in the shapes of polyhedra [119]. The templating functions of DNA have been demonstrated in recent experiments in which a linear arrangement of nanoparticles, such as gold nanoclusters or streptavidin proteins, was achieved on the surface of the DNA molecule [120]. It is becoming evident that DNA can be regarded as a "backbone" for the fabrication of information-processing devices, chemical and biological sensors, and molecular transistors at the nanometer-size scale [121, 122].

By taking advantage of DNA self-assembly capabilities [123], one can envision that single DNA and/or self-assembled DNA constructs can be used as a "skeleton" for metallic or even superconducting networks of wires. The approach could lead to the creation of complex metallic networks having the smallest dimensions of the order of the diameter of DNA. The key to practical realizations of DNA molecular templating lies in the possibility of creating a homogeneous metal coating on single molecules, which transforms the molecules into thin metallic wires. In the first such attempts, a wet-chemistry approach was used to metal-coat DNA [124–127]. This approach tends to yield rather granular wires, which typically exhibit very high electrical resistance at low temperatures. These two problems (granularity and very high resistance at cryogenic temperatures) are in fact related to each other. If the wire is composed of weakly connected metallic grains, electrons tend to localize on these grains due to the Coulomb blockade effect [9, 128, 129]. The Coulomb repulsion between the electrons leads to an impossibility of electrons to jump from one grain of metal to another, thus leading to a strong increase of the electrical resistance at low temperatures.

Metal-coating of DNA using physical (e.g., sputter-deposition) rather than chemical methods gives more homogeneous wires which show high conductivity and even superconductivity [105]. This section describes in some detail the process of the DNA deposition on the substrates having a trench for the purpose of subsequent coating with a metal. The deposition begins with a solution of λ-DNA (which is 16 µm long) in water, having a concentration of $\sim$ 500 µg/ml. One can purchase such solution from Promega, for example. The stock solution of λ-DNA is too concentrated and needs to be diluted with deionized (DI) water down to $\sim$ 3 µg/ml in order to ensure that the average distance between molecules crossing the trench is many microns [111]. Whenever one transfers a solution containing DNA through a micropipette, it is recommended that one widens the pipette output to $\sim$1 mm in diameter to ensure the easy passage of long DNA molecules. A 4 µl drop of the DNA solution is placed on the surface of the Si chip containing the trench. The chip is then placed in a desiccator to dry under vacuum. As the stock solution contains some buffer salts, which dry as crystals on the surface, it is necessary to rinse the chip in DI water after the DNA deposition. The λ-DNA molecules become strongly fixed to the surface upon drying due to van der Waals forces, and they do not wash away when the chip is rinsed in DI water, while most of the salt deposits dissolve almost completely. After the final complete drying of the sample, the end result is that many DNA molecules cross the trench, the average distance between them being in the range of 20–200 µm, which is precisely what is needed for performing further steps of MT. An important fact is that the molecules crossing the trench are pulled straight during the drying process, and always dry nearly perpendicular to the trench. Possibly due to the van der Waals attraction, the DNA has a lower energy

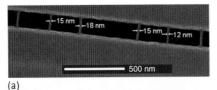

(a)

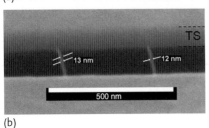

(b)

Figure 4.5 SEM micrograph of nanowires made using DNA templates [111]. The wires appear morphologically homogeneous, without noticeable granularity. The apparent width of the wires is indicated on each image. The actual width of the metallic core is smaller than the indicated number, for example, due to the surface oxidation (by ~ 2.5 nm on each side of the wire, in the case of MoGe wires), due to a carbon coating of the wires during SEM imaging, which contributes to a larger apparent width, and due to SEM resolution limitations which smear the image and make the wire look somewhat wider. (a) Top view of a sample having a trench (black) crossed by six nanowires templated by DNA molecules. (b) SEM image of a tilted sample which shows that DNA crosses the trench at the level of the bottom surface of SiN film. The two horizontal dashed lines, marked with letters "TS," show the thickness of the SiN film. The surface of SiN between the lines is called "tilted side" (TS).

per unit length when it is bound to the SiN membrane compared to the suspended state of the molecule. Thus, the length of the suspended segment is minimized, leading to the result that all suspended molecules are straight and roughly perpendicular to the trench sides (Figure 4.5a). After the sample is dried completely, it is ready for metal deposition. A thin metallic film is deposited over DNA molecules, thus converting them into thin wires suspended across the trench.

4.3
Significance of the So-Called "White Spots"

The term "white spots" is used to describe the small and bright regions visible on SEM micrographs at the ends of many, but not all, of the nanowires. White spots occur near the points where the wire connects to the electrodes. For example, on Figure 4.3, one of the white spots is located between the arrows A and B. The reason for the appearance of these bright regions at the ends of the wire is that the segment of wire between points A and B is suspended above the tilted side of a trench. These sides of the trench are also covered by metal, as illustrated in Figure 4.1a, because the sputtering is a more or less an isotropic deposition process. Thus, in the SEM images, the brightness due to the metalized side of the trench adds to the brightness of the wire. So, the presence of the white spots is an indication that the wire is straight and coplanar with the leads, i.e., it stays above the tilted region of the SiN surface. The white spots can be used as a guidance in the wire selection process in MT, before the photolithography and etching steps are performed to remove all wires except the chosen one. Amongst many wires formed across the trench after the sputtering process, it is necessary to select one among those which make good electrical connections to the electrodes and which are straight and coplanar with the electrodes. It turns out that in some cases, the wire might *not* be coplanar if the templating molecule sticks to SiN too much and bends into the trench, as is illustrated in Figure 4.1b. In such cases, the film electrodes would not be connected to the wire directly, but through the tilted regions on the inner sides of the trench. They are marked as "TS" in Figure 4.1 because these sides of the trench are tilted with respect to the horizontal top surface of the SiN film. A way to distinguish between these configurations is to examine SEM micrographs of the wire and check if white spots are visible. White spots do occur if the molecule is straight (Figures 4.3 and 4.1a), and they do not occur if the molecule dives down into the trench and crosses the gap at the level of the bottom surface of the SiN film (Figures 4.5b and 4.1b).

A typical apparent width for the tilted side is ~ 100 nm. The tilted sides of the trench might not be as well coated with metal as the top surface of the SiN film. Therefore, the tilted side can exhibit suppressed superconducting characteristics, for example, a lower T_c, leading to an additional transition or a step on the $R(T)$ curves of the sample. Such problems can be avoided or minimized if the angle θ (defined in Figure 4.1) of the tilted side region is much smaller than $90°$. Yet, even if $\theta = 90°$, the tilted regions will be coated with MoGe to some extent since the sputter deposition process is not directional, but rather isotropic in nature. Another measure to avoid complications is to make the thickness of the deposited metal large enough so the film thickness on the tilted region is large compared to the thickness at which the T_c of the chosen type of superconducting film is significantly suppressed. For example, for $Mo_{79}Ge_{21}$, the film should be thicker than ~ 5 nm in order to make sure that T_c is close to the bulk T_c of the material. Thus, it is important to know which of the two configurations, shown schematically in Figure 4.1a or b, is realized in any given device. Presence or absence of white spots is a good indicator in the cases when the angle θ is less than $90°$.

Fluorinated carbon nanotubes are more robust than DNA and remain straight when placed over up to ~ 300 nm wide trenches. Regular carbon nanotubes are even more rigid and remain straight when suspended, even on a scale of a few microns. DNA molecules, on the other hand, are flexible, and so they almost never show white spots, even if the trench is as narrow as ~ 100 nm. Absence of white spots suggests that the DNA molecule dives into the trench before crossing it. This expectation is directly confirmed by tilting the sample during SEM imaging. For example, in Figure 4.5b, the DNA molecules appear suspended near the bottom of the trench.

So, when DNA molecules are used as templates, it is necessary to sputter thicker superconducting films in order to ensure that the tilted side regions, which are electrically measured in series with the wire, have a critical current much larger than the studied nanowire. Since in MT the metal is deposited by sputtering, which is not a directional process, coating of the tilted sides of the trench is quite possible, as it just requires longer deposition times and thicker films. On the other hand, if the focus of the study is a superconductor-insulator transition, which means that extremely thin wires need to be measured, then using DNA molecules as templates is not justified and carbon nanotubes have to be employed in the MT procedure. White spots also do not occur when the wires are long, even if fluorotubes are used as the template. For example, if the trench is $\sim$ 500 nm wide, white spots usually are not observed. Thus, if the metallic films are not thick enough, such long samples frequently show multiple resistive transitions, probably due to the weakened superconductivity in the tilted region of the SiN surface leading to the nanowire. Under weakened superconductivity, we understand a reduced T_c compared to the film electrodes and a low critical current, comparable to the critical current of the nanowires. If the molecule is not straight, it could be a problem in the cases in which the films must be thin. For example, studies of a superconductor-insulator transition in thin wires [130] mostly involve short wires. Empirically, this means shorter than $\sim$ 200 nm. Longer wires, made using MT with fluorotubes, exhibit signs of inhomogeneity, possible related to the tilted side regions acting as weak links. For short wires, that is, those shorter than $\sim$ 200 nm, the MT technique works reliably and reproducibly. Reproducibility is one of the key factors that is used to conclude that the wires are homogeneous and have no weak links, because, if present, weak links are random and thus make $R(T)$ curves randomly shaped.

5
Experimental Methods

5.1
Sample Installation

After the fabrication process is finished, the sample is mounted for electrical measurements. The mounting procedure is very "dangerous" for the wires, as this is when many of them are typically burned by static electricity.

To ensure the successful mounting of the wire, the person making the mounting should be electrically grounded, for example, using an antistatic wristband and a seat that is sprayed with an antistatic solution. The Si chip is mounted onto a plastic chip carrier having six metallic pins (Figure 5.1) which are nonmagnetic. Electrical connections between the pins of the chip carrier and the contact pads on the Si chip are made using 50 µm diameter gold wire. First, four or five gold wires, each $\sim$ 1 cm long, are soldered (or attached using silver paint) to desired pins of the chip carrier. Then, a small piece of double-sided sticky carbon tape (which is typically used for mounting samples in SEM) is placed in the center of the chip carrier, and the sample is placed over the carbon tape. The tape serves the purpose of fixing the Si chip on the surface of the chip carrier. One needs to make sure that the tape does not touch any of the pins of the chip carrier since the tape, typically used to mount samples onto SEM holders, is electrically conducting. Next, the free ends of the gold wires are connected to the corresponding contact pad of the sample using $\sim$ 250 µm diameter indium spheres. This is done as follows. One sphere is placed on a contact pad and pressed from the top using the flat-backed surface of a stainless steel drill bit, or using the flat end of a metal lead of a common commercial resistor. The In particle must stick to the contact pad. Then, the corresponding gold wire is placed over the In dot and pressed upon again. Finally, another In dot is placed over the gold wire and pressed one more time. The second In dot is needed in order to reduce electrical resistance of the contact and make the connection reliable enough that it can withstand the process of cooling the sample down to cryogenic temperatures. With some training, such a connection process allows one to connect thin gold wires to a thin-film MoGe contact pad of the sample without using a soldering iron or an ultrasonic bonder, either of which might bring an unwanted voltage to the sample and thus burn or damage the wire.

Superconductivity in Nanowires, First Edition. Alexey Bezryadin.
© 2013 WILEY-VCH Verlag GmbH & Co. KGaA. Published 2013 by WILEY-VCH Verlag GmbH & Co. KGaA.

Figure 5.1 A photograph of the chip carrier having a mounted Si chip. The patterned MoGe electrodes and contact pads are visible on the chip. The chip, which is 4.8 mm × 4.8 mm in size, is connected to the plastic chip carrier using a vacuum sticky tape. At least four contact pads are typically connected to the chip carrier pins by gold wires. Indium "dots" (i.e., small spheres) (visible on the picture as brighter round-like regions on the chip surface) are used to connect each gold wire to the contact pad. Usually, an In sphere is placed on the contact pad and pressed upon with, for example, the flat end of a thin drill bit. It sticks to MoGe strongly enough to hold the gold wire. Then, the thin gold wire is pressed into the indium dot slightly. Finally, one more In dot is placed above the gold wire and squeezed against the first dot. The nanowire itself is of course too small to be visible on this large scale image [111].

Once all the pads have been connected to the pins of the chip carrier, the chip carrier is inserted into a matching socket positioned on the "cold finger" of the cryostat. Since the chip carrier is made of plastic, which does not conduct heat well, some arrangement has to be made to cool down the sample to the base temperature of the cryostat. In a ^{4}He cryostat, which operates with liquid helium isotope ^{4}He, the temperature can be reduced down to ~ 1.3 K by pumping on the liquid helium. In such systems, the Faraday cage, in which the sample is located, is filled with a low pressure He gas. The thermal conductivity of the gas does not depend strongly on the pressure, and thus the choice of the pressure is not important. It can be, for example 1 Torr. The ^{4}He in this case acts as the thermal exchange gas or a "thermalization agent." It provides a thermal link between the sample and the reservoir of He outside the Faraday cage. The gas allows the sample to cool down to the bath temperature, that is, to the temperature of the liquid ^{4}He which is pumped upon. Since the heat contact between the sample and the helium bath is achieved through the gas which does not have a particularly high thermal conductivity, the temperature of the sample can be controlled using a resistor connected to the Faraday cage and warmed up by Joule heating.

If a ^{3}He cryostat is used, thermal exchange gas is not usually present. The thermalization of the sample is achieved exclusively through the leads connected to its contact pads. In the considered example, the sample gets cold through the gold wires connected to the contact pads (Figure 5.1). The gold wires become cold due to their connection to the pins of the chip carrier. The pins in turn are inserted into the matching socket. The pins of the socket are soldered to the leads, which are thin resistive wires carrying electrical signals to and from the sample and the room-temperature part of the cryostat ("top" of the cryostat). These leads are thermalized by winding them over a copper rod (Figure 5.2), that is, the "cold finger" of the cryostat. The cold finger is a part of the cryostat which is refrigerated to the base temperature, that is, to the lowest temperature available in the used refrigerator. The thermal connection between the leads and the cold finger must be made strong so the leads acquire the base temperature of the cryostat. At the same time, there should be no electrical connection between the leads and the cold finger at all. Thus, one has to use wires which are covered with varnish or any other plastic that provides reliable electrical insulation. To achieve the desired thermal contact, the leads are wound many times (say 100 or 1000 times) around the cold finger and coated using Stycast epoxy glue (from *Lakeshore Cryotronics*), having small, $\sim$ 100 μm diameter, Cu spheres mixed into it. The role of the Cu particles is not only to improve thermal conductivity of the glue but, more importantly, to impose a strong damping on any electromagnetic (EM) waves propagating along the leads to the sample. The damping is achieved by means of Joule heating, occurring due to so-called Eddy or Foucault currents which are induced in the Cu par-

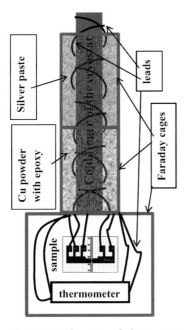

Figure 5.2 Schematics of a low temperature electromagnetic (EM) noise filter. It has two stages, namely, the Cu powder and silver paste filters. The filters cut the noise and ensure good thermalization of the leads connected to the sample. Most of the microwave photons propagating along the leads are absorbed by Cu and Ag particle in the filters. The cold finger of the cryostat (the Cu rod in the center) provides the low temperature. The Faraday cages, separate for each filter but the same for the sample and the thermometer, are shown as red boxes.

ticles by the propagating waves [33]. The unwanted electromagnetic waves, which represent EM noise, originate at the top of the cryostat due to black body radiation. Alternatively, the thermalization of the leads and the noise filtering can be achieved by coating the leads, which are always wound around the cold finger, by a glue filled with Ag particles, typically marketed as "silver paste." Not all types of sliver paste are equal. One has to choose a sliver paste which penetrates well between the wires and which is stable mechanically, even at low temperatures. The leads must be made of a resistive thin wire, the length of which under the Cu-filled Stycast should be at least 0.3 m (the lower the temperature is, the longer the thermalization length should be). The resistance of the leads is usually chosen in the range of 0.1–1 kΩ. For example, for making the leads, one can use 50 μm diameter, nylon-coated nickel-chrome alloy wires, "Stablohm 800A" from California Fine Wire Co., Grover Beach, CA, USA. These wires might be somewhat hard to handle since they are thin and the insulation is thin as well, so it might break if the wire is not handled with extreme care. A good effect of thermalization and filtration can also be achieved with slightly thicker wires. For example, if the lowest temperature is 0.3 K, an adequate filtering can be achieved by using one meter of constantan wire (Cu(55%)Ni(45%) alloy having resistance 60 Ω/m and having the diameter of 100 μm) isolated using a layer of varnish coating plus a thin cotton insulation. The presence of the cotton insulation greatly reduces the risk of breakage of the insulation and corresponding unwanted electrical connections between the leads and the cold finger, and just between the leads themselves. The cotton insulation must be thin enough that the epoxy can leak through it and come into contact with the wire to ensure thermalization. Otherwise, the wire will remain hot, bringing too much heat load to the sample, and it would not cool down to the base temperature. The wire can be embedded into a mixture of copper powder (325 mesh, *Alfa Aesar*) and the epoxy Stycast #1226 from *Emerson and Cuming*. The filter design, involving a combination of Cu powder filled epoxy stage and the silver paste stage, provides improved filtering characteristics. To reduce the EM pickup by the leads, they have to be arranged in twisted pairs and at all stages have to be enclosed in a metallic shield, that is, a Faraday cage. If twisted pairs are not available, each wire wound N times clockwise around the cold finger (Figure 5.2) should be wound also N times counterclockwise in order to reduce any potential EM pickup.

It is important to ensure that the leads are made from a thin wire, typically a resistive and nonmagnetic alloy with low heat conductivity. Before thermalizing the leads to the ^{3}He pot, it is necessary to thermalize the leads at every higher-temperature stage of the refrigerator. In particular, the leads should be wound on a Cu rod and coated with a low-temperature-compatible epoxy, and the Cu rod should be thermally connected, using, for example, a nonmagnetic brass bolt, to a part of the cryostat maintained at 4 K. The same should be done at the 1 K pot stage of the ^{3}He cryostat.

The thermometer, a calibrated RuO (or Cernox) resistor (from *LakeShore Cryotronics Inc.*), can be mounted in the same way as the sample, that is, on a separate chip carrier that is analogous to the chip carrier of the sample. Under such conditions, the thermometer is also thermalized through the leads which are thermalized in the same way as the leads connected to the sample. The sample and thermometer chip carriers can be placed into the same large socket, near each other, to make sure they are exposed to very similar conditions and thus attain the same temperature.

5.2
Electronic Transport Measurements

In this chapter, let us discuss a typical setup that can be used to study quantitatively superconducting wires of the type shown in Figure 4.3. Assume that the sample and the contact pads are shaped as is illustrated in Figure 4.2. In this figure, the contact pad marked V_f is not used, except for measuring the properties of the thin film electrodes connected to the nanowire, separately from the nanowire itself. This setup can be used, for example, to observe a superconductor-insulator quantum phase transition on short wires [100], or, to collect evidence supporting the hypothesis that Little's phase slips can pass across the wire by means of quantum tunneling [59, 116, 131].

A typical setup circuit diagram is shown in Figure 5.3. To measure the sample resistance, one needs to apply a known current between the contacts "$+I$" and "$-I$" of Figure 4.2 and measure the voltage on the contacts "$+V$" and "$-V$." Then, in the linear regime, that is, if the voltage is proportional to the current, the resistance equals the measured voltage divided by the known current. In reality, applying a DC current and measuring the corresponding DC voltage is not the most accurate approach. Possible reasons causing errors in such DC methods include contact voltages on the leads occurring due to thermal gradients, and offset voltages and offset currents on the signal sources and amplifiers. In addition to this, so-called "$1/f$" noise becomes especially strong at low frequencies, so that DC measurements appear quite noisy. Thus, an AC probe current is typically used.

The frequency of the AC probe current has to be very low compared to the internal response frequencies of the sample (which is easy to achieve) and lower than the cut-off frequency of the leads, which is typically in the range of 0.1–1 kHz. The cut-off frequency of the leads can be that low due to the fact that the leads pass through various noise-reducing devices, for example, the Cu-powder filter (Figure 5.2) and the π-filters[1] (see Figure 5.3). Also, the frequency should not be commensurate with the frequency of the power line (60 Hz in the US and 50 Hz in Europe). A possible choice of the measurement frequency is 12.7 Hz. Other choices might be good also, but one needs to experimentally verify that the frequency is such that its further reduction does not change the measured value of the sample resistance R. For example, if the frequency is reduced to 1 Hz, the measured resistance of the sample must remain the same as was measured at 12.7 Hz within the noise level of the setup. If the result is different, then both measurements might be incorrect. In such cases, one needs to search for the frequency level at

---

1) For example, BLP-1.9, purchased from www.minicircuits.com, accessed: 9 August 2012.

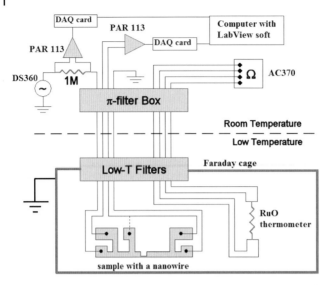

Figure 5.3 Diagram of a typical circuit setup that can be installed within, for example, a ^{3}He refrigerator, which provides temperatures down to 0.3 K [132]. The sample is current-biased using a low-noise, low-distortion function generator (DS360) connected to the current contact pads of the sample through a well calibrated, so-called series resistor, which is connected in series with the sample. Its typical value is 1 MΩ, which is much larger than the sample resistance in the normal state. The voltages on the voltage contact pads of the sample and on the series resistor are measured using low noise, battery-operated preamps (e.g., PAR 113). To eliminate electromagnetic (EM) noise and interference on the sample, it is necessary to use π-filters on the top of the cryostat and cold "low-T filters" near the sample. A detailed design of the low-T filters is shown in Figure 5.2. In the illustration, the low-T filter includes two stages, namely, the Cu powder filters and the silver paste filter (Figure 5.2). The sample, the filters and the thermometer must be enclosed in a grounded Faraday cage to further reduce the EM noise. A ruthenium oxide thermometer is placed near the sample. The thermometer is basically a calibrated resistor. It is attached to the same type of leads and in the exact way as the sample. Since both the thermometer and the sample are cooled through the leads, it can be expected that their temperature will be equal with high precision. The voltages on the preamps are digitized using data acquisition (DAQ) cards or digital voltmeters and submitted to a computer for numerical analysis. No DAQ card or a digital voltmeter should be attached directly to the leads connected to the sample since EM noise would be too large.

which further reduction of the excitation frequency does not lead to a noticeable change of resistance. For instance, it might happen that the measured value of R is the same when measured at 0.1 and 1 Hz. Then, 0.5 Hz excitation signal can be considered acceptable and the measurement is correct. Since the measurement frequency is so low that the result does not depend on it, one can reasonably expect that the result would be the same as if measured at a DC current (if the noise could be eliminated somehow). Thus, such low frequency measurements are commonly called "DC measurements," in contrast with true AC experiments in which the frequency is so high that the superconducting response of the sample itself is frequency-dependent. We will not discuss such high frequency experiments, but will concentrate on DC transport properties of superconducting nanowires, which are measured at frequencies sufficiently low so that further reduction of the frequency does not alter the results.

As was discussed above, AC measurements provide a higher precision, even if DC transport properties are being studied. The question to be discussed is the amplitude of the AC probe current. One common way to characterize the sample is to measure its linear or "zero-bias" resistance, that is, the resistance defined as $R = V/I$ and measured under such small values of V and I that the $V(I)$ curve is linear. The $V(I)$ curve is usually linear at low bias and becomes nonlinear when the applied current becomes larger than the current noise induced by the thermal fluctuations. In the commercial resistor, the range of the linearity is much larger than this conservative estimate. For superconducting samples of the type of Figure 4.3, the linear part of the $V(I)$ curve typically extends to the current of the order of 10 nA. Thus, the amplitude of the AC excitation current needed to produce a linear $V(I)$ curve has to be 10 nA or less.

While doing measurements, one has to verify that it is conducted in the linear regime. Otherwise, the obtained resistance could not be considered linear or zero-bias resistance, and would be incorrect, in general. Being in the linear regime practically means that a reduction of the voltage or current by some factor, for example, a factor of 3, does not change the resulting R value. If it does, the measurement is still nonlinear and one needs to reduce the probe current further.

Practically, the linear (or ohmic) resistance of the sample can be determined from the slope of the best linear fit to the linear segment of the $V(I)$ curve. A LabView program collects an array of voltages on the sample, V_i, where i is the point number. At the same time, the corresponding array of voltages, $V_{ser,i}$, is measured on the series resistor. Both signals change in time following a sine function. The current is then computed as $I_i = V_{ser,i}/R_s$. Then, V_i is plotted versus I_i. The resulting curve must be linear if one wants to obtain the linear resistance. The $V(I)$ curve is then fitted with a straight line. LabView provides convenient options for finding the best linear fit to such ensemble of the data points. The obtained slope is equivalent to the linear resistance of the sample, R. It is called "zero-bias," contrary to the fact that it is not measured at exactly zero-bias, because for a linear $V(I)$ curve, it is true that $R \equiv \lim_{I \to 0}(V/I) = V/I = dV/dI$.

We will always use the term "resistance" in the meaning of the linear or zero-bias resistance, measured at low bias, at which the $V(I)$ curve is linear. It should be distinguished from the "differential resistance" which is defined as $R_d = dV/dI$, which is the slope of the $V(I)$ curve and which can be measured at low or at high bias, including nonlinear segments of the $V(I)$ curve. In some cases, the resistance can also be defined as V/I at high bias currents, at which the $V(I)$ curve is already nonlinear. With such a definition, the resistance is a function of the bias current or bias voltage. We will not use such a definition. In all cases, when we discuss the resistance of the sample R, we will mean the resistance measured using a sufficiently low bias current so that the V–I curve is linear. Thus, R is independent of I or V, that is, $dR/dI = 0$ and $dR/dV = 0$. On the other hand, the differential resistance is a function of the bias current in most mesoscopic samples, that is, in general, $dR_d/dI \neq 0$ and $dR_d/dV \neq 0$.

The measurement of the voltage must be done using a low noise battery-operated preamplifier, called "preamp" for brevity. For example, either PAR 113 (from Princeton Applied Research) or SR 560 (from Stanford Research Systems) can be used. One of possible test circuits is shown in Figure 5.3. The AC current bias is obtained using a high precision, low noise, function generator which acts as an AC voltage source (e.g., Stanford Research Systems, DS 360 can be used). The voltage source is connected to the sample through a standard $R_s = 1\,M\Omega$ resistor which is much larger than the resistance of the measured sample. This series resistor defines the current in the sample. It can also be called "standard resistor" to emphasize its constant value which is well calibrated. The voltage on the standard resistor is also measured using a low-noise preamp. The bias current is then calculated using the Ohm's law, as explained above. For achieving a good precision of the current measurement, it is a good idea to purchase a very stable high precision series resistor, which value is known and is guaranteed by the company to a high precision (e.g., allowed deviations from the nominal value are less than 0.1%). During the measurement, the resistor must be installed in a Faraday cage (i.e., a sealed metallic box). All connections, including those leading to the box containing the resistor R_s and to the sample, must be done by well shielded cables. The typical choice is the so-called BNC cable. Banana-type connections must not be used since they are not well shielded and will pick up a strong electromagnetic noise. If the shielding is not done well, the noise and interference would alter the measured properties of the sample, strongly weakening or even completely suppressing its superconducting characteristics, for example, the measured critical current value.

To convert the voltage on the output of the preamps (PAR 113 preamps in Figure 5.3) into the digital format required for the computer, one can use either a digital analogue converter (DAQ) card (e.g., National Instruments, PCI-6030E or PCI-MIO-16XE-10 card) or a Keithley voltmeter. Note that such digital devices should never be connected directly the leads. Even though the leads are passing through EM filters, the noise created by a digital device can be too strong and can alter the apparent behavior of the nanowire. This is one of the reasons that the preamps are necessary.

Some of the more traditional measuring devices, for example, digital voltmeters, may generate high frequency (10–100 kHz) voltage spikes, up to $\sim 100\,mV$ in amplitude, on their *input* terminals. Such voltage or current noise, if it reaches the sample, can induce additional phase slips in the nanowire and can make the wire appear more resistive than it actually is. This is why the majority of digital devices should never be connected to the sample leads. Incorrect grounding of the measurement circuit may also cause the occurrence of the resistive "tails" in the resistance versus temperature plots. Such tails can be mistakenly understood as QPS-related resistive tails. The entire electrical circuit must be grounded (connected to the ground of the power line for example) just at one spot. Grounding the measurement circuit at two locations can produce strong EM interference and infringe the measurements. Usually, if the measured resistance changes as the grounding point is changed, it is a sign that the noise has some significant effect on the measured resistance.

6
Resistance of Nanowires Made of Superconducting Materials

6.1
Basic Properties

Although it may sound paradoxical, the resistance of nanometer-scale wires made of any superconducting metal or alloy can be larger than zero, even far below the critical temperature. Such resistance is caused by thermal (and possibly quantum) fluctuations. A basic picture is that each segment of the wire has, by virtue of fluctuations, a nonzero probability to become normal (i.e., nonsuperconducting) for a short time. Such fluctuating normal regions break the continuity of the supercurrent, especially if the typical size of such fluctuating normal regions is comparable or larger than the wire diameter. In this chapter, we discuss the basic experimental facts and the model describing such fluctuations, and the resistance of superconducting wires due to such fluctuations.

A typical resistance versus temperature, $R(T)$, curve of a thin $Mo_{79}Ge_{21}$ nanowire produced by sputter-deposition of only 5 nm of the MoGe alloy on a suspended carbon nanotube is shown in Figure 6.1. There are two curves on the figure. The top one shows the resistance of the entire sample, namely, the nanowire and the film electrodes connected in a series with the wire. The bottom curve corresponds to the resistance of the film electrodes only. The sample geometry is shown in Figure 4.2. During the measurements, the bias current was applied between the contact pads marked $+I$ and $-I$. The top curve was obtained while the voltage was measured on the contact pads $+V$ and $-V$, while the bottom curve was obtained by measuring voltage on the electrodes V_f and $+V$. Both curves show a resistive transition, that is, a sharp drop of the resistance, at ~ 5.5 K, which is the critical temperature of the film electrodes, $T_{c,film}$[1]. The film critical temperature is indicated by the black vertical line in Figure 6.1. Below this temperature, the bottom curve remains at the level of zero resistance, naturally, because the film electrodes are superconducting. The top curve at $T < T_{c,film}$ remains close to the level of $10\,k\Omega$. The interpretation of this data is that *below the critical temperature of the electrodes*, the total sample resistance measured on the contacts $+V$ and $-V$ (top curve) equals the resistance of

---

1) Note that the resistance drop at T_c is never discontinuous. An observation of a discontinuous, jump-wise drop should be considered a sign that the measurement is done incorrectly.

Superconductivity in Nanowires, First Edition. Alexey Bezryadin.
© 2013 WILEY-VCH Verlag GmbH & Co. KGaA. Published 2013 by WILEY-VCH Verlag GmbH & Co. KGaA.

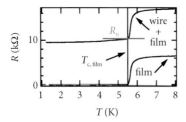

Figure 6.1 Typical resistance versus temperature curves measured on ultrathin $Mo_{79}Ge_{21}$ nanowire, produced by sputter-deposition of only 5 nm of the MoGe alloy on a suspended carbon nanotube. The top curve is the resistance of the entire sample. The bottom curve is only the resistance of the film electrodes. Both curves show a resistive transition at ~ 5.5 K (marked by the vertical solid black line), which is the film superconducting transition. Below the film transition, the electrodes have zero resistance, as is clear from the bottom curve. The nanowire, on the other hand, does not enter the superconducting state in this example. The resistance of the wire remains near $10\,k\Omega$ (see the top curve), as the temperature is reduced. The normal state resistance of the wire, R_n, is indicated by the horizontal grey line. In this example, the resistance of the wire remains close to its normal resistance at all temperatures tested. Such nanowires can be called "normal." Yet, traditionally, such wires are termed "insulating" because their resistance always shows a slight upturn under cooling (not visible in this large-scale graph).

the nanowire because the electrodes contribute zero resistance. Such a conclusion is corroborated by the fact that the sample resistance R_{n1} measured at $T \ll T_c$ using a bias current higher than the critical current of the wire and lower than the critical current of the electrodes (not shown) equals the zero-bias resistance of the sample R_n (measured at low bias currents and at the temperature slightly below the critical temperature of the film electrodes). In other words, the fact that $R_{n1} = R_n$ provides evidence that the interpretation of R_n as the normal-state wire resistance is correct, at least approximately.

The geometry of the electrodes and contact pads shown in Figure 6.1 can be called "pseudo-four-probe." The pseudo-four-probe geometry provides accurate results for the wire resistance when the electrodes become superconducting. The geometric definition of the "normal-state" or simply "normal" resistance, R_n, of the wire is shown by the red horizontal line. It is defined as the sample resistance measured at a temperature slightly lower than $T_{c,film}$. In such definitions, it is assumed that the critical temperature of the wire is considerably lower than $T_{c,film}$, that is, at T slightly lower than $T_{c,film}$ the wire is still normal. This definition of R_n is exactly valid if the wire is long compared to the coherence length, so the proximity effect [65–68] contribution is not significant. Additionally, if the electronic mean free path is much shorter than the wire, the electronic transport is diffusive. Both of these conditions are valid for the MoGe wires of the type of Figure 4.3. The elastic mean free path in amorphous MoGe is very short, about 3 or 4 Å, which is roughly the interatomic distance [151]. This is much shorter than the wire length, which is of the order of 100 nm. The coherence length is typically also much smaller than the typical wire length. This is because in the optimized MoGe alloy, the coherence length is in the range ~ 5–15 nm. Under "optimized alloy," we understand the alloy having relative concentrations of Mo and Ge such that the critical temperature is the maximum, which are 79 and 21% correspondingly. In all cases considered, the alloy composition was close to the optimum composition. The $R(T)$ in Figure 6.1 correspond to a wire, which is not superconducting. As will be discussed in detail later, wires made of a superconducting alloy do not always behave as superconductors. Figure 6.1 shows one such example.

Now, consider an example of a superconducting wire (Figure 6.2). Here, two resistive transitions are observed: The first one (at $T_{c,\text{film}} \approx 4.55$ K) manifests the onset of superconductivity in the electrodes. Speaking precisely, this transition might not take place precisely at $T_{c,\text{film}}$ but, rather, at the T_{KT}, which is the temperature of the so-called Kosterlitz–Thouless (KT) transition, which is slightly lower (typically by a small percentage for amorphous films of the types discussed here) than $T_{c,\text{film}}$, as was discussed for example in [102].

The resistance of the film above T_{KT} is given by the following expression, derived by Halperin and Nelson [137], and Hebard and Fiory [138] (HF)

$$R_{\text{HNHF}}(T) = 10.8 b_{\text{KT}} R_{n,\text{film}} \exp\left(-2\sqrt{\frac{b_{\text{KT}}(T_{c,\text{film}} - T_{\text{KT}})}{T - T_{\text{KT}}}}\right) \quad (6.1)$$

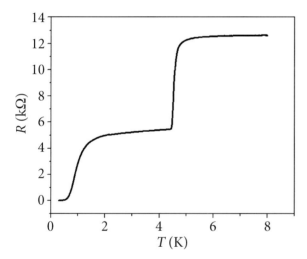

Figure 6.2 Typical resistance versus temperature curve of a nanowire produced by the sputter-deposition of 6.5 nm of the MoGe alloy on a suspended carbon nanotube. The first transition, the middle of which is located at $T \approx 4.55$ K, is due to the film converting into the superconducting state. The best estimate for the normal state resistance of the wire is $R_n = 5.44$ kΩ, which is the sample resistance measured at 4.33 K, that is, at a temperature slightly lower than T_{KT}. The T_{KT} is the temperature at which the film acquires zero resistance. The second transition, apparent at ≈ 0.95 K, is in fact a crossover representing Arrhenius-type freezing out of the rate of phase slips in the nanowire. The interpretation of the two transitions is based on the fact that in the samples discussed throughout this book, the nanowire is connected in series with thin-film electrodes, as is shown in Figure 4.2, for example.

Here, b_{KT} is a nonuniversal constant.

The resistance of the film is indeed zero below T_{KT}, even in theory. A detailed discussion of the two-dimensional (2D), that is, thin film, superconductor physics is beyond the scope of this book. We will focus only on the understanding of thin wires, which are one-dimensional (1D) or quasi-one-dimensional (quasi-1D), that is, the wire diameter is either much smaller or up to four times larger than the coherence length. Basically, the 1D physics occurs when vortex cores do not "fit" into the nanowire because it is not sufficiently wide, and thus superconducting vortices can be excluded from the considerations. On the contrary, the KT physics is due to the interaction of vortices and antivortices generated by thermal fluctuations.

The second transition in Figure 6.2 is observed at $T \approx 0.95$ K. This resistive transition shows the development of superconductivity in the nanowire. The resistance of the wire appears to be zero at $T \approx 0.3$ K. As will be discussed below in detail, the second transition is not a true phase transition, but a crossover. This means that the resistance of the wire is never exactly zero. In fact, it follows the Arrhenius activation law at low temperatures, namely, $R(T) \sim \exp(-\Delta F(0)/k_B T)$, where $\Delta F(0)$ is the energy needed to create a phase slip (i.e., a normal spot) in the wire at zero temperature. This expression is true if quantum phase slips can be neglected and in the limit of low temperatures, at which the barrier for phase slips can be assumed constant, that is, when $\Delta F(T) \approx \Delta F(0)$. In the example curve of Figure 6.2, the resistance appears to be zero at low temperatures (close to 0.3 K) only because it becomes smaller than the resolution limit of the setup used to measure the resistance. In order to understand the temperature dependence of the wire resistance, the concept of Little's phase slips (LPS) is needed, which will be discussed in what follows.

The homogeneity of the nanowire and the absence of weak links and/or constrictions in it is difficult to assert based on SEM images only. Thus, many other indicators are used to confirm homogeneity and uniformity of nanowires. The shape of $R(T)$ curves is a good indicator of the homogeneity. If the sample shows exactly two resistive transitions, the first corresponding to the film electrodes and the second corresponding to the nanowire, then it can be classified as homogeneous. On the other hand, if the lower-temperature transition shows subtransitions or bumps of any sort, the wire probably has weak links. Each constriction of a weak link has its own barrier for LPS and thus contributes its own resistive transition to the collective $R(T)$ curve.

Another factor that helps to confirm homogeneity of a nanowire is the presence of just one jump at the critical current on its voltage-current curve. If there are a few jumps and steps, then the wire probably has weak links, and each weak link has its own critical current.

One more factor that helps to conclude that the wire is uniform is its residual resistance ration $\rho_{300} \equiv R(300\,\text{K})/R(10\,\text{K})$, where $R(10\,\text{K})$ is the value of the wire resistance at $T = 10$ K, which is a low temperature, but still above the critical temperature, and $R(300\,\text{K})$ is the resistance at room temperature. If this ratio is about 0.9, then the wire is homogeneous. Note that the ratio is always smaller than unity for optimized amorphous MoGe due to localization effects which become stronger at lower temperatures. If it is much lower than 0.9, then there are weak links or breaks in the wire which become highly resistive at low temperature due to Coulomb blockade. Furthermore, if the wire would be granular, a gate voltage would change its resistance. Such is not observed. Thus, the wires produced by sputter coating of nanotubes with amorphous MoGe are confirmed to be homogeneous.

6.2
Little's Phase Slips

The notion of a phase slip was proposed by William Little in 1967 [17], and is thus named the Little's phase slip (LPS). This concept is the key for understanding most of the properties of thin superconducting wires. The LPS concept was proposed to understand the mechanism of the supercurrent decay in thin wires and to justify his earlier hypothesis of a superconducting macromolecule [139]. The picture of a phase slip event is based on the assumption that the order parameter in a thin superconducting wire is defined locally as well as globally [140, 141]. If a supercurrent flows along the wire, the order parameter is $\psi(x) = \sqrt{n_s}\exp(-ikx)$, where the wavevector is $k = I_s m/\hbar A_{cs} e n_s$ and A_{cs} is the wire cross-section, I_s is the supercurrent in the wire and n_s is the density of condensed electronic pairs or superpairs. Following Little, such a complex order parameter can be plotted as a spiral in the Argand plane extended by the spatial direction (horizontal x-axis), representing the position along the wire (Figure 6.3). The radius of the spiral represents the amplitude of the order parameter which equals $\sqrt{n_s}$. The number of turns in the spiral is $n = \phi/2\pi$, where ϕ is the superconducting order parameter phase difference between the end of the wire ($x = l$) and the beginning of the wire ($x = 0$). If the wire is closed into a loop (i.e., the point $x = l$ coincides with $x = 0$), then due to single-valuedness of the order parameter, n is an integer and represents the number of phase vortices trapped in the loop. In any case, the number of turns n of the order parameter spiral is proportional to the supercurrent in the wire. The decay of the supercurrent occurs if and only if the spiral loses one or more turns.[2] Such loss can happen if the order parameter amplitude is reduced to zero at some point on the wire for a short time. In other words, the number of turns in the order parameter spiral can change (assuming that the spiral is fixed at the ends, that is, there is no voltage applied to the wire) only if n_s becomes zero at least at one point on the wire. Such a loss of a turn caused by thermal or quantum fluctuations,

---

2) One more reason which effectively leads to a decrease of the supercurrent is a reduction of the spiral radius with the number of its turns per unit length being constant. Such a decrease of the radius can only occur if the external conditions are changed. For example, the radius would become smaller if the temperature is increased or if a nonsuperconducting, that is, normal, metal is deposited on the surface of the wire. While discussing the phenomenon of LPS, we assume that the external conditions are not changing and that the temperature is constant. Under such, quite general assumptions, the loss of turns of the order parameter is the only mechanism by which the supercurrent can decay in time.

which can temporarily and locally bring n_s to zero, is called Little's phase slip (LPS) or simply a phase slip. It is illustrated in Figure 6.3 where the order parameter spiral makes a transition from a configuration making one turn ($n = 1$) around the x-axis to a configuration having zero turns ($n = 0$). We emphasize again that the difference between the two configurations of the order parameter is local. The only difference is that the state of the order parameter marked $n = 1$ makes a single revolution around the x-axis, while the $n = 0$ state does not.

The LPS, that is, the crossing of the x-axis by the order parameter curve, is a short lasting event which happens on a timescale of order $\hbar/\Delta \sim 10^{-12}$ s (for this rough estimate, we assume typical numbers for the critical temperature and the energy gap, namely, $T_c = 2$ K and, by a BCS expression, $\Delta = 1.76 k_B T_c = 0.3$ meV). The LPS is a local even which does not depend on the state of the order parameter at the ends of the wire (assuming the wire is much longer than the coherence length), and therefore it is not influenced by the electrodes which could be connected to the ends of the wire. One exception from this locality rule is the case when the phase slips happen by quantum-coherent quantum tunneling rather than by thermal activation. Then, the flow of energy from the supercurrent, released through the LPS events, to photon and/or plasmon modes in the electrodes can suppress, fully or partially, the tunneling of LPS [28]. On the other hand, the probability of a thermal activation of a phase slip (TAPS) is only dependent on the energy (ΔF) needed to suppress the order parameter to zero at a spot on the wire. This probability or the rate of TAPS Γ is controlled by the Arrhenius activation exponent as

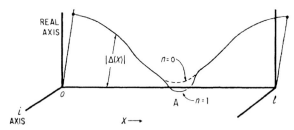

Figure 6.3 Little's phase slip helix [17]. The diagram shows a variation of the complex order parameter as a function of the position along a thin wire, which begins at "0" and ends at "l." The continuous line corresponds to the case when the superconducting phase difference between the ends of the wire is about 2π, meaning that there is a nonzero net supercurrent flowing through the wire. The dashed curve represents the case in which the phase difference is zero and the net supercurrent between the ends of the wire is zero. The transition from the state "$n = 1$" to the state "$n = 0$" takes place when the order parameter is suppressed to zero at some spot on the wire and the spiral loses a turn by crossing the horizontal x-axis. Such an event of crossing the horizontal axis by the order parameter curve is the Little's phase slip (LPS). It is the only mechanism by which a DC supercurrent can decay in a superconducting wire (at a fixed temperature below T_c). Note that the LPS cannot happen and the supercurrent cannot decay unless the order parameter amplitude (marked as $|\Delta(x)|$) reaches zero at some point on the wire, at least for a short time, to allow the order-parameter curve to cross the x-axis. Such a conclusion follows from the consideration of the topology of the present diagram.

$\Gamma = \Omega \exp(-\Delta F/k_B T)$, where Ω is the so-called "attempt frequency" [142] which will be discussed later.

The LPS is the only mechanism by which a DC supercurrent in a thin wire can diminish. Thus, the resistance of a superconducting wire is proportional to the rate of phase slips, $R \sim \Gamma$. The resistance is only strictly zero if there are no phase slips at all. Practically, the resistance of a wire can be extremely small, but never exactly zero in the mathematical sense. This follows from the fact that the temperature of a real sample is never exactly zero and that the Arrhenius factor is larger than zero if the temperature is larger than zero (note also that $\Omega > 0$ if $T > 0$). Since each LPS produces a 2π change of the phase difference, it is possible to use Gor'kov's relation (1.5) to calculate the voltage and therefore the resistance of the wire, provided the rate of LPS is known. Such an approach was adopted in [143], although the attempt frequency was not evaluated correctly. Later, the attempt frequency was computed for the case of gapless superconductivity in [142]. Such a combined theoretical model of TAPS is also known as the LAMH model, after the names of the contributors: Langer, Ambegaokar, McCumber, and Halperin.

Early experiments on superconducting, 0.5 μm diameter, tin whiskers, conducted by Lukens et al. [22] and Newbower et al. [23], confirmed the TAPS theory [17, 142, 143, 171]. Yet, the attempt frequency has a very weak effect on the shape of $R(T)$ curves. Thus, the predictions for the value and the temperature dependence of the attempt frequency [142] could not be confirmed experimentally so far. In fact, the prediction of the attempt frequency [142] might not be applicable in most practical cases since it was derived from the time-dependent Ginzburg–Landau equation, which is only valid when the superconductivity is gapless (B.I. Halperin, private communication, 2005). The applicability range of the LAMH model, estimated in [144], is limited to the temperature interval $0.9T_c < T < 0.94T_c$. This limitation originates from the limited applicability of the calculated attempt frequency. Yet, the Arrhenius exponential factor, which is the dominant part of any theory of TAPS, is correct at all temperatures. Recently, a structure of TAPS in a clean, one-dimensional superconductor, in which superconductivity occurs only within one or several identical conducting channels, was calculated by Zharov et al. [145] for all temperatures. Another recent development is a generalization of the TAPS theory that includes two-wire devices and the quantum interference between the corresponding current paths in magnetic fields [105, 111, 146] and in presence of a phase gradient in the electrodes created by a supercurrent [110, 111].

Qualitatively speaking, the occurrence of LPS provides a nonzero electrical resistance to an otherwise superconducting wire in the following way. Consider the following stationary state. Assume there is a small voltage applied to the ends of the wire. This voltage, due to the Gor'kov phase rotation equation (1.4) and (1.5), increases the phase difference between the ends of the wire: Because the voltage is applied, the electrochemical potentials are different at the ends of the wire and so the phase of the complex order parameter evolves differently with time at the ends of the wire. As the absolute value of the phase difference increases, the magnitude of the supercurrent also increases (up to a certain limit, when the current starts to weaken the condensate density, n_s). This is not unexpected of course since the applied voltage creates an electric field in the wire which acts with a force on the BCS condensate of the electrically charged electrons. This force accelerates the condensate, thus increasing the current. The LPS, on the other hand, occurs inside the wire, at random spots, and acts effectively as a friction force by decelerating the flow of the condensate. This happens because the LPS occurring stochastically at all points along the wire tend to unwind the phase, and thus tend to reduce the supercurrent. The net current would be constant on average (i.e., a dynamic equilibrium would be established) if the voltage applied is such that it winds the phase as much as phase slips unwind the phase, per unit time. The voltage required to maintain such dynamic equilibrium must obviously be greater than zero, unless the net number of phase slippage per unit time is zero. As will be discussed in more detail, such is always the case if the applied current is greater than zero (here, we always know that the current is DC). Thus, a finite resistance occurs. To calculate the resistance quantitatively, there are currently two "competing" TAPS models: the Little's fit and the LAMH model. They are discussed in the following sections.

6.3
Little's Fit

The Little's fit is a simple semiphenomenological formula that provides reasonably good fits to the experimental $R(T)$ curves of nanowires. The formula can be justified not only near T_c, but down to low temperatures. In this section, we consider the phenomenological Little's fit as well as discuss its significance and usefulness.

Initially, we discuss a limiting case in which the wire is so short and thin that if the order parameter is zero at one spot on the wire, then it is close to zero everywhere on the wire. Such a case is realized if all dimensions of the wire are smaller than the coherence length. Suppose also that the wire cannot switch between normal and superconducting states instantaneously. On the contrary, we assume that the time needed for the higher-energy normal state to relax to the equilibrium superconducting state is finite and we denote this relaxation time by τ_{LPS}. In other words, τ_{LPS} is the duration of a phase slip. Let Γ be the rate of LPS, that is, the number of phase slips per second. Each LPS is an event when the wire becomes normal. Thus, Γ equals the average number of times the wire jumps into the normal state each second and τ_{LPS} equals the time the wire spends in a single normal state fluctuation. The probability to find the wire in the normal state, P_n, equals the average total time the wire spends in the normal state within a second. For example, if the wire spends 0.14 s in the normal state during each second (on average), then the probability of being normal is 0.14 and so on. The probability of being normal is, on the other hand, the number of times the wire enters the normal state each second times the duration of each normal-state fluctuation, that is, $P_n = \Gamma \tau_{LPS}$. Now, consider equation (2.28). Suppose in the fluctuation state "x" there is the normal state of the wire. Taking into account the above discussion and (2.28), the probability of being normal can be written in two ways, that is,

$$P_n = \Gamma \tau_{LPS} = \exp\left(-\frac{\Delta F}{T}\right) \tag{6.2}$$

where ΔF is the increase of the wire's free energy as it switches to the normal state. As was discussed in detail in Section 6.2, the change of the free energy equals the minimum work needed to be done in order to drive the wire normal. From (6.2), one gets the expression for the rate of phase slips, namely,

$$\Gamma = (\tau_{LPS})^{-1} \exp\left(-\frac{\Delta F}{k_B T}\right) \tag{6.3}$$

On the other hand, using the Arrhenius activation law for the thermal activation rate, the rate of phase slips can be also written as [21]

$$\Gamma = \Omega \exp\left(-\frac{\Delta F}{k_B T}\right) \tag{6.4}$$

where $\Omega \sim \tau_{LPS}^{-1}$ is the attempt frequency, which, generally speaking, depends on temperature and the bias current. Yet, in many cases, one can assume the attempt frequency to be a constant and still get a good agreement with the experiment because the exponential factor changes much more rapidly with the temperature and with the bias current than the attempt frequency could. This is of course only true if phase slips are rare, that is, if $\Delta F/k_B T \ll 1$.

To estimate the wire average resistance, we use the Little's hypothesis [17], that is, during the time when the wire is normal, its resistance equals its normal resistance (R_n), and that when the wire is superconducting its resistance (R_s) is zero ($R_s = 0$). The usual DC transport measurement probes the average resistance, not the instantaneous resistance because the duration of an LPS is many orders shorter than the time constant of a typical DC transport measurement setup. The average resistance R is then the normal resistance times the probability of the wire to be in the normal state (P_n) plus the probability of being in the superconducting state, $(1 - P_n)$, times the resistance in the superconducting state (R_s), that is,

$$R = R_n P_n + R_s(1 - P_n) \tag{6.5}$$

It is assumed here that the considered short and narrow wire can only exist in one of the two states: the superconducting state, characterized by the order parameter amplitude being close to its equilibrium value, or the normal state, in which case the order parameter is either zero or close to zero everywhere in the wire. Using (6.2) for P_n and the Little's assumptions discussed above, the average wire resistance given in (6.5) can be transformed into

$$R = R_n \exp\left(-\frac{\Delta F}{T}\right) \tag{6.6}$$

This expression for the temperature dependence of the wire resistance will be referred to as the "Little's fit." The strength of this formula is that it does not require the knowledge of the attempt frequency or the duration of phase slips which are hard-to-derive quantities. The Little's fit (LF) for resistance is simply the Arrhenius activation exponential factor multiplied by the normal resistance of the wire. So far, we have provided some justification of the LF validity in the case when the wire is shorter than the coherence length, and thus the wire can access, by means of thermal fluctuations, only two distinct macrostates, namely, the normal state and the superconducting state. Let us now consider the more general case of a wire, which is longer or even much longer than the coherence length. We will see that exactly the same formula for the wire resistance is applicable.

If the wire is long, the size of the normal state fluctuations can be smaller or even much smaller than the wire length L. The analysis of the wire free energy shows that the lowest energy needed to produce a normal spot on the wire is realized when the normal spot size x_ξ is of the order of the coherence length ξ, that is, $x_\xi \sim \xi$. The minimum work needed to produce a normal spot will be denoted by ΔF. If the length of the wire is small, $L < \xi$, then ΔF would simply be the volume of the wire multiplied by the superconducting condensation energy density. For longer wires, the normal spot associated with the phase slip is of the order of the coherence length, that is, it is shorter than the wire. Thus, in longer wires, the energy needed to produce a phase slip is $\Delta F = (H_c^2/8\pi) A_{cs} x_\xi$. This expression is the definition of the size x_ξ of the normal core of the Little's phase slip. Here, A_{cs} is the wire cross-section area and $H_c^2/8$ is the condensation energy density (in Gaussian units).

To compute the resistance of a long wire, imagine that it is composed of independent segments of length x_ξ connected in a series. Since the size of the segments is, by our choice, equal to the size of the most probable normal-state fluctuation, therefore if one segment goes normal, the others may remain superconducting. Thus, the number of such uncorrelated fluctuating segments in the wire is $N_x = L/x_\xi$. Each segment can be either normal, characterized by its normal state resistance $R_{x,n} = R_n(x_\xi/L)$, or superconducting, in which case its resistance is zero. Thus, each such short segment is analogous to a short wire, and thus its time-average resistance can be estimated using the Little's fit, derived above, as $R_x(T) = R_{x,n} \exp(-\Delta F/k_B T)$. In this argument, the length of the segment is chosen to be equal to the size of a single Little's phase slip. Thus, the energy increase ΔF of the system corresponding to one segment fluctuating into the normal state equals the energy of a single Little's phase slip in a long wire. Note that if the size of the imagined segments would be chosen to be much smaller than the size of a single phase slip x_ξ, then typical fluctuations would involve multiple neighbor segments going normal. On the other hand, if the segment size is taken larger than x_ξ, then in a typical fluctuation, only a part of the segment, but not the entire segment, would go normal. As usual, the "typical fluctuation" means the most probable fluctuation, that is, the fluctuation requiring the least amount of work on the system. Note also that here we only consider the so-called strong fluctuation, that is, those which bring the order parameter to zero at least at one place on the wire because only such strong fluctuations allow the phase to slip and thus can cause some resistance to the flow of the supercurrent.

6.3 Little's Fit

Now, since the segments are statistically independent, the total resistance of the wire is the number of segments N_x multiplied by the resistance of each segment $R_x(T)$. Therefore,

$$R(T) = N_x R_x(T) = \left(\frac{L}{x_\xi}\right) R_n \left(\frac{x_\xi}{L}\right) \exp\left(-\frac{\Delta F}{k_B T}\right) = R_n \exp\left(-\frac{\Delta F}{k_B T}\right)$$

Thus, we arrive at the same formula as was obtained for short wires, (6.6). The only difference is that for a short wire $\Delta F = (B_c^2/2\mu_0) A_{cs} L$ (in SI units). However, for a wire longer than the size of the phase slip, one needs to use $\Delta F = (B_c^2/2\mu_0) A_{cs} x_\xi$ (in SI units), where $B_c^2/2\mu_0$ is the condensation energy density in SI units and the size of the phase slip is $x_\xi = 8\sqrt{2}/3\xi$, [19]. The vacuum permeability is $\mu_0 = 1.257 \times 10^{-6}$ N/A². In particular, at temperatures near zero, we have the equality

$$\Delta F(0) = \left(\frac{B_c(0)^2}{2\mu_0}\right) A_{cs} x_\xi(0) \tag{6.7}$$

Of course, thus obtained Little's fit (6.6) is only approximate. Its usage is justified by the absence of a more exact formula applicable in a wide temperature interval. Note that the LAMH model is only applicable in a narrow temperature interval near T_c. The validity of the Little's fit is justified by the fact that the Arrhenius activation law is correct at any low temperature, not just near T_c, and it dominates the temperature dependence of the $R(T)$ curves.

Just as an example, let us estimate what would be the apparent experimental critical temperature $T_{c,exp}$ of a long nanowire. The apparent critical temperature can be defined in as many ways, in principle. One can define it as the temperature at which the resistance drops to the 10% level of the normal resistance. Many define it as the temperature at which the resistance drops by factor 2. Here, for the sake of simplicity of the estimate, we defined $T_{c,exp}$ as such temperature at which the resistance of the wire is $e = 2.718$ times lower than the normal resistance, that is, $R(T_{c,exp}) = R_n/e$. Using the LF, one then obtains $R_n e^{-1} = R_n \exp(-\Delta F/k_B T)$ and therefore $\Delta F = k_B T_{c,exp}$.

One can show (see [19]) that the barrier for phase slips is proportional to the critical current of the wire and can be expressed as

$$\Delta F(T) = \sqrt{6}\frac{\hbar I_c(T)}{2e} \tag{6.8}$$

The critical current temperature dependence is given by the Bardeen formula (see the discussion of (11.1) for more details) [20]

$$I_c(T) = I_c(0) \left(1 - \frac{T^2}{T_c^2}\right)^{3/2} \tag{6.9}$$

which is valid at all temperatures, not only near T_c. Here, the T_c is the true critical temperature of the wire, that is, the temperature at which the barrier for phase slips goes to zero, which is the same as the temperature at which the parameter of the GL theory is $a(T_c) = 0$.

Now, let us combine (6.7), which is valid in SI units, with (6.8) and (11.1), which are valid in the SI as well as in Gaussian units, to obtain the formula for the barrier height as

$$\Delta F(T) = \left[\frac{B_c^2(0)}{2\mu_0}\right] A_{cs} x_\xi(0) \left(1 - \frac{T^2}{T_c^2}\right)^{3/2} \tag{6.10}$$

Assume for this example that the parameters are the same as in bulk optimized MoGe (see Appendix A). Therefore, $T_c = 7.36$ K and the thermodynamic critical field at zero temperature is $B_c(0) = 55$ mT. Also, assume that the wire diameter is 5 nm and its zero-temperature coherence length is $\xi(0) = 5$ nm. The equation we now get is $k_B T_{c,\exp} = (B_c(0)^2/2\mu_0) A_{cs} x_\xi(0)(1 - T_{c,\exp}^2/T_c^2)^{3/2}$. The size of the phase slip at zero temperature can be estimated as $x_\xi(0) = 18.8$ nm, assuming that the same relation between the coherence length and the phase slip core size holds at zero temperature near T_c. With these assumptions, the equation can be solved numerically, and the result is $T_{c,\exp} = 6$ K. This example illustrates the fact that one cannot determine the true critical temperature simply by measuring the $R(T)$ and looking for a point where the resistance drops by a large factor, since $T_{c,\exp} \neq T_c$.

Following the same approach, let us see if TAPS along can explain the experimentally observed transition to a nonsuperconducting state in very thin wires. Assume that the wire diameter is so small that on average, only three atoms fit in its cross-section (two Mo and one Ge). Such a diameter is about 1 nm. The estimated apparent critical temperature for such a wire is $T_{c,\exp} = 1.2$ K. Thus, even such thin wire would appear superconducting if TAPS is the only mechanism suppressing superconductivity in the wire. Nanowires, which are about ~ 5 nm in diameter, exhibit a complete suppression of superconductivity. Thus, other mechanisms, for example, MQT, surface spins, or enhanced Coulomb repulsion in low-dimensional disordered samples, need to be involved in the explanation of the experimentally observed SIT. Finally, let us emphasize that the introduced above "experimental critical temperature" $T_{c,\exp}$, which is introduced through the $R(T)/R_n$ ratio, depends on the wire diameter as much as on the true T_c of the wire. Therefore, we will not use it. Instead, we will always attempt to find the true critical temperature, for example, by fitting the $R(T)$ curve with the Little's fit and extracting the T_c as the best fitting parameter.

Let us now return to the discussion of the prefactor multiplying the Arrhenius exponent. The interpretation for the prefactor Ω in (6.4) is that it is the so-called attempt frequency, that is, the frequency with which the system "attempts" to overcome the phase-slip barrier. Within this interpretation, the rate of phase slips is understood as follows. Due to the thermal energy present in the wire and in the environment and due to the equipartition principle, the order parameter amplitude fluctuates in time. The characteristic frequency of these amplitude fluctuations then equals Ω. This means that the amplitude is not constant in time, but it either increases or decreases with time. In most cases, the deviations from the equilibrium value are small, but sometimes the fluctuation is strong, so strong that the order parameter amplitude reaches zero at some spot on the wire and an LPS occurs. The attempt frequency Ω represents the characteristic frequency of all fluctuations (weak and strong), while Γ is the rate of the strong (i.e., those which reach zero) fluctuations only. The attempt frequency Ω equals the number of times per second the sign of the rate of change of the order parameter is altered. In other words, Ω is the number of times per second the order parameter begins to decrease. Only a small fraction of such "attempts" of the order parameter to reach zero is successful. Since the fluctuations leading to a phase slip correspond to the usable energy increase ΔF (or ΔG if the current is fixed), the fraction of such successful attempts is $\exp[-\Delta F/k_B T]$ (or $\exp[-\Delta G/k_B T]$ if the current is fixed). Thus, the rate of successes (when the order parameter "manages" to reach zero) is equal to the product of the number of attempts (when the order parameter begins to decrease) and the exponential factor defining the fraction of successful attempts. Thus, one justifies (6.4) (in this equation, ΔF needs to be replaced with ΔG if the current rather than the phase difference is fixed). As we have argued above, the Little's fit argument allows us to remove the attempt frequency Ω from the final expression for the wire resistance. This is helpful since the exact derivation of Ω, applicable in a wide temperature interval, has not yet been achieved.

It is instructive to note, for comparison, that in SIS junctions, the attempt frequency is known exactly, down to zero temperature. According to Kramers escape rate theory [21], the attempt frequency for escape from a metastable minimum equals the frequency of small oscillations near the metastable equilibrium. Thus, the attempt frequency of an undamped SIS Josephson junction, Ω_{SIS}, equals the so-called plasma frequency divided by 2π, namely, $\Omega_{SIS} = \omega_p/2\pi$. The plasma frequency is the angular frequency of small supercurrent oscillations in the JJ. Such periodic oscillations occur if, initially, a small charge is moved from one electrode of the junction to the other and then the system is allowed to evolve without further external interference (see the section on Stewart–McCumber model). The supercurrent in this case will flow back and force, in exact analogy with a harmonic oscillator, having the frequency $\omega_p/2\pi = (1/2\pi)\sqrt{2eI_c/\hbar C}$ (in SI units). Here, I_c is the critical current of the junction and C is the capacitance formed between the electrodes of the junction. This expression for the frequency of small oscillations is derived from the Stewart–McCumber (SM) model (see the corresponding section for details). The plasma frequency can be reduced if the junction is biased with a constant current I. The corresponding expression is $\omega_p(I) = \omega_p(0)(1 - I^2/I_c^2)^{1/4}$, where $\omega_p(0) = \sqrt{2eI_c/\hbar C}$. Thus, due to the Kramers theory, the bias-current-dependent attempt frequency is $\Omega_{SIS}(I) = (1/2\pi)\omega_p(I)$. This formula is valid only if the damping is negligibly low, meaning that the quality factor corresponding to the considered small oscillations of the supercurrent is $Q = \omega_p(I)R_n C \gg 1$, where R_n is the effective normal resistance shunting the junction.

Let us make an example estimate of the attempt frequency for a typical JJ having a critical current of 1 μA and the size of 1 μm square. The tunnel junction capacitance, assuming that the dielectric barrier is 1 nm thick and its dielectric constant is $\epsilon = 9$, is $C = 80$ fF. Assume zero-bias current. The corresponding attempt frequency is 31 GHz. Such low frequency motivates, among other things, the application of SIS junctions in qubit designs, for example, phase qubits, which take advantage of the possibility of tuning the frequency through the bias current. The calculation of the attempt frequency is much harder in the case of nanowires. A computation was done by McCumber and Halperin (MH) [142, 149] for temperatures near T_c, assuming that the superconductivity is approximately gapless. The result is

$$\Omega_{\text{MH}} = 0.397 \left[\frac{k_B(T_c - T)}{\hbar} \right] \left(\frac{L}{\xi} \right) \sqrt{\frac{\Delta F}{k_B T}} \tag{6.11}$$

In the discussion above, we introduced the experimental critical temperature $T_{c,\text{exp}}$ as the temperature at which the resistance drops by a factor e with respect to the normal-state resistance. There, we derived a simple equation to such as $\Delta F = k_B T_{c,\text{exp}}$. Using this definition, we can write an estimate for the attempt frequency at this temperature for the purpose of providing a concrete example: $\Omega_{\text{MH}} = 0.397[k_B(T_c - T_{c,\text{exp}})/\hbar](L/\xi) = 1.4\,\text{THz}$. The expression provided within the LAMH model for the resistance of a nanowire is

$$R(T) = \left(\frac{\pi \hbar^2}{e^2 k_B T} \right) \Omega_{\text{MH}} \exp\left(-\frac{\Delta F}{k_B T} \right)$$

Using the above example conditions, the prefactor at $T = T_{c,\text{exp}}$ is estimated as $(\pi\hbar^2/e^2 k_B T_{c,\text{exp}})\Omega_{\text{MH}} = 22.4\,\text{k}\Omega$. In the Little's fit, the prefactor is taken to be equal to simply the normal resistance of the wire. Remembering that the effect of the prefactor is much weaker than the effect of the exponential factor, one concludes that the Little's fit expression is generally in agreement with the LAMH model since the normal resistance of typical nanowires is indeed a few kiloOhms.

Let us now discuss factors which limit the applicability of the Little's fit (6.6). The Arrhenius law might not be applicable, or might need nontrivial modifications, if the conditions are such that phase slips in the wire overlap in time and thus experience an interaction. The interaction occurs because each phase slip suppresses the order parameter to zero, and therefore less energy is required for the creation of another phase slip which is close to the first one in space and in time. So, for example, the following phase slip would experience a reduced barrier if it arrives sooner than the time needed for the order parameter to grow and reach its equilibrium value after the passage of the first phase slip through the wire. Thus, if ΔF in (6.6) is understood as the energy of a single phase slip, without any regard for a possible attraction between LPSs, then (6.6) is valid only if $1/\Gamma > \tau_{\text{LPS}}$, which, according to (6.2), can be written as $\exp(-\Delta F/T) < 1$. The latter condition is equivalent to $R < R_n$. In practice, to ensure that the interaction between LPSs is negligible, one should require that R is significantly lower than R_n.

Another reason why the Little's fit might disagree with the data close to T_c is the fact that some fraction of the total current injected into the nanowire can be carried by the normal electrons (bogoliubons). To take this dissipative current into consideration, the formula for the resistance might be written as

$$\frac{1}{R(T)} = \frac{C_{\text{ns}}}{R_n} + \frac{1 - C_{\text{ns}}}{R_n \exp(-\Delta F/T)} \tag{6.12}$$

where the first term represents the conductance of normal quasiparticle excitations, the second term represents the conductance of the BCS condensate, and C_{ns} is a phenomenological fitting parameter defining relative contributions of normal electrons to the total conductance. The regular Little's fit can be obtained by choosing $C_{ns} = 0$.

6.4
LAMH Model of Phase Slippage at Low Bias Currents

The model was advanced by Langer and Ambegaokar (LA) and further improved by McCumber and Halperin (MH). The normal resistance of the wire is not explicitly included in the LAMH model [142, 143], as was the case with the Little's fit. The effective resistance is calculated by considering the time evolution of the phase of the complex superconducting order parameter. Let $\phi(x, t)$ be the local value of the phase and $\Delta\phi$ be the phase difference between the ends of the wire. The local value of the supercurrent density is $j_s = (e\hbar/m)|\psi|^2[\partial\phi(x,t)/\partial x]$, where x is the coordinate along the wire. It is assumed, as always throughout the book, that the wire is thin enough so that the phase and the amplitude of the order parameter $|\psi|$ are constant within the cross-section of the wire and depend only on x. It is also assumed that the total current $I_s = A_{cs} j_s$ is small enough so that the magnetic field is negligibly small. Therefore, the vector potential has been put to zero. Under the condition of a homogeneous current, the phase difference and the phase gradient are related as $\Delta\phi = L(\partial\phi(x,t)/\partial x)$. Therefore, the supercurrent in the wire of length L is proportional to the phase difference and can be expressed as

$$I_s = \left(\frac{A_{cs} e\hbar}{m}\right)|\psi|^2 \left(\frac{\Delta\phi}{L}\right) \tag{6.13}$$

Our goal here is to analyze constant-current (DC) charge transport and find the linear resistance corresponding to DC bias conditions. Assume that a small constant voltage V is present between the ends of the wire. The voltage accelerates the condensate and the supercurrent I_s increases correspondingly. The LAMH model is applicable when the temperature is high enough and the voltage is low enough so that a dynamic equilibrium can be reached. The dynamic equilibrium state, which is also called the stationary regime, is such that the acceleration of the condensate by the applied voltage is compensated, on average, by the stochastic occurrence of Little's phase slips which decelerate the condensate. The phase slips must be frequent enough so that the current does not reach the depairing current. In such a stationary regime, the supercurrent fluctuates in time, yet its averaged value remains constant, provided that the averaging is done over a timescale much longer than the time interval between phase slips, $1/\Gamma$. In typical transport experiments, only the average current $\langle I \rangle$ is known since the fluctuations due to individual phase slips are too fast and too small to be detected using a typical amp-meter setup. The linear resistance is then defined simply as $R = V/\langle I \rangle$, provided that Ohm's law (i.e., the current is proportional to the voltage) holds. The LAMH model neglects the current of bogoliubons, I_n, and assumes that the total current equals the supercurrent, that is, $I = I_s$ and $I_n = 0$.

Since the LAMH regime corresponds to a constant average current, the phase difference must be, on average, constant also, that is, $d\langle\Delta\phi\rangle/dt = 0$. The averaging $\langle\ldots\rangle$ has to be done over a timescale much longer than $1/\Gamma$, but shorter than the typical timescale of the measurement. Two processes contributing to the change of the phase difference are (1) the voltage-driven phase rotation (1.5), and (2) the phase difference relaxation (unwinding), occurring through Little's phase slips. Taking into account both contributions, the evolution of such an averaged phase difference can be expressed as

$$\frac{d\langle\Delta\phi\rangle}{dt} = \frac{2eV}{\hbar} + 2\pi\Gamma_{net} \qquad (6.14)$$

where Γ_{net} is the net phase slippage rate. To make sense of the signs in this equation, let us assume for a moment that phase slips are not present. Then, the phase difference grows more and more negative with time if the voltage is positive because the charge of the electron is negative. However, this is not a contradiction since according to (6.13), the current is positive if the phase difference is negative (again, because the electronic charge e is a negative number). If a strong fluctuation occurs on the wire and the order parameter goes to zero, $\Delta\phi$ can either increase or decrease by 2π. Topologically, a phase slip can only change the number of revolutions of the spiral in the Little's helix by an integer number, as is illustrated in Figure 6.3. The effect of phase slips is to prevent the supercurrent to grow too large. Thus, let us agree that the LPS that decreases the absolute value of the phase difference by 2π will be called the "phase slip" (PS) and the event that increases $|\Delta\phi|$ by 2π will be termed "antiphase-slip" (APS). Of course, $\Delta\phi$ can also remain unchanged, but such events are neglected as causing no resistance within the LAMH model. In addition, the order parameter can, in principle, change by any integer number larger than unity times the 2π value. Such events can be termed multiquanta phase slips. Although the LAMH model neglects such events as insignificant, recent experiments by Belkin et al. showed that slippage of phase by 4π can have a higher probability than the slippage by 2π [148] at high values of the supercurrent. Here, we discuss the linear, low-bias transport, so these multiquanta events can be neglected without making the model invalid.

From the discussion above, it follows that for the stationary charge transport, when the supercurrent and correspondingly the phase difference averaged values are not changing over long periods of time, the dynamic equilibrium condition is (Langer and Ambegaokar [143] have used this type of condition)

$$2\pi\Gamma_{net} = -\frac{2eV}{\hbar} \qquad (6.15)$$

The net rate of the phase slippage is $\Gamma_{net} = \Gamma_\uparrow - \Gamma_\downarrow$, where $\Gamma_\uparrow$ is the rate of LPS increasing the phase difference by 2π and $\Gamma_\downarrow$ is the number of LPS per second, which decrease the phase difference by 2π. In order to calculate the resistance $R = V/I_s$, one needs to determine the value of the supercurrent I_s at which the dynamic equilibrium condition (6.15) is satisfied, that is, at which the phase rotation due to the applied voltage is compensated, on average, by the phase unwinding due to LPS.

Within the terms we have introduced above, it is possible to argue that a thin wire cannot be a perfect superconductor at any temperature greater than zero. Suppose the voltage on the wire is zero, $V = 0$. In this case, the dynamic equilibrium condition (6.15) is satisfied if the rate of phase slips equals the rate of antiphase-slips, $\Gamma_\uparrow = \Gamma_\downarrow$. As will be explained below, the rates of PS and APS are equal only if the supercurrent is zero; the rates become different if the supercurrent deviates from zero. Thus, one concludes that the supercurrent must be zero on average ($\langle I_s \rangle = 0$) under the condition that $V = 0$ in order to satisfy (6.15). Therefore, a thin wire made of a superconducting material, that might have in general a nonzero order parameter at low temperatures, still is not a perfect superconducting at any nonzero temperature because if $V = 0$ one gets that $I_s = 0$. Under a "perfect superconductor," we understand a sample that can carry a constant supercurrent, of the order of the depairing current, indefinitely, under zero voltage applied.

To find the rates of PS and APS, one needs to know the associated energy barriers. Let us first consider the case of a zero supercurrent. The barrier equals the difference of the free energy in the state having a zero order parameter at one point on the wire and the equilibrium state characterized by a constant order parameter. The solution of the GL equation having a constant order parameter is simply $\psi(x) = \psi_0 = \sqrt{-\alpha(T)/\beta}$, where x-axis is parallel to the wire. The solution characterized by a zero order parameter at one point on the wire was given by de Gennes [2] and has the following form

$$\psi(x) = \psi_0 \tanh\left[\frac{x - x_0}{2\xi(T)}\right] \tag{6.16}$$

The corresponding phase difference between the ends of the wire is $\Delta\phi = \pi$. This function is shown in Figure 6.4.

To find the LPS barrier, one needs to calculate Helmholtz free energy functional F_0 for the homogeneous solution and the free energy F_{LPS} for the de Gennes solution, given in (6.16). The integration in the free energy functional (2.7) can be done over the volume of the superconductor only since the magnetic field is negligible at low currents. The Helmholtz energy is used here since the current is assumed zero. This way, one gets the energy barrier for a phase slips $\Delta F(T) = F_{LPS} - F_0 = V_{LPS}(H_c^2/8\pi)$ where H_c is the temperature-dependent thermodynamic critical field, $H_c^2/8\pi$ is the condensation energy density, $V_{LPS} = (8\sqrt{2}/3)A_{cs}\xi(T)$ is the effective volume of the phase slip core, A_{cs} is the cross-section area of the wire, and $\xi(T) = \xi(0)/\sqrt{1 - T/T_c}$ is the temperature-dependent coherence length. Thus the expression for the phase slip energy barrier, valid at negligible bias currents and at temperatures close to T_c, is

$$\Delta F = \Delta F(T) = \Delta F_{LA} = \frac{8\sqrt{2}}{3} A_{cs} \xi(0) \frac{H_c(0)^2}{8\pi} \left(1 - \frac{T}{T_c}\right)^{3/2} \tag{6.17}$$

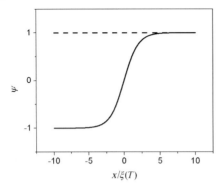

Figure 6.4 The order parameter as a function of the position x along the wire normalized by the coherence length $\xi(T)$. A homogeneous solution of the GL equation, corresponding to zero supercurrent, is shown by the dashed line. It corresponds to the minimum of the free energy of the wire. The continuous curve represents a solution of the GL equation corresponding to a phase slip, namely, $\psi(x) = \psi_0 \tanh[(x - x_0)/2\xi(T)]$. It appeared in the de Gennes book [2] in 1966, so we will refer to it as the de Gennes solution (of the GL equation) or de Gennes state. This solution reaches zero at one point along the wire, namely, at $x = x_0$. In the example, $x_0 = 0$. The de Gennes solution defines the saddle-point barrier for phase slips at $I_s = 0$. Note an important property of the solution, namely, that the order parameter reverses the sign as the coordinate goes through the point of the zero order parameter. Thus, the phase difference corresponding to the case when the core of a phase slip is positioned on the wire is either 2π or -2π, because $\exp(i\pi) = \exp(-i\pi) = -1$. Whether it is 2π or -2π depends on whether the phase difference was increasing or decreasing before the de Gennes state is reached.

In this model, it is assumed that at $T_c - T \ll T_c$, the coherence length and the thermodynamic critical field can be written as $\xi(T) = \xi(0)\sqrt{1 - T/T_c}$ and $H_c(T) = H_c(0)(1 - T/T_c)$. These two formulas provide formal definitions for the material constants $\xi(0)$ and $H_c(0)$. They might differ somewhat from the corresponding quantities measured at $T \approx 0$. Note that the barrier ΔF decreases with temperature as $(1 - T/T_c)^{3/2}$. If it is desirable to extend the expression for the barrier to lower temperatures, it is necessary to choose a more general formula for the critical field, which is $H_c(T) = \text{const} * (1 - T^2/T_c^2)$. No matter which form for $H_c(T)$ is used, it is always true that $\Delta F(T) \propto (1 - T/T_c)^{3/2}$ for $T \to T_c$.

If $I_s = 0$, the rates of thermal activation of phase slip, $\Gamma_\uparrow$, and antiphase slips, $\Gamma_\downarrow$, are equal. In such cases, both are called the rate of thermally activated phase slip (TAPS) and denoted by Γ_taps. Both are given by the Arrhenius thermal activation activation law

$$\Gamma_\text{taps} = \Gamma_\uparrow = \Gamma_\downarrow = \Omega \exp\left(-\frac{\Delta F(T)}{k_B T}\right) \tag{6.18}$$

where the attempt frequency is

$$\Omega = \Omega_\text{MH} = \frac{\sqrt{3}}{2\pi^{3/2}} \frac{1}{\tau_r} \frac{L}{\xi(T)} \sqrt{\frac{\Delta F}{k_B T}} \tag{6.19}$$

The attempt frequency was derived by McCumber and Halperin [142, 149] from a simple version of the so-called time-dependent GL equation (TDGL). Here, $\tau_r = \pi\hbar/8k_B(T_c - T)$ is the order parameter relaxation time. Notice that the attempt frequency goes to zero at $T = T_c$.

Recently, this result was corrected by Golubev and Zaikin [171], who showed that the factor $(1/\tau_r)$ has to be removed from (6.19). Yet, in any formulation, the attempt frequency formally goes to zero at $T = T_c$. Thus, the LAMH model predicts that the frequency of phase slips should approach zero as $T \to T_c$. Such a prediction is unphysical since in practice the phase slip rate increases as one approaches T_c. Thus, the LAMH model is not applicable very near T_c. It is only applicable if $\Delta F(T) \gg k_B T$. Note that the attempt frequency based on TDGL is also not correct far from T_c. This is because TDGL is designed for the cases when the superconducting gap is negligible, that is, smaller than the thermal energy, or if the superconductivity is gapless. Thus, the range of applicability of the model is quite narrow [144]. Yet, the exponential Arrhenius factor, which is more important since it varies with temperature much faster than the prefactor (the attempt frequency), is correct for any temperature, as long as the rate of thermal activation is concerned. Thus, the expressions of LAMH still can be used to fit the data reasonably well.

In order to calculate the resistance of a thin wire, assume that the bias current is greater than zero ($I_s > 0$), but much smaller than the depairing current ($I_s \ll I_c(T)$). It is also assumed that the wire is very long, so that a phase slip changes the phase gradient by an amount much smaller than the applied current I_s. As was originally stated by Anderson and Dayem (AD) (Eq. (3) in [11]) in 1964, the presence of the current reduces the barrier for phase slips and increases the barrier for antiphase-slips. The AD expression for the current-dependent barrier height for phase slips can be generalized as (for $e < 0$)

$$\Delta F_{PS} = \Delta F + \frac{\pi\hbar I_s}{2e} \qquad (6.20)$$

The barrier for antiphase-slips is increased due to the current (assuming $I_s > 0$ and $e < 0$):

$$\Delta F_{APS} = \Delta F - \frac{\pi\hbar I_s}{2e} \qquad (6.21)$$

The convention we choose is that the phase slips reduce the supercurrent magnitude and the antiphase-slips increase the current magnitude. It is clear that the net fluctuation effect is to reduce the current magnitude. The barrier correction $\pi\hbar I_s/2e$ will be discussed in detail below. Briefly, it can be justified as follows: In general, the barrier for a phase slip equals the work required to create a phase slip core on the wire when there is no current (ΔF), less the amount of work provided by the external current source. The work by the source is, as usual, equals the charge passing from one end of the wire to the other multiplied by the difference of potential (i.e., the voltage) maintained between the ends of the wire. If the voltage changes in time, then the total work needs to be expressed as an integral. Assuming the current is fixed and equals I_s and using the Gor'kov phase equation, the work can be written as

$$W = \int I_s V dt = \left(\frac{\hbar}{2e}\right) \int I_s d\Delta\phi = (\Delta\phi_{LPS} - \Delta\phi_{min})\left(\frac{\hbar I_s}{2e}\right) \quad (6.22)$$

where $\Delta\phi_{min}$ is the phase difference corresponding to a state having a homogeneous order parameter (Figure 6.4, dashed line) and $\Delta\phi_{LPS}$ is the neighbor state having a zero order parameter at one point on the wire. In the limit of an infinitely long wire, the de Gennes solution (Figure 6.4) describes the state having one point at which $\psi = 0$. The additional phase difference corresponding to this solution is π. In a real experiment, the wire is not infinitely long and the current is not infinitely weak. Thus, the phase path the systems needs to travel to reach the top of the barrier for a phase slip only equals π approximately: $\Delta\phi_{LPS} - \Delta\phi_{min} \approx \pi$.

This Anderson–Dayem work (6.22) changes the Helmholtz free energy and leads to a reduction of the barrier for phase slips. In other words, the modified barrier is $\Delta F \pm \pi\hbar I_s/2e$, where the smaller barrier corresponds to the case when the phase slip tends to reduce the supercurrent magnitude and the larger barrier corresponds to the case when the phase slip tends to increase the supercurrent magnitude (we call such phase slips antiphase-slips). Thus, we have proven (6.20).

Using the expression for the PS and APS barriers and assuming that the attempt frequency is the same for both of them, one gets the net rate of the phase slippage as

$$\Gamma_{net} = \Gamma_{PS} - \Gamma_{APS} = \Omega\left[\exp\left(-\frac{\Delta F_{PS}}{k_B T}\right) - \exp\left(-\frac{\Delta F_{APS}}{k_B T}\right)\right] \quad (6.23)$$

Now, we take into account that the barriers for PS and APS are slightly different. Thus, the net rate of phase slippage can be written as

$$\Gamma_{net} = \Omega \exp\left(-\frac{\Delta F}{k_B T}\right)\left[\exp\left(-\frac{\pi\hbar I_s}{2ek_B T}\right) - \exp\left(\frac{\pi\hbar I_s}{2ek_B T}\right)\right] \quad (6.24)$$

which can be simplified using the hyperbolic sine function $\sinh(x) = (1/2)(e^x - e^{-x})$ as follows

$$\Gamma_{net} = -2\Omega \exp\left(-\frac{\Delta F}{k_B T}\right)\sinh\left(\frac{\pi\hbar I_s}{2ek_B T}\right) \quad (6.25)$$

Now, let us take into account (6.15) for the steady current and constant bias. Then, we get the expression for the voltage on the wire caused by the presence of the current I_s and due to phase slips.

$$-\frac{eV}{\pi\hbar} = -2\Omega \exp\left(-\frac{\Delta F}{k_B T}\right)\sinh\left(\frac{\pi\hbar I_s}{2ek_B T}\right) \quad (6.26)$$

and finally the expression for the voltage is

$$V = \left(\frac{2\pi\hbar\Omega}{e}\right)\exp\left(-\frac{\Delta F}{k_B T}\right)\sinh\left(\frac{\pi\hbar I_s}{2ek_B T}\right) \quad (6.27)$$

In the limit of small bias currents $|I_s| \ll I_0 = |e|k_B T/\pi\hbar$, the hyperbolic sine can be replaced by the first term in its Taylor expansion as $\sinh(x) \approx x$. Here, I_0 is the current at which the V–I curve becomes nonlinear. The first term in the Taylor expression is linearly proportional to the current:

$$V = I_s \left(\frac{\pi^2 \hbar^2}{e^2 k_B T} \right) \Omega \exp\left(-\frac{\Delta F}{k_B T}\right) \tag{6.28}$$

Assume that the current and the voltage are both small, and thus the V–I curve is linear. Then, using Ohm's law, the resistance is simply $R_{\text{LAMH}} = V/I_s$, which can be written as

$$R_{\text{LAMH}} = \frac{\pi^2 \hbar^2}{e^2} \frac{\Omega}{k_B T} \exp\left(-\frac{\Delta F}{k_B T}\right) = \frac{\pi^2 \hbar^2}{e^2} \frac{\Gamma_{\text{taps}}}{k_B T} = R_q \frac{h \Gamma_{\text{taps}}}{k_B T} \tag{6.29}$$

Substituting the explicit expression for the rate of TAPS into this expression, we get the LAMH resistance as

$$R_{\text{LAMH}} = \sqrt{\frac{3}{\pi}} \frac{\pi \hbar^2}{2 e^2 k_B T} \frac{L}{\xi(T)} \frac{1}{\tau_r} \sqrt{\frac{\Delta F}{k_B T}} \exp\left(-\frac{\Delta F}{k_B T}\right) \tag{6.30}$$

Various useful expressions for the barrier $\Delta F = \Delta F(T)$ are discussed in Section 9.1. In the formula above, all the coefficients, including $\sqrt{3/\pi}$, are written to ensure an easy comparison with the formulas used in the literature. The presence of these coefficients should not be taken as an indication that the formula for R_{LAMH} is exact. Modern theory, due to Golubev and Zaikin, shows, for example, that the coefficient $1/\tau_r$ should be removed from the preexponential factor [171].

Interestingly, the LAMH resistance does not depend, by any explicit manner, on the normal resistance of the wire. This is due to the implicit assumption that if the order parameter reaches zero, the phase can only slip by 2π, but not by 4π or $2n\pi$, with the integer n being larger than unity. There is some indirect evidence that such an assumption might not always be correct and phase slips having $n > 1$ can happen [148].

Note that the result we obtain for the resistance is a factor 2π larger than the one given in the Tinkham book [1]. This is because Tinkham uses the angular frequency as the attempt frequency Ω and the angular frequency is larger than the attempt rate (number of attempts per second) exactly by the factor 2π. We follow the notation of the original reference [142] and assume that Ω is the attempt rate (i.e., simply the number of attempts per second). It is interesting that the prefactor in the expression for the resistance is listed differently in various publications, for example, in [149, 150]

The LAMH model neglects the presence of Bogoliubov quasiparticles (BQP or bogoliubons). Although the charge of BQP is zero when the energy of BQP exactly equals the superconducting gap energy, the charge becomes distinct from zero as the energy deviates from Δ. Thus, BQP contribute to the current if $V > 0$. The computed quantity R_{LAMH} does not take into account bogoliubons, but only the condensate which dissipates energy through phase slips. To take into account the bogoliubons, Tinkam and coauthors suggested a phenomenological formula for the wire resistance as follows

$$R^{-1} = R_{\text{LAMH}}^{-1} + R_n \tag{6.31}$$

This formula is not exact since the contribution of BQP is modeled simply as a constant parallel normal resistor, having resistance R_n being equal to the normal resistance of the wire. Yet Tinkham's formula gives a better fit to the data [23]. The exact theory, which would include BQP is not available yet.

6.5
Comparing LAMH and Little's Fit

In previous sections, we have discussed two possible theoretical models predicting TAPS-based $R(T)$ curves, namely, the phenomenological Little's fit, which explicitly involves the normal resistance of the wire, and the LAMH model, which is based purely on the dissipation of energy of the moving condensate through phase slips and does not take normal electrons into consideration. In this section, we present a comparative analysis of these two models. The comparison is based on the results published in [135]. In this work, a set of $Mo_{79}Ge_{21}$ bridges has been measured at low bias currents (linear regime) and at high bias currents (nonlinear regime). The conclusion was that the both LAMH and Little models provide good fits to the data. Yet, the LAMH fit requires an unreasonably high critical temperature, which is used as a fitting parameter in both models, in order to fit the data.

Let us illustrate this conclusion by examining the sample B2 of [135] which can be classified as a Dayem bridge. The fabrication of the bridges in this reference was done similar to the molecular templating of nanowires. The only difference is that for the bridges, no molecule is used, but, instead, a SiN bridge acts as a template for the superconducting alloy deposition. Therefore, such bridges are usually a few times wider than nanowires produced on the surface of carbon nanotubes. This sample B2 was 28 nm wide, which is about four times wider than the zero-temperature coherence length. It is still narrower than the limit of $4.4\xi(T)$, at which vortices become energetically favorable excitations for the bridge, according to [18]. Therefore, the dissipation in the bridge is mostly due to Little-type phase slips. Thus, the sample considered can be used to compare the applicabilities of the LAMH and the Little's models. The $R(T)$ curves are shown in Figure 6.5. The open circles show the low-bias resistance of the bridge. The solid black circles represent the resistance determined by an indirect procedure, namely, by an extrapolation of the high bias (nonlinear) dV/dI vs. I measurements to zero-bias. The Little's fit is shown as a solid red curve and the LAMH fit is shown as a dashed blue curve. At low temperatures, where the resistance drops rapidly, both fits agree with the data. This agreement is observed over a range of eleven orders of magnitude of the resistance if the indirectly determined points are counted.

However, there is a significant difference between the two fits, namely, in the reasonableness of the corresponding critical temperature, which is used as a fit-

6.5 Comparing LAMH and Little's Fit

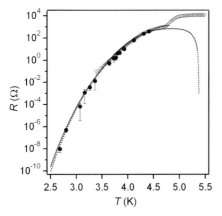

Figure 6.5 Resistance versus temperature plot obtained on a narrow Dayem bridge of width less than $4.4\xi(T)$ (sample B2 of [135]). Open circles show the linear resistance of the sample measured using a bias current lower than the nonlinearity onset current $I_0 = \frac{4ek_BT}{h} = 13$ nA/K and much lower than the critical current of the bridge. The filled black circles are obtained indirectly, by conducting differential resistance measurements at high bias currents and then by performing an extrapolation procedure in order to obtain the linear or low-bias resistance. The solid (red) and the dashed (blue) curves give the best fits generated by the Little (6.6) and the LAMH (6.30) formulas correspondingly. The critical temperature of the bridge was used as a fitting parameter. The best fits were obtained by taking $T_c = 4.81$ K for the Little fit and $T_c = 5.38 K$ for the LAMH fit. The first value, corresponding to the Little's fit, is realistic because the critical temperature of the film is 4.91 K and the bridge is expected to have the same critical temperature or slightly lower. The value of T_c required for the LAMH fit is significantly larger than the T_c of the electrodes, which is physically unrealistic because the T_c of the MoGe nanoscale samples of any sort always decreases as the sample dimensions are reduced. Please find a color version of this figure in the color plates.

ting parameter. The best Little's fit is obtained using $T_{c,\text{bridge,Little}} = 4.81$ K and the best LAMH fit is obtained using $T_{c,\text{bridge,LAMH}} = 5.38$ K. By an independent direct measurement, it was determined that the critical temperature of the film electrodes is $T_{c,\text{film}} = 4.91$ K in this sample. Thus, in order for the LAMH expression to fit the data, it is necessary to choose $T_{c,\text{bridge,LAMH}}$ significantly larger than $T_{c,\text{film}}$. Physically speaking, such a choice is meaningless because in the samples studied, the bridge is not a separate object but a constriction in the film. It is made in the same sputtering run and has the same thickness as the film electrodes into which the bridge extends seamlessly. All existing publications on MoGe samples indicate that for this material, the critical temperature of the sample decreases as the dimensions of the sample shrink (see e.g., [102, 151–153]). Thus, it can be confidently expected that the Dayem bride of sample B2 should have an equal or a lower critical temperature compared to the thin film electrodes of the same thickness. Thus, the result obtained using the LAMH fit, namely, that the critical temperature of the bridge is higher than that one of the electrodes, is not correct.

The conclusion is that for accurate determination of the critical temperature of a narrow weak link, the Little's fit is more reliable. The LAMH gives the wrong results for the T_c because the attempt frequency goes to zero at T_c (see (6.11)). This unphysical result is obtained since the attempt frequency is equated to the frequency of phase difference oscillations in the local minimum of the free energy. The minimum naturally becomes more and more shallow as $T \to T_c$, also causing the phase-dependent part of the free energy to approach zero. On the other hand, at such high temperatures, the attempt frequency is also determined by strong Brownian diffusion of the effective phase particle representing each ξ-segment of the wire. This fact is neglected in [142]. Thus, within the LAMH model, a rather wide region exists near T_c where the model is not applicable. This region of temperature is created by the unphysical assumption that $\Omega \to 0$ as $T \to T_c$. This existence of this region within the model, which does not have analogous temperature regime in real samples, might be the reason why the LAMH model tends to overestimate the critical temperature of the weak link samples. In some sense, it needs to put the T_c higher up to gain space for the regime in which the resistance drops with increasing temperature. Certainly, such a regime in the LAMH model, characterized by $dR/dT < 0$, has no counterpart in reality. The Little's fit appears to be a more accurate predictor of the critical temperature for weak links, such as bridges and nanowires, since the fit does not have any such unphysical regime at high temperature. Another reason to prefer the Little's fit is that it only relies on the Arrhenius law, which does not break down at temperatures far from T_c, as the TDGL theory does.

7
Golubev and Zaikin Theory of Thermally Activated Phase Slips

In this section, we will follow the original Golubev and Zaikin publication [171] and use theoretical units in which $k_B = 1$ and $\hbar = 1$. In such systems, the frequency and the temperature are measured in the units of energy. Such choice of units is quite usual in theoretical physics papers since it makes expressions shorter. The transition to more practical SI units can be made by replacing T with $k_B T$, $I/2e$ with $\hbar I/2e$, and Ω with $\hbar\Omega$ in the mathematical expressions representing energy. In the expressions representing time through energy, a reduced Planck's constant also needs to be restored in order to convert from the theoretical to SI units. For example, if some time scale is written as $1/\Delta$, and from the context it is clear that Δ is some energy scale, then the expression should be changed to $\hbar/\Delta$ and so on.

The most accurate analysis of thermally activated phase slips to date is due to Golubev and Zaikin (GZ) [171]. They analyze the case when the wire is biased with the constant current I, which equals the supercurrent in the wire $I_s = I = \text{const}$. Thus, their analysis deals with Gibbs (not Helmholtz) free energy. With the aid of an effective action approach, combined with GLAG theory, they calculated the TAPS rate, which turns out to exceed the rate found by MH by the factor $(1 - T/T_c)^{-1}$. The inaccuracy of the MH approach is due to the fact that their analysis is based on the TDGL theory, which is inaccurate by itself and is hardly applicable below T_c. At the same time, the GZ analysis confirms the ΔF value used in the LAMH model and in the Little's fit approximate formula (6.17), at least in the limit of $T_c - T \ll T_c$ and near-zero bias currents.

In the GZ theory, the general expression for the rates of the phase slips, Γ_+, and antiphase-slips, Γ_-, is the usual Arrhenius expression

$$\Gamma_\pm = \Omega_{GZ\pm} \exp\left(-\frac{\Delta G_{GZ\pm}}{T}\right) \tag{7.1}$$

Here, $\Omega_{GZ\pm} = \Omega_{GZ\pm}(I, T)$ and $\Delta G_{GZ\pm} = \Delta G_{GZ\pm}(I, T)$ represent, correspondingly, the attempt frequency and the barrier height, as derived from the GZ microscopic theory. The index "+" corresponds to phase slips and the index "−" corresponds to antiphase-slips, which, in general, have different barrier heights and also different attempt frequencies. Remember that phase slips are those fluctuations which change the phase difference by 2π and reduce the magnitude of the supercurrent in the wire. The antiphase-slips have the opposite effect with respect to phase slips

Superconductivity in Nanowires, First Edition. Alexey Bezryadin.
© 2013 WILEY-VCH Verlag GmbH & Co. KGaA. Published 2013 by WILEY-VCH Verlag GmbH & Co. KGaA.

so they increase the magnitude of the supercurrent. Thus, Ω_{GZ-} and Ω_{GZ+} represent the attempt frequency for the antiphase-slips and phase slips correspondingly, and ΔG_{GZ-} and ΔG_{GZ+} represent the barrier height for antiphase-slips and phase slips correspondingly.

The GZ analysis of the barrier begins with Gibbs energy $G[\Delta(x)]$ for the superconducting nanowire. This energy is a functional of the superconducting order parameter $\Delta(x)$, normalized to be proportional to the superconducting energy gap. In the general case, $\Delta(x)$ is a function of the position x along the wire. The coordinates of the ends of the wire are taken to be $-L/2$ and $L/2$. The effect of the electrodes on the order parameter is neglected in this analysis. Thus, the wire has to be much longer than the coherence length for the analysis to be valid. The Gibbs energy functional involves the Ginzburg–Landau part, written in the Gor'kov format, and the term proportional to the bias current which describes the work on the condensate, and correspondingly the change of its usable energy produced by the bias current I_s. The expression for the Gibbs functional is

$$G[\Delta(x)] = A_{cs} N_0 \int_{-L/2}^{L/2} dx \left(\frac{\pi D}{8T} \left|\frac{\partial \Delta}{\partial x}\right|^2 + \frac{T - T_c}{T_c}|\Delta|^2 + \frac{7\zeta(3)}{16\pi^2 T^2}|\Delta|^4 \right)$$

$$- \frac{I}{2e}\left[\phi\left(\frac{L}{2}\right) - \phi\left(-\frac{L}{2}\right)\right]$$

where $D = l_e v_F/3$ is the diffusion constant, A_{cs} is the wire's cross-section area, v_F is the Fermi velocity, l_e is the electronic mean free path, $N_0 = mk_F/2\pi^2$ (or $N_0 = mk_F/2\pi^2\hbar^2$ in SI or Gaussian units) is the density of states of one spin at the Fermi surface, k_F is the Fermi wavevector, m is the electronic mass, $\phi(x)$ is the phase of the order parameter, which takes the form $\Delta = \Delta(x) = |\Delta(x)|e^{i\phi(x)}$, with $i^2 = -1$. The Riemann zeta function is $\zeta(3) = 1.202$.

The extrema (called "saddle points") of this functional are obtained through the usual minimization procedure and are given by the GL equation

$$-\frac{\pi D}{8T}\frac{\partial^2 \Delta}{\partial x^2} + \frac{T - T_c}{T_c}\Delta + \frac{7\zeta(3)}{8\pi^2 T^2}|\Delta|^2\Delta = 0 \tag{7.2}$$

Within these notations, the supercurrent is $I_s = (\pi e N_0 D A_{cs}/2T)|\Delta|^2 \nabla \phi$.

Suppose $I_s = 0$. Then, the solution for the order parameter corresponding to the minimum of G is $\Delta_0(T) = \sqrt{8\pi^2 T_c(T_c - T)/7\zeta(3)}$. The corresponding phase slip free energy barrier $\Delta F(T)$ is determined by this solution and the solution that has one zero of the order parameter somewhere along the wire, shown graphically in Figure 6.4. The expression for this solution, which corresponds to the maximum of $G[\Delta]$, is $\Delta(x) = \Delta_0 \tanh[(x - x_0)/2\xi(T)]$, where the coherence length in the currently used notations is $\xi(T) = \sqrt{\pi D/4(T_c - T)}$. The energy difference of such solutions gives the barrier for phase slips at zero bias current (for $I = I_s = 0$), which is

$$\Delta F = \Delta F_{0,GZ} = \Delta G_{0,GZ} = \frac{16\pi^2}{21\zeta(3)} A_{cs} N_0 (\pi D)^{1/2}(T_c - T)^{3/2} \tag{7.3}$$

Suppose now $0 \leq |I_s| \leq I_c$. Then, the barrier is defined by two solutions of the GL equation. The first one corresponds to the minimum of G and can be written as $\Delta_m = |\Delta_m| \exp i\phi_m(x)$. The corresponding amplitude and the phase are

$$|\Delta_m| = \Delta_0(T) \sqrt{\frac{1 + 2\cos\nu}{3}}, \quad \phi_m(x) = \frac{2x T I_s}{\pi e N_0 D A_{cs} |\Delta_m|^2} \quad (7.4)$$

The parameter in the solution is

$$\nu = \frac{\pi}{3}\theta\left(|I_s| - \frac{I_c}{\sqrt{2}}\right) + \frac{1}{3}\arctan\frac{2|I_s|\sqrt{1 - (|I_s|/I_c)^2}}{I_c[1 - 2(I_s/I_c)^2]} \quad (7.5)$$

with the critical current being

$$I_c = \frac{16\sqrt{6}\pi^{5/2}}{63\zeta(3)} e N_0 \sqrt{D} A_{cs} (T_c - T)^{3/2} \quad (7.6)$$

The second solution needed to compute the barrier is $\Delta_s = |\Delta_s|\exp[i\phi_s(x)]$, corresponding to the saddle point of G for the given values of the bias current I_s. It is defined through

$$\frac{|\Delta_s(x)|}{\Delta_0(T)} = \sqrt{\frac{1 + 2\cos\nu}{3} - \frac{2\cos\nu - 1}{\cosh^2\left\{[x/\xi(T)]\sqrt{2\cos\nu - 1}\right\}}} \quad (7.7)$$

which has a minimum at the center of the wire, namely, at $x = 0$. Note that the order parameter amplitude reaches zero at the energy saddle point only if $I_s = 0$, otherwise the order parameter remains above zero in order to allow constant nonzero I_s at all points of the wire (remember that the problem is solved under the assumption that the wire is biased with a constant current). The phase of the order parameter changes along the wire as

$$\phi_s = \frac{2T I_s}{\pi e N_0 D A_{cs}} \int_{-L/2}^{x} \frac{dx'}{|\Delta_s(x')|^2} \quad (7.8)$$

Gibbs energy barrier for a phase slip is

$$\Delta G_{GZ+} = G[\Delta_s(x)] - G[\Delta_m(x)]$$

$$= \Delta F_{0,GZ}\left[\sqrt{2\cos\nu - 1} - \sqrt{\frac{2}{3}}\frac{I_s}{I_c}\arctan\left(\frac{\sqrt{3}}{2}\sqrt{\frac{2\cos\nu - 1}{1 - \cos\nu}}\right)\right] \quad (7.9)$$

The expression for the barrier for antiphase-slip, Δ_{GZ-}, follows from the fact that the "distance" between the two barriers (the one for phase slips and the one for antiphase-slips) in the phase space is exactly π. By integrating the work done by the external source, which is the only factor leading to the difference between the two barriers, one gets (for $I_s > 0$)

$$\Delta G_{GZ-} = \Delta G_{GZ+} + \frac{I_s \pi}{|e|} \qquad (7.10)$$

Without going into detail, we reproduce the GZ attempt frequencies, also called preexponents. The result of the theory, valid with high accuracy only if $\Delta G_{GZ\pm} \gg T$ and $T_c - T \ll T_c$, is

$$\Omega_{GZ\pm} = a_q \kappa(I_s) T_c \frac{L}{\xi(T)} \sqrt{\frac{\Delta G_{GZ\pm}}{T}} \qquad (7.11)$$

where $\kappa(I_s)$ is a function that slowly changes in the range $5.53 < \kappa(I_s) < 8.74$ as the current changes from near zero to near the critical current, and a_q is another unknown constant of order unity. Physically, it represents the ratio of the crossover temperature T_q and the critical temperature of the wire T_c, that is, $a_q = T_q/T_c$. The crossover temperature T_q is the temperature below which the quantum phase slips are more frequent than thermal phase slips. On the other hand, above T_q, TAPS dominate QPS. According to the GZ theory, this factor a_q is somewhat smaller than one. Usually, the value of the attempt frequency has an extremely weak impact on the experimental results. Therefore, it is not possible to determine the attempt frequency precisely. Thus, when fitting experimental results, it should always be acceptable to simplify the expression and make the replacement $a_q \kappa(I_s) \to 1$. Thus, one can use the following approximate result

$$\Omega_{GZ\pm} = T_c \frac{L}{\xi(T)} \sqrt{\frac{\Delta G_{GZ\pm}}{T}} \qquad (7.12)$$

Note that the GZ theory differs from the MH prediction: The GZ result for the attempt frequency does not contain the order parameter relaxation time τ_r, which is present in the MH attempt frequency, (6.19). The difference is due to the fact that the TDGL, used in the MH theory, is not correct below T_c, as was discussed in [182].

The complete expression for the thermal phase slip rate is given by (7.1), using (7.9) and (7.11).

V–I curves, linear resistance and the phase slip noise – The voltage is related to the rate of phase slips Γ through Gor'kov phase equation. The phase slippage rate is $d\phi/dt = 2\pi \Gamma$, so the voltage is $V = \hbar(d\phi/dt)/2e = h\Gamma/2e$. In a way similar to the LA analysis, GZ obtained the following expression for the voltage $V = [2\pi \Gamma(0)/e]\sinh(\pi I_s/2eT)$. It is valid for weak currents, that is, $I_s \ll I_c$. Here, $\Gamma(0)$ is the phase slip rate corresponding to zero bias current.

The linear or zero-bias resistance, by definitions, is $R(T) = (\partial V/\partial I)_{I=0}$. This general definition can be extended to superconducting wires by making a replacement $I \to I_s$ in the above formula. If $V = 0$ for $I_s \neq 0$, then, obviously, $R = 0$ as well. However, as we have discussed above, for a superconducting wire at nonzero temperature, it is true that $V(I_s) > 0$ if $I_s > 0$ due to the occurrence of Little's phase slips. Thus, the resistance is greater than zero. According to the expression for the voltage given above, we have

$$R(T) = \frac{2\pi R_q \Gamma(0)}{T} \qquad (7.13)$$

The factor 2π reflects our convention that our Γ's and Ω's are measured in units of s^{-1} and not in units of rad/s. Here, $R_q = \pi/2e^2$ and the rate of phase slips at zero or negligible bias current is

$$\Gamma(0) = \frac{a_q \kappa(I_s) T_c L}{\xi(T)} \sqrt{\frac{\Delta F}{T}} \exp\left(-\frac{\Delta F}{T}\right) \approx \frac{L T_c}{\xi(T)} \sqrt{\frac{\Delta F}{T}} \exp\left(-\frac{\Delta F}{T}\right) \qquad (7.14)$$

The barrier height should be taken from (7.3). A combination of (7.13) and (7.14) leads to the final result for the linear resistance as [171]

$$R(T) = 8 a_q \sqrt{6\pi} R_q \left(\frac{T_c}{T}\right) \left[\frac{L}{\xi(T)}\right] \sqrt{\frac{\Delta F}{T}} \exp\left(-\frac{\Delta F}{T}\right)$$

If we now remember that $a_q \sim 1/2$ and change to SI units, though just for now, then the GZ linear resistance becomes

$$R(T) \approx 17 \frac{T_c}{T} R_q \frac{L}{\xi(T)} \sqrt{\frac{\Delta F}{k_B T}} \exp\left(-\frac{\Delta F}{k_B T}\right)$$

If the heating effects due to phase slips can be neglected, meaning the wire temperature is fixed and equals the bath temperature, then the thermal phase slips are independent and therefore obey Poissonian statistics. Hence, it can be shown [171] that the voltage noise, defined as $S_V = \int dt \langle \delta V(t) \delta V(0) \rangle$, is $S_V = [4\pi^2 \Gamma(0)/e^2] \cosh(\pi I_s/2eT)$. Here, the phase slip rate $\Gamma(0)$ decreases exponentially with cooling, according to (7.1), in which one needs to put $I = 0$. According to this, the voltage noise decreases rapidly with cooling.

It is instructive to compare the GZ zero-bias barrier height for phase slips, (7.3)

$$\Delta F_{0,GZ} = \frac{16\pi^{5/2}}{21\zeta(3)\sqrt{3}} A_{cs} N_0 (l_e v_F)^{1/2} T_c^{3/2} (1 - T/T_c)^{3/2}$$

$$\approx 6.4 A_{cs} N_0 \sqrt{l_e v_F} T_c^{3/2} \left(\frac{T_c - T}{T_c}\right)^{3/2} \qquad (7.15)$$

and the corresponding LAMH barrier height, defined in [143]

$$\Delta F_{LA} = \left(\frac{8\sqrt{2}}{3}\right) E_{cond} A_{cs} \xi(T) \approx 3.771 E_{cond} A_{cs} \xi(T) \qquad (7.16)$$

In writing the GZ expression, it is taken into account that $D = l_e v_F/3$. The coherence length near T_c is

$$\xi(T) = 0.85 \sqrt{l_e \xi_0} / \sqrt{1 - T/T_c} \qquad (7.17)$$

where the clean-limit coherence length at zero temperature is defined as $\xi_0 = v_F/\pi\Delta$ and the gap is $\Delta = 1.76 T_c$ in the theoretical units used here.

The condensation energy density, according to [143], is

$$E_{cond} \equiv \frac{H_c^2}{8\pi} = \left[\frac{4\pi^2}{7\zeta(3)}\right] N_0 (T_c - T)^2 = 4.692 N_0 (T_c - T)^2 \tag{7.18}$$

Combining (7.16), (7.16), and (7.16), we can transform the LAMH barrier height expression to

$$\Delta F_{LA} = 6.396 A_{cs} N_0 \sqrt{l_e v_F} T_c^{3/2} \left(1 - \frac{T}{T_c}\right)^{3/2} \tag{7.19}$$

Now, comparing (7.19) with (7.15), we conclude that $\Delta F_{LA} = \Delta F_{0,GZ}$. In other words, the Golubev and Zaikin microscopic theory of TAPS predicts exactly the same barrier as the LA model does, at least in the limit of $I \approx 0$. Both results are exact only near T_c. On the other hand, for both theories to be valid, the temperature must be sufficiently low, so the phase slip barrier is much larger than the energy of thermal fluctuations. In both theories the wire is assumed sufficiently long so the effect of the electrodes is negligible. In particular Coulomb charging of the electrodes plays no role, unlike the SM model of JJ.

8
Stochastic Premature Switching and Kurkijärvi Theory

Understanding the significance and physical meaning of stochastic premature switching in superconducting devices was achieved via the development of the Kurkijärvi theory. Kurkijärvi's original publication presents an analysis of the statistics of premature switching events, or vortex entry events, in RF SQUIDs [179]. Later, his approach was adapted to the analysis of switching events in current-biased Josephson junctions [33, 80, 81] and many other systems such as magnetic nanoparticles [35] and, more recently, superconducting nanowires [131]. This Kurkijärvi approach is very productive as it led to the discovery of macroscopic quantum tunneling (MQT) [33, 80, 81]. The phenomenon of MQT is of fundamental importance since it demonstrates that the laws of quantum mechanics are applicable to macroscopic systems and objects, including their macroscopic degrees of freedom, for example, the position or the velocity of the center of mass or the total current in a conducting system. The MQT also led to the development of the concept of quantum computers and to the creation of qubits. Here, we discuss the Kurkijärvi analysis of the switching statistics using the example of superconducting nanowires.

8.1
Stochastic Switching Revealed by V–I Characteristics

Thus far, we have only discussed the linear charge transport regime in which the current is weak. The weakness of the current means that if it is made even weaker, the measured resistance value remains unchanged. In this section, we discuss phenomena occurring at high bias currents which are revealed through measurements of voltage versus current, V–I, curves.

An example of the $V(I)$ curve of a thin, homogeneous and short wire is shown in Figure 8.1. This typical curve exhibits a pronounced hysteresis. The plot shows that the nanowire sample can be found in two distinct states: The superconducting state (SS) is observed when the bias current is sufficiently low. The SS is characterized by zero voltage at sufficiently low temperatures. As the bias current is increased, the sample shows a jump-wise transition into the normal state, which is characterized by significant Joule heating and an elevated temperature of the wire.

Superconductivity in Nanowires, First Edition. Alexey Bezryadin.
© 2013 WILEY-VCH Verlag GmbH & Co. KGaA. Published 2013 by WILEY-VCH Verlag GmbH & Co. KGaA.

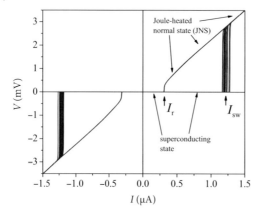

Figure 8.1 Typical voltage–current, V–I, characteristics of a $Mo_{79}Ge_{21}$ nanowire produced by molecular templating. Data points are connected by lines. Many V–I curves are plotted on top of each other in order to show reproducibility. All curves are measured under the same conditions, by sweeping the bias current as $I_b = I_a \sin(2\pi f_b t)$, with the amplitude I_a chosen somewhat higher than the depairing current and the frequency being $f_b \sim 10\,\text{Hz}$. The measurement was done at a temperature much lower than the critical temperature of the wire. As the current increases, the voltage remains zero, or below the noise "floor" of the DC measurement setup. Then, as the current is increased further, a sharp jump to the normal state is observed. The value of the current at which the jump happens is called the switching current I_{sw}. The switching current is a stochastic quantity, meaning that each new sweep of the current gives a different I_{sw} value. The fluctuations are of the order of $\Delta I_{sw} \sim 100\,\text{nm}$. The average switching current is $I_{sw} \approx 1.1\,\mu\text{m}$. After the wire switches to the normal state, the Joule heating prevents it from returning back to the superconducting state, until the current is reduced to the so-called retrapping current I_r. As the graph clearly shows, the retrapping current is not stochastic. In other words, the retrapping always occurs at the same current.

This resistive state will be referred to as the Joule heated normal state (JNS). The jump from SS to JNS occurs at the switching current I_{sw}, which is stochastic, that is, it changes from one measurement to another. This stochasticity is illustrated in Figure 8.1, in which many V–I curves are plotted on top of each other. Since the curves are measured under the same conditions, they are identical and fall on top of each other in Figure 8.1. The only exception is the region of the switching current. There, the jumps from SS to JNS are shown by almost vertical lines (which occur because the data points are connected by straight lines) and these lines occur at somewhat different values of I_{sw} for each new bias current sweep. Provided that the leads, which bring the bias current to the wire and which transfer the voltage on the wire to the room-temperature amplifier, are well-filtered, so that the electromagnetic noise reaching the wire is negligible, the fluctuations of the switching current reflect either thermal or quantum fluctuations *inherent to the nanowire itself*. Thus, by studying the statistics of I_{sw}, one can learn about the internal thermal and quantum fluctuations in the wire and understand their effect on measurable properties of the nanowire devices.

The transition from JNS to SS is observed at a much lower current called the retrapping current or return current, and denoted by I_r. The value of I_r is low because as the switching to JNS happens, the wire Joule heating increases from zero to about $R_n I_{sw}^2$. This amount of heating power is typically sufficient to increase the wire temperature by many times its critical temperature. Thus, to bring the wire to the normal state, the bias current needs to reach a much lower value compared to I_{sw}. This is why I_r is a few times smaller than I_{sw}. In the example of Figure 8.1, the ratio is $I_{sw}/I_r \approx 4$. Note that the region of the V–I curve near the retrapping current is not significantly broadened. This means that all V–I curves fall on top of each other. In other words, I_r is reproducible from one measurement to the next one (within the precision of the setup) and thus it is not stochastic. This fact is to be contrasted with I_{sw}, which is not reproducible from one measurement to the next one, and thus is considered stochastic quantity. As will be discussed in detail in subsequent section, the stochasticity of I_{sw} is due to the fact that it is determined by one variable, namely, the phase difference ϕ, which is subject to thermal and quantum fluctuations. The absence of pronounced stochasticity in I_r is due to the fact that the JNS → SS transition happens from the normal state, in which case the number of degrees of freedom of the wire is huge (equals the number of electrons). The relative fluctuation of any quantity is proportional to $1/\sqrt{N_{df}}$, where N_{df} is the number of the independent degrees of freedoms or, simply speaking, the number of particles which influence the quantity. For the retrapping current, this number is the total number of electrons in the wire. Due to averaging of this large number of independently fluctuating degrees of freedom, the fluctuation of a macroscopic quantity, namely, of the current value I_r at which the temperature drops below the critical temperature, is very small and cannot be observed.

The variation of the V–I curves with temperature is illustrated in Figure 8.2. Consider the curve measured at 1.75 K. The switching current (I_{sw} = 365 nA) is again marked by an arrow (Figure 8.2a). As the temperature is increased, the switching current decreases rapidly, while the return current remains constant. Thus, the hysteresis disappears at an elevated temperature which is still lower than the critical temperature of the wire. In the example shown, the hysteresis disappears at about 2.2 K.

The sharp voltage jump occurring on the V–I curve at $I = I_{sw}$ is called the switching event. The wire switches from the superconducting regime to the normal regime. The superconductivity disappears in the wire at the switching event. An important issue to understand is the relationship between the easily observable switching events and the Little's phase slips. It is tempting to to say that each LPS causes a switching event. Is this true? The analysis shows that it is not always true, but it can be true if the temperature is sufficiently low. First, let us consider cases where this is not true. Such examples are shown in Figure 8.2b. This figure shows three V–I curves taken from Figure 8.2a, namely, those measured at $T = 1.75$ K, $T = 1.9$ K, and $T = 2.1$ K. The only difference is that in Figure 8.2b, the voltage scale is much finer, namely, one division of the vertical axis in Figure 8.2b is 5 µV, while one division in Figure 8.2a is 1 mV. Thus, (b) allows us to see with a high resolution the shape of the behavior of the voltage at currents slightly lower than

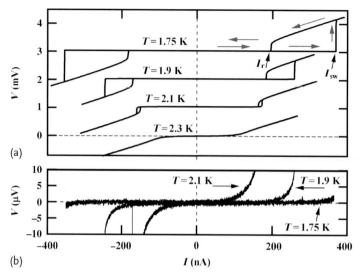

Figure 8.2 V–I curves of a $Mo_{79}Ge_{21}$ nanowire at various temperatures [183]. The data points are connected by lines (black). The horizontal scale is the same for both figures. The curves are measured by sweeping the bias current as $I_b = I_a \sin(2\pi f_b t)$, having the amplitude I_a chosen somewhat higher than the depairing current and the frequency being $f_b \sim 10\,Hz$. (a) The current and voltage change in time during the current sweep, as shown by arrows. The curve measured at 2.3 K is not shifted, while all other curves are shifted upward (for clarity) as follows: the 2.1 K curve – by 1 mV, the 1.9 K curve – by 2 mV, and the 1.75 K – by 3 mV. At low temperatures, the V–I curves exhibit a pronounced hysteresis, which shrinks by increasing the temperature and finally disappears at about 2.2 K for this sample. The switching current (I_{sw}) and the retrapping current (I_r) are shown explicitly by black arrows for the curve measured at 1.75 K. (b) A zoomed-in view on some of the V–I curves presented in (a). The figure illustrates the "voltage tails" occurring at a low voltage scale and at currents slightly lower than the switching current. This tails indicate that the switching is caused by multiple phase slips (multiphase slip switching process). Such tails are well-pronounced at higher temperature, but quickly freeze-out as the temperature is decreased. In the three examples shown, the tail is very strong at 2.1 K, less strong at 1.9 K, and barely distinguishable from the noise floor at 1.75 K.

the switching current. The observation is interesting: it turns out that the voltage starts to increase quite noticeably even before the switching even takes place. Since the voltage is proportional to the rate of LPS, we have to conclude that some significant number of phase slips happens before the switching event takes places. Thus, in the cases presented in Figure 8.2b, it is not possible to make a one-to-one correspondence between individual LPS events and the observable switching events. Note that detecting individual LPS is not possible using the DC measurement setup discussed here since a single LPS causes a very small voltage change, during a very short time. For example, the voltage detected in the tail of the 1.9 K curve, just before switching to the JNS, is of the order of 10 μV (of Figure 8.2b). According to the Gor'kov equation, this corresponds to about five billion phase slips per second on average. The setup upper limit frequency is typically about a hundred or a thousand hertz, which is so low in part due to the Cu powder filters used to eliminate any possible microwave noise. Thus, the rate of phase slips is much higher than the highest frequency that can be measured using a typical DC transport setup, for example, with the one illustrated in Figure 5.3. Therefore, the only hope of detecting single phase slips using a DC transport setup is to find some triggering mechanism in which easily observable events would have a one-to-one correspondence with LPS.

It turns out that such mechanism does exist. In fact, at low temperatures, the switching events in suspended nanowires are triggered by single phase slips and, most importantly, each phase slip switches the wire into the normal state. Such triggering happens because a single phase slip releases enough heat to increase the temperature of the wire above its critical temperature, for a given value of the bias current [116, 131, 178].

8.2
"Geiger Counter" for Little's Phase Slips

In a Geiger counter, a single ionized atom causes an avalanche of ionization events so that a current pulse occurs. Thus, a microscopic event (i.e., ionization of a single atom of the gas by an elementary particle flying through the chamber of the counter) causes a macroscopic event (an avalanche of ionizations leading to a large current pulse), which then can be detected. Thus, by measuring the rate of current pulses, one can draw conclusions about the rate of elementary particles in the environment.

In a similar way, a suspended nanowire can be used as a detector of LPS [59, 116, 131, 178]. The analysis of the statistics of the switching event is done under the condition that the bias current is swept up and down rather fast, as the V–I curves are measured. The term "rather fast" means much faster than the rate of LPS at zero-bias for the given temperature. Typically, the bias current is changed as $I_b = I_a \sin(2\pi f_b t)$, having the amplitude I_a chosen somewhat higher than the depairing current and the frequency being $f_b \sim 10$ Hz. The directions of the current and the voltage variation with time are indicated by arrows in Figure 8.2a. We are going to argue that at low temperatures, the heat released by even a single phase slip is sufficient to overheat the wire above its current-dependent critical temperature and thus cause the switch visible on the V–I curve. The main condition for this argument to work is the current sweeping condition. If the current is not changing, whatever its value is, one can get an LPS sooner or later. However, the switch to the normal state can only happen if the LPS occurs near the depairing current. Thus, the sweep must be fast enough so that the first LPS occurs when the bias current is already near the depairing current.

Another assumption here is that, as usual, we neglect the normal current when the wire is in the S-state, that is, we assume that before the switching happens, all current injected into the wire flows as supercurrent. After the switching, all injected current flows as normal current and thus it causes Joule heating of the wire, preventing it from returning back to the superconducting state.

The amount of energy released by one phase slip equals the work done by the current, which is given by (6.22). There, the phase change should be doubled since to reach the top of the barrier, the phase difference should change by π, and to reach the next local minimum of the free energy, corresponding to the completion of the phase slip, the phase should change by 2π. Thus, the energy released by one phase slip is [11, 51]

$$Q_{\text{LPS}} = \left(\frac{\hbar}{2e}\right) \int_0^{2\pi} I_s \, d\Delta\phi = \frac{h I_s}{2|e|} = \phi_0 I_s \tag{8.1}$$

where I_s is the supercurrent in the wire, $h = 2\pi\hbar$ is the Planck's constant, and $\phi_0 = h/2e = 2.07 \times 10^{-15}$ Wb (in SI units) is the superconducting flux quantum. Note that by the convention we have accepted, phase slips dissipate the energy of the supercurrent into heat and antiphase-slips convert the energy of thermal fluctuations into the kinetic energy of the supercurrent. Such convention is valid for $I_s \geq 0$. At high currents (i.e., when the current is close to the depairing current of the wire), the PS are much more frequent than APS, and thus the existence of antiphase-slips can be neglected completely when the switching events are analyzed.

Our intention now is to argue that if the temperature is low enough, even a single LPS can increase the temperature of the wire above its critical temperature and thus cause the wire to become normal. According to (8.1), each LPS generates heat of the amount $Q_{\text{LPS}} = 2 \times 10^{-21}$ J per each microamp of current flowing through the wire. Let us assume that all this heat is released inside the wire. Is this sufficient enough to convert the wire into the normal state? The answer depends on the heat capacity of the wire, c_{wire}, which, in turn, depends on the dimensions of the nanowire and the specific heat of the material from which the wire is made. However, the answer also depends on how much lower the bias current I_s is compared to the depairing current $I_c(T)$. If the bias current is very near the depairing current at the moment when the phase slip happens, then obviously, even a small amount of heat can raise the temperature above the critical temperature. As the wire enters the normal state, the Joule heating becomes very strong, so the temperature of the wire jumps up even higher, which, in turn, causes more heating and so on. Thus, the wire enters the normal state and remains trapped in the normal state.

Remember that the critical current is defined by the Bardeen formula, (11.1). If the current is fixed at the I_s value, the formula can be inverted. In other words, we can find the temperature $T_c(I_s)$ at which the given current I_s would be the depairing current. Such temperature can be appropriately named the current-dependent critical temperature. Note that $T_c(I_c) = 0$ since if the current equals the critical current (we use the term "critical current" interchangeably with the "depairing current"), the wire is normal even at $T = 0$. The inversion of the Bardeen formula gives

$$T_c(I_s) = T_c(0) \left\{ 1 - \left[\frac{I_s}{I_c(0)} \right]^{2/3} \right\}^{1/2} \tag{8.2}$$

One phase slip would trigger a switch to the normal state if $\Delta T_{\text{LPS}} > T_c(I_s) - T$, where $\Delta T_{\text{LPS}} = Q_{\text{LPS}}/c_{\text{wire}} = \phi_0 I_s/c_{\text{wire}}$ is the temperature increase due to the heat released by one phase slip and c_{wire} is the heat capacity of the wire, averaged over the temperature range between T and $T + \Delta T_{\text{LPS}}$. Thus, the minimal condition for single LPS to supply sufficient heat and make the wire normal is (assuming $I_s > 0$)

$$T > T_c(I_s) - \frac{\phi_0 I_s}{c_{\text{wire}}} \tag{8.3}$$

Note that the left side of this equation is always positive since $T > 0$ by the definition of temperature as being proportional to the kinetic energy of atoms forming the wire. Yet, the right side of the inequality becomes negative if the supercurrent is approaching the depairing current ($I_s \to I_c$). To see that the right side of the inequality can be negative, notice that the magnitude of the negative term, $-\phi_0 I_s/c_{\text{wire}}$, increases linearly by increasing the bias current I_s. On the other hand, the positive term $T_c(I_s)$ decreases with increasing I_s since the current suppresses the current-dependent critical temperature of the wire (8.2). Therefore, the inequality cited above will be satisfied and single phase slips will be switching the wire to the normal state with necessity if $T_c(I_s) < \phi_0 I_s/c_{\text{wire}}$. This sufficient condition for one-to-one correspondence between single (microscopic) phase slips and the switching events, which have macroscopic consequences, can be written as

$$\frac{\phi_0 I_s}{c_{\text{wire}}} > T_c \left[1 - \left(\frac{I_s}{I_c} \right)^{2/3} \right]^{1/2} \tag{8.4}$$

Here and everywhere, $T_c(0) \equiv T_c$ is the critical temperature of the wire at zero-bias current. The right side of the inequality above approaches zero as $I_s \to I_c$. Thus, the inequality is satisfied if the supercurrent is close enough to the critical current. What can prevent the supercurrent from approaching the critical current close enough to satisfy the above inequality? The answer is the premature switching to the normal state. The premature switching is caused by LPS (assuming that all other noise sources are eliminated by filtering the leads and by screening the sample). Therefore, a single LPS can switch the wire if the LPS occurs when the bias current is very near the critical current. Such a situation happens at very low temperature, when phase slips are not present at low currents and start to occur with a noticeable rate only when the bias current is very near the depairing current. Thus, if the temperature is lower than some sample-specific temperature, the LPS would not happen until the current is increased to a value very near I_c, in which case a single LPS, as it finally happens, would release enough heat to push the temperature a bit higher and switch the wire to the normal state. The lower the temperature, the easier it is to satisfy the above inequality since the first LPS happens close and closer to the critical current as the temperature is made lower. A more detailed theoretical analysis of this process is given in [178]. The phase diagram in this paper shows that for a typical nanowire, there is a region of low temperatures in which the observable switches are in one-to-one correspondence with single phase slips. The existence of this region allows one to study the rate of phase slips as a function of the current and the temperature by studying the rate of easily observable switching events.

A sufficient condition for a single phase slip to generate enough heat to switch the wire into JNS is given by (8.4). It can be better understood if it is analyzed graphically. The equation can be rewritten as

$$\left(\frac{\phi_0 I_c}{c_{\text{wire}} T_c}\right) i_s > \left[1 - (i_s)^{2/3}\right]^{1/2} \qquad (8.5)$$

where $i_s \equiv I_s/I_c$ is the normalized supercurrent in the wire. The inequality can also be written as $Y_l(i_s) > Y_r(i_s)$, where $Y_l(i_s) = (\phi_0 I_c/c_{\text{wire}} T_c) i_s$, which depends on the parameters of the wire and $Y_r(i_s) = [1 - (i_s)^{2/3}]^{1/2}$, which is a universal function. Let us consider an example and estimate the factor $\phi_0 I_c/c_{\text{wire}} T_c$ for a realistic wire. Typically, the critical current is $I_c \sim 3\,\mu\text{A}$ and the critical temperature of a MoGe wire is $T_c \sim 3\,\text{K}$. The heat capacity of the wire is $c_{\text{wire}} = L\pi r^2 \rho_w c_{3\text{K}} \sim 5 \times 10^{-21}\,\text{J/K}$, where $L \sim 100\,\text{nm}$ is the wire length, $r \sim 5\,\text{nm}$ is the wire diameter, $\rho_w \sim 10^4\,\text{kg/m}^3$ is the wire density (which is taken here equal to the density of molybdenum, which is the main component in a typical MoGe wire), and $c_{3\text{K}} \sim 0.065\,\text{J/(K kg)}$ is the relevant specific heat (taken as the specific heat of Mo at 3 K). Note that the specific heat of Mo is 250 J/(K kg) at room temperature. The specific heat of Mo drops by a factor of ~ 3850 as the temperature is reduced from 300 to 3 K. With these typical numbers, we get the coefficient in front of i_s in the function $Y_l(i_s)$ as $\phi_0 I_c/c_{\text{wire}} T_c = 0.4$. The corresponding plots are shown in Figure 8.3. Since the function $Y_l(i_s)$ is positive and increases linearly from zero and the function $Y_r(i_s)$ is positive and decreases monotonously, reaching zero at $i_s = 1$, these functions always cross. The crossing point is denoted by i^*. In the example shown, $i^* = 0.84$. It corresponds to a supercurrent $I^* = 0.84 I_c$. The condition of interest, $Y_l(i_s) > Y_r(i_s)$, is satisfied for any $I_s > i^* I_c$. Thus, if a single phase slip happens at $I_s > i^* I_c$, it releases enough heat to increase the wire temperature above its current-dependent critical temperature. Thus, each phase slip switches the wire to the normal state with necessity. If, on the other hand, a phase slip happens at $I_s < i^* I_c$, then the switching to the normal state can happen only if one or a few more phase slips follow the first one within a short period of time, so short that the wire cannot cool down. In other words, for $I_s > i^* I_c$, all switching events are initiated by single phase slips, while for $I_s < i^* I_c$, the majority of switching events would correspond to a collective overheating effect of many phase slips occurring within a short interval of time, shorter than the cooling time constant of the wire [131, 178]. The value of the supercurrent at which the switching takes place is denoted by I_{sw}. It is a stochastic quantity which changes from measurement to measurement. Yet, the distribution of the switching currents shifts to higher values as the temperature is reduced. At sufficiently low temperature, the distribution

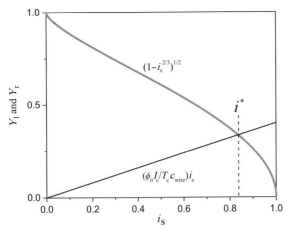

Figure 8.3 Graphical solution of (8.5). The downward curve is the function $Y_r(i_s) = [1 - (i_s)^{2/3}]^{1/2}$ which approaches zero as the normalized supercurrent $i_s = I_s/I_c$ approaches unity. The straight line represents the function $Y_l(i_s) = (\phi_0 I_c/c_{wire} T_c) i_s$ which increases linearly with increasing i_s. The graph illustrates that these two curves always cross. The crossing point is denoted by $i^* \equiv I_s^*/I_c$. In the example shown, $I_s^* \approx 0.84 I_c$. This means that if a single phase slip happens at $I_s > 0.84 I_c$, then the heat released by the LPS is definitely sufficient to heat the wire above its current-dependent critical temperature. Therefore, the observable switching events are in one-to-one correspondence with single phase slips if the corresponding switching currents are near the depairing current, namely, if $I_{sw} > 0.84 I_c$ in this example.

approaches I_c so closely that practically all switching events in a large ensemble of measurements satisfy the condition $I_{sw} > I^*$. In this case the entire distribution can be considered as a collection of single phase slip events, detected through the switching mechanism overheating.

8.3
Measuring Switching Current Distributions

The switching statistics can be characterized quantitatively by measuring the probability density for the switching current. By definition, if the switching probability density is $P(I)$, then $P(I)dI$ equals the probability that, as the bias current is increased from zero, the wire switches from the superconducting to the normal state in the interval of currents from I to $I + dI$. The probability density graph is also called the distribution of the switching currents. To understand the probability density concept, imagine that the bias current is swept from zero up to $I_{max} > I_c$, and then down back to zero. Assume that such a current cycling is repeated N times and the switching current is recorded for each cycle. Then, according to the definition of the probability density, the number of times the switching is detected in the interval between I and $I + dI$ is $NP(I)dI$, assuming N is very large.

The value of the bias current at which the switch takes place is the switching current and denoted by I_{sw}. The distribution current can also be written as I_{sw} because at the moment of the switching, even the bias current equals the switching current, $I = I_{sw}$, by definition.

To provide more concrete example of how the distribution $P(I_{sw})$ of the switching current I_{sw} can be measured, we follow [131]. The temperature is kept constant and the current is swept as $I = I_{max} \sin(2\pi f t)$, where f is the frequency of the current bias. The typical choice of the amplitude is $I_{max} = 1.3 I_c$. The current biasing is achieved technically as explained in Figure 5.3. The frequency needs to be chosen much lower than the frequency cut-off of the setup used. To understand this, consider a concrete example. Suppose one point on the V–I curve can be measured using a National Instruments DAQ card within $\tau_{DAQ} = 100\,\mu s$. Suppose the distribution width is $\sigma = 10\,nA$, and the critical current is $I_c = 0.7\,\mu A$, so that the amplitude can be chosen as $I_{max} = 1\,\mu A$. Suppose our desire is to obtain $N_\sigma = 10$ significant points on the distribution $P(I_{sw})$. The time it takes the current to sweep though the region where the switching occurs frequently is $\tau_\sigma = (\sigma/I_{max})(1/f)(1/4)$. If N_σ points are needed, then each point should be reliably measured within the time interval which equals $\tau_1 = (\sigma/I_{max})(1/4 f N_\sigma)$. Thus, the condition for a correct measurement is $\tau_1 > \tau_{DAQ}$

To measure the switching current I_{sw} at which the wire abruptly switches to the normal state, one needs to detect the accompanying voltage jump. It is customary to use a National Instruments digitizer card controlled using LabVIEW programming language. The LabVIEW program receives data from the card which alternatively measures the voltage on the sample and the voltage on the series resistor (which is then converted into current using Ohm's law). The data acquisition frequency is typically in the range of 10–100 kHz. The data are stored in the card memory and then a big array, representing one cycle of the V–I curve, is sent to the computer assigned to control the experiment. A typical measurement circuit diagram is shown in Figure 5.3, where the digitizer cards are marked as DAQ cards, and DAQ means "data acquisition." LabVIEW can be programed to record the instantaneous value of the bias current exactly (or almost exactly) at the moment when the voltage on the sample exceeds some preset threshold. Typically, the threshold is set ten times larger than the voltage noise level. The threshold is still much lower than the voltage observed after the switch, which equals $R_n I_{sw}$. The reference noise level is estimated at currents below the switching current when the sample is still superconducting. The jump, taking place at I_{sw}, happens on a timescale of nanoseconds. Therefore, it appears as a vertical or almost vertical line on the V–I plot since the V–I curve is typically measured at a much lower frequency scale ($\sim 10\,Hz$). Therefore, the exact choice of the voltage threshold is not important, that is, this choice does not change the measured value of I_{sw} and the corresponding dispersion σ. Yet, the voltage threshold should be much less than $R_n I_{sw}$ in order to avoid distortions.

For the statistical analysis, the measurement of the I_{sw} must be repeated many times, typically $N = 10\,000$ times. The reason for this is that a typical goal of such experiments is to determine the standard deviation σ, defined as $\sigma^2 = N^{-1} \sum_1^N (I_{sw,i} - \langle I_{sw} \rangle)^2$, white i is the number of the measurement, $I_{sw,i}$ is the result of the ith measurement, and $\langle I_{sw} \rangle$ is the switching current averaged over all measurements. At large N, σ is not dependent on N, though at small N, the resulting sigma might be subject to statistical errors. A practical way to estimate the required N is to repeat the measurement of clusters of N points a few times and to determine σ for each cluster. The difference between the results should be smaller than the desired accuracy of the σ measurement.

To get a detailed picture of thermal and quantum fluctuations in thin wires, it is useful to repeat such measurements at various temperatures. For example, in [131], large arrays of switching currents, $N = 10\,000$ points each, have been repeatedly acquired at equally spaced temperatures, between 0.3 and 2.3 K. The corresponding distributions are shown in Figure 8.4. The curves are proportional to the probability density functions. Each curve is normalized such that the summation over all bins, bin size being 4 nA, gives the total of 10 000 switching events.

In the example shown, an increase of the standard deviation of the switching current is observed as the temperature is lowered. This fact is made explicit in the Figure 8.4 insert. It is also quite obvious from the distributions themselves (Figure 8.4) since σ is the measure of the width of the distribution curves and the curves are seen to broaden as the temperature is reduced. Note also that the width is inversely proportional to the height of the distribution since their product, or, to be more exact, the area under each curve, equals the total number of acquired points, which is 10 000 in the example discussed (sometimes, a different normalization is chosen, such that the area equals unity).

The fact that the distribution curve for the switching currents broadens with cooling is opposite to the Kurkijärvi prediction which states that $\sigma \propto T^{2/3}$ [179]. The explanation is as follows: The Kurkijärvi theory is developed for the case when (1) each detected event is a single Little's phase slip, (2) all phase slips are detected, and (3) phase slips are caused only by thermal fluctuations. Although such conditions are quite common, see, for example, [39], they are not satisfied in the case of thin metallic wires discussed here.

At low temperatures, which for the example shown means $T < 1$ K, a different scenario is realized. At such temperatures, as it was argued above, there is one-to-one correspondence between individual phase slips and the switching events. This means that each phase slip causes a switching, and therefore the measure σ represents the dispersion of the phase slips. Such a dispersion is expected to be described by the Kurkijärvi theory, yet the results of Figure 8.4 do not confirm this: instead of decreasing according to $T^{2/3}$ law, σ is observed to increase with cooling. This is explained by the occurrence of quantum phase slips (QPS). Ideally, the QPS rate is supposed to be independent of temperature. Therefore, σ, if it is due to QPS, should be independent of temperature. The data presented in Figure 8.4 show some slight increase of σ with cooling (for $T < 1$ K). This increase is probably caused by a slight increase of the superconducting energy gap with cooling, which takes place even at low temperatures if the T_c of the wire is not much high than the temperature at which σ is measured. A more detailed analysis of the QPS switching statistics will be presented later. In the next section, we provide a more formal analysis of the switching current standard deviation and the switching rate. The foundation for the analysis was laid down by Kurkijärvi [179], and thus we will refer to the techniques involved in the analysis as the Kurkijärvi theory.

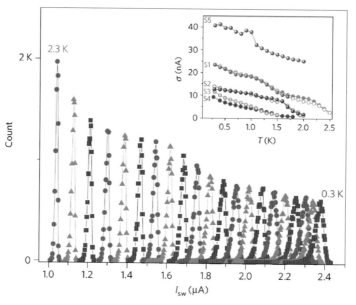

Figure 8.4 Examples of distributions of the switching current corresponding to different temperatures [131], all taken on the same sample (S1). The horizontal axis is divided into consequent segments, 4 nA each, which are called bins. The total number of measurements of the I_{sw} is $N = 10\,000$ at each temperature. The vertical axis corresponds to the number of counts in each bin. In other words, the "Count"-axis gives the number of switching events for each bias current bin. Note that the current bins are not explicitly marked on the horizontal I_{sw}-axis. Inset: The standard deviation σ vs. T for five different nanowires, including sample S1. For samples S1 and S2, the measurements have been repeated (and the results are shown by different symbols) to verify the reproducibility of the temperature dependence $\sigma(T)$. In these examples, it is observed that $d\sigma/dT < 0$, which is unusual. This is explained in terms of multiple phase slips [131]. The detailed analysis of [131] showed that below about 1 K, the switching events are caused by individual phase slips and each phase slip causes a switching. Therefore, the fact that at $T < 1$ K, no decrease of σ is seen as the temperature is decreased provides evidence that the phase slips in these samples occur by quantum tunneling, that is, they are quantum phase slips (QPS). Some increase of σ at $T < 1$ K could be due to the fact that the barrier for phase slips increases slightly with cooling. Note that if the switching at low temperatures would be caused by thermally activated phase slips (TAPS) then σ would decrease with cooling as $\sigma \sim T^{2/3}$. This is not observed even at the lowest temperature tested. Therefore it is concluded that QPS is the dominant mechanism causing the switching events. Please find a color version of this figure in the color plates.

8.4
Kurkijärvi–Fulton–Dunkleberger (KFD) Transformation

In this section, how to extract switching rates from distributions of the current at which the switching happens is explained. In the previous section, we have seen how the probability distribution of the switching current can be obtained experimentally. Here, we look into the statistical analysis of such distributions based on the Kurkijärvi theory (KT) [179] which was generalized to the current-biasing experiments by Fulton and Dunkleberger [180] and Garg [181]. Generally speaking, the Kurkijärvi theory provides a quantitative statistical analysis of stochastic premature switching events. Although the theory can be applied to the analysis of different physical systems, we formulate it in the form suitable for the analysis of the switching currents in thin wires. Thus, our example system is a current-biased superconducting wire. Therefore, bias current is assumed to be slowly increasing, starting from zero. As the current is increased, the wire approaches the critical state which is reached at $I = I_c$ in which the normal state of the wire has a lower Gibbs energy than the superconducting state and in which the barrier separating the two states becomes zero. Thus, the wire must switch to the normal state if the current has reached the $I = I_c$ level and if the wire is still superconducting. At lower currents, the energy barrier separating the normal state from the superconducting state is larger than zero. However, due to thermal or quantum fluctuations, the wire can overcome the barrier prematurely and switch to the normal state at $I < I_c$. Typically, these premature switchings happen very near I_c. Their statistical analysis, to be discussed in detail below, provides information about the strength of various fluctuations in the wire. The stochasticity of a switching process signifies an observation that even when the temperature and the sweeping protocol of the control parameter (i.e., the bias current) are kept unchanged, the switching event occurs at a different value of the control parameter in each new measurement run.

Experimentally, to measure the switching current, one initially applies zero current to a nanowire to ensure that it reaches the equilibrium superconducting state. The temperature of the wire, T, is assumed lower than the critical temperature. Then, the bias current I is increased linearly in time. The current increase must be so slow that any possible nonequilibrium gradients in the wire induced by the current sweep have enough time to decay. We assume, as always, that the entire amount of the bias current flows as the supercurrent (i.e., $I_s = I$).[1] The sweep

---

1) This is reasonable since the condensate has zero resistance, at least for a slowly changing bias current and when the phase slips are absent, while the fluid of normal electrons (bogoliubons) in the wire possesses, naturally, a finite electrical resistance (since normal electrons are not superconducting of course). For the theory presented here to be valid, we only need the condition $I_s = I$ to hold until the system reaches the maximum of the Gibbs barrier, which happens during a phase slip event, as the phase difference on the wire increases by a small amount only, namely, significantly less than π near the critical current [64, 143]. A detailed theoretical analysis shows that the system can reach the maximum of Gibbs energy in a course of a fluctuation, and maintain a constant supercurrent [64, 143]. In such fluctuations, the order parameter magnitude diminishes while the phase gradient increases in such a way that the supercurrent remains unchanged.

speed of the bias current, defined as $v_I = dI/dt$, is either constant or at least positive during the measurement of the I_{sw}. At some value of the bias current, called the switching current I_{sw}, the wire switches to the normal state. This switch is easily detectable since it causes a strong voltage jump on the wire. Soon after the jump is detected, the bias current is reduced to zero in order to bring it back to the superconducting state. Then, the measurement is repeated. The total number N of such measurements needs to be large for the accuracy of the statistical analysis (typically, $N = 10\,000$).

Generally speaking, the wire can be only in two distinct states, the superconducting (before the switching takes place) and normal (after switching). The probability density (also called the probability distribution) for the switching event to occur as the bias current reaches value I is denoted as $P(I)$. To understand the meaning of this distribution, assume that instead of repeating the measurement N times, we have an ensemble of N identical wires and conduct the same switching experiment in each of them. According to the ergodic hypothesis of the thermodynamic, repeating the measurement N times on the same system or making the measurements just once, but on N identical systems, generally gives the same results. In our thought experiment at $t = 0$, we have $I = 0$ and all wires are superconducting. Then, the current is increased in all wires by the same exact monotonic function $I = I(t)$. The number of wires in the superconducting state N_1 decreases with time (and with increasing current) and approach zero as the current approaches the critical current of the wire I_c. The number of wires in the normal state (i.e., the number of switched wires), N_2, increases with time (and with current) and reaches N at $I = I_c$. Note that the current must increase with time in the same, well-defined manner, for all systems. It can, for example, be linear as $I = v_I t$, but can also be chosen sinusoidal or any other smooth monotonic function.

It is possible now to define the distribution $P(I)$. The meaning of it is that the probability for the wire to switch in the bias current interval dI is $P(I)dI$. Assume that the number of switched wires at current I is $N_2(I)$. The number of switched wires in the considered ensemble increases to $N_2(I + dI)$ as the current reaches $I + dI$. Then, the number of wires which experience a switching event in the current interval dI is $dN_2 = N_2(I + dI) - N_2(I)$. Thus, the probability (by the definition of probability) for any particular wire to switch in the considered current interval dI is

$$P(I)dI = \frac{dN_2}{N} \tag{8.6}$$

where N is the total number of the wires in the ensemble. Since the total number is fixed, we have $N_1 + N_2 = N = \text{const}$ and therefore $dN_1 = -dN_2$. Thus, from the equation above, it follows also that

$$dN_1 = -NP(I)dI \tag{8.7}$$

This equation can be integrated over current, from zero to I, leading to

$$N_1(I) - N_1(0) = -N \int_0^I P(I')dI' \tag{8.8}$$

Since at the beginning of the current sweep all wires are not switched, then $N_1(0) = N$, and thus one gets

$$N_1(I) = N\left[1 - \int_0^I P(I')dI'\right] \tag{8.9}$$

Following Kurkijärvi, we can also introduce the probability $W(I)$ for the wire not to switch up to the current I as the number of unswitched wires divided by the total number of them, that is, $W(I) = N_1(I)/N$. Then, (8.7) can be rewritten as

$$P(I) = -\frac{dW}{dI} \tag{8.10}$$

Note that if thermal and quantum fluctuations would be absent (practically, this is not possible of course), then all wires would switch to the normal state at $I = I_c$ and so the probability density would be $P(I) = \delta(I - I_c)$, where $\delta(x)$ is the Dirac delta function. This conclusion, as well as all other conclusions in this section, is exactly correct only if the number of wires in the thermodynamic ensemble is infinite or if the number of measurements of I_{sw} on one wire is infinite (if $N \to \infty$). Similarly, if the fluctuations are present but the sweep speed v_I is infinitely high, the wires would not have time to switch prematurely and so all would switch at $I = I_c$, where they must switch since the energy of the normal state becomes lower than the energy of the superconducting state, and the energy barrier separating these two states becomes zero. In such hypothetical cases, again, $P(I) = \delta(I - I_c)$. These conclusions are important since one can understand, based on these facts, that in the experimentally accessible cases, the maximum of the switching distribution should move to the $I = I_c$ point as the sweep speed is increased or if the temperature is decreased.

The next step in our analysis is to introduce the switching rate $\Gamma(I)$, which is more fundamental than the experimentally measured $P(I)$. An approximate formula for the switching rate is given by (6.3) for the case of negligibly low bias currents. The formula is written for the premature switching events caused by thermal fluctuations. In the context of our present discussion, the rate has to be generalized to include the dependence of the barrier on the bias current. Generally, is can be written as $\Gamma(T, I) = \Omega(T, I)\exp[-\Delta G(T, I)/k_B T]$, where $\Omega(T, I)$ is the attempt frequency and $\Delta G(T, I)$ is the Gibbs barrier for the switching. The switching rate is the inverse ($\Gamma(I) = 1/\tau(I)$) of the lifetime $\tau(I)$ (Kurkijärvi notation) of the system in the metastable (superconducting) state under a constant bias current I.

The reason Γ is considered more fundamental is that it does not depend on the sweep speed v_I, while the distribution $P(I)$ is sweep-speed dependent (the maximum of the distribution P(I) shifts to higher current if v_I is increased). The rate Γ

is defined such that Γdt is the probability for the sample to switch in the time interval dt, that is, to switch in the time interval between t and $t + dt$, assuming that the wire is not switched at the beginning of this time interval, that is, the wire is in the superconducting state at time t.

Therefore, in the thermodynamic ensemble, the number of wires switching during the time interval dt is $dN_2 = N_1 \Gamma dt$, where N_1 is the number of unswitched wires, that is, those which still have a possibility to switch.[2] Thus, we arrive at the second important equation describing the probability flow

$$dN_1 = -N_1 \Gamma dt \tag{8.11}$$

where, as everywhere, $\Gamma = \Gamma(T, I)$ where $T = $ const and $I = I(t)$ is a certain monotonically increasing function of time. If (8.11) is divided by N, it gives

$$dW = -W\Gamma dt \tag{8.12}$$

This can be solved by rewriting as $dW/W = -\Gamma\,dt$, and integrating to get $\ln[W(t)] - \ln[W(0)] = -\int_0^t \Gamma(I(t'))dt'$. Since at $t = 0$ we have $W(0) = N_1(0)/N = N/N = 1$, the solution becomes

$$W(t) = \exp\left[-\int_0^t \Gamma(I(t'))dt'\right] \tag{8.13}$$

which means that the number of unswitched wires drops exponentially as

$$N_1(t) = N \exp\left[-\int_0^t \Gamma(I(t'))dt'\right] \tag{8.14}$$

which means, among other things, that if the current would be fixed, then the number of unswitched wires in the ensemble would drop as $N_1(t) = N\exp(-\Gamma t)$, where $\Gamma = \Gamma(I, T)$ depends, in general, on the set value of the bias current and the temperature.

To find the expression for the switching rate through the distribution $P(I)$, we have to remember that the current depends on time in a deterministic way according to $dI = v_I dt$. This expression, if put into (8.7), leads to $dN_1 = -NP(I)v_I dt$. Now, we combine this with (8.11) to get $-N_1 \Gamma dt = -NP(I)v_I dt$, where dt cancels and we obtain an important but simple equation which does not contain time explicitly:

$$N_1 \Gamma(I) = NP(I)v_I \tag{8.15}$$

This equation will be called the master equation.

---

2) A wire, as it switches, enters the normal state which is stable and cannot return back to the superconducting state due to the Joule heating released under a constant current bias. The wire can only be switched back to the supercurrent state if the bias current is substantially reduced.

The master equation can be further transformed using (8.9) in

$$\Gamma(I) = P(I)v_I \left[1 - \int_0^I P(I')dI' \right]^{-1} \tag{8.16}$$

This equation will be called the Kurkijärvi–Fulton–Dunkleberger (KFD) transformation. It provides a link between the sweep-speed dependent distribution function $P(I)$ measured in experiments, to the sweep-speed independent switching rate which could be computed theoretically, at least in principle.

Since the integral of the probability density must be normalized to unity, that is, $\int_0^{I_c} P(I)dI = 1$, the following equality is true, namely, $\int_I^{I_c} P(I')dI' = 1 - \int_0^I P(I')dI'$. Therefore, the KFD transformation can be written in a different form, namely, as

$$\Gamma(I) = P(I)v_I \left[\int_I^{I_c} P(I')dI' \right]^{-1} = -v_I \frac{d}{dI}\left[\ln \int_I^{I_c} P(I')dI' \right] \tag{8.17}$$

The last equality is obtained due to the well-known integral differentiation rule as $d(\int_I^{I_c} P(I')dI')/dI = -P(I)$.

The data obtained in experiments are discrete. To make a convenient comparison with experiments, one can define bias current bins following [180]. Let $I_{sw,m}$ be the greatest switching current detected in a series of measurements. Then, the bin #1 is $I_{sw,m} - \Delta I < I \le I_{sw,m}$, the bin #2 is $I_{sw,m} - 2\Delta I < I \le I_{sw,m} - \Delta I$ and so on. The kth bin is defined on the interval $I_{sw,m} - k\Delta I < I \le I_{sw,m} - (k-1)\Delta I$, where k is an integer and $k > 0$. Here, ΔI is the chosen bin size, which should be much smaller than the width of the distribution, but much larger than the resolution of the ammeter employed to measure the switching current. Each bin can be matched with the corresponding mean current, as $I_1 = I_{sw,m} - \Delta I/2$, $I_2 = I_{sw,m} - 3\Delta I/2$, $I_3 = I_{sw,m} - 5\Delta I/2$, and so on. The mean current of the kth bin is $I_k = I_{sw,m} - (k-1/2)\Delta I$. The mean bin current represents the center of the bin. The maximum bin number, k_m, is chosen such that in the entire array of data, there no switching current events having the switching current less than $I_{sw,m} - k_m \Delta I$.

The probability to switch at a current within bin k can be computed as

$$P_k \equiv P(I_k) = \frac{1}{\Delta I} \int_{I_k - \Delta I/2}^{I_k + \Delta I/2} P(I)dI \tag{8.18}$$

Remember that P_k can be obtained from experimentally obtained switching current arrays. It is defined as the number of switching events N_k such that the switching current belongs to the bin k, that is, $I_k - \Delta I/2 < I_{sw} \le I_k + \Delta I/2$. Once N_k is determined (usually this is done using a computer program written in LabVIEW), the distribution function is computed as $P_k = N_k/N$.

As soon as the array P_k is known, one can compute the discrete-set switching rates $\Gamma_k \equiv \Gamma(I_k)$. According to (8.17), we have $\Gamma(I_k) = -v_I d(\ln Z)/dI|_{I=I_k}$, where

$Z = Z(I) \equiv \int_I^{I_c} P(I')dI'$. To obtain a fully discrete KFD formula, we have to replace the continuous variable I with its discrete analogue, I_k. This can be done, approximately, by making the following replacement

$$\left[\frac{d(\ln Z)}{dI}\right]_{I=I_k} \to -\frac{\ln Z_k - \ln Z_{k-1}}{\Delta I} = (\Delta I)^{-1} \ln \frac{Z_{k-1}}{Z_k}$$

Here, we use the notation $Z_k \equiv Z(I_k)$. Therefore, for the discrete switching rate, we obtain

$$\Gamma(I_k) = \frac{v_I}{\Delta I} \ln\left(\frac{Z_k}{Z_{k-1}}\right)$$

which is valid for $k > 1$. The integral Z can be presented as a sum of short-segment integrals as

$$Z_k = \int_{I_k}^{I_c} P(I')dI' = \sum_{l=1}^{l=k} \int_{I_l - \Delta I/2}^{I_l + \Delta I/2} P(I')dI'$$

where l is an integer. Using the definition of the discrete probability distribution (8.18), one then gets $Z_k = \Delta I \sum_{l=1}^{l=k} P_l$. Thus, we arrive at the expression for the discrete KFD transformation

$$\Gamma(I_k) = \frac{v_I}{\Delta I} \ln\left(\frac{\sum_{l=1}^{l=k} P_l}{\sum_{l=1}^{l=k-1} P_l}\right) = \frac{v_I}{\Delta I} \ln\left(1 + \frac{P(I_k)}{\sum_{l=1}^{l=k-1} P(I_l)}\right) \quad (8.19)$$

If ΔI_{sw} is the interval of currents obtained experimentally in which $P(I)$ is above zero, then the formula above works well only for small bin sizes, $\Delta I \ll \Delta I_{sw}$. Of course, if the resolution of the apparatus used to measure the switching current is δI, then it is necessary to choose $\Delta I \gg \delta I$.

If necessary, the accuracy of the KFD expression can be further improved. Notice that the derivative of $\ln Z$ was computed on the interval between I_k and I_{k-1}, that is, on the interval from $I_k = I_{sw,m} - (k - 1/2)\Delta I$ to $I_k = I_{sw,m} - (k - 3/2)\Delta I$. The center of this interval is $I_k = I_{sw,m} - (k - 1)\Delta I = I_k + \Delta I/2$, which is the higher limit of the kth interval. Thus, a slightly more precise expression is

$$\Gamma\left(I_k + \frac{\Delta I}{2}\right) = \frac{v_I}{\Delta I} \ln\left(\frac{\sum_{l=1}^{l=k} P_l}{\sum_{l=1}^{l=k-1} P_l}\right) = \frac{v_I}{\Delta I} \ln\left(1 + \frac{P(I_k)}{\sum_{l=1}^{l=k-1} P(I_l)}\right)$$

Presumably, ΔI is small enough so the difference discussed above provides only a small correction to the form of the function $\Gamma(I)$.

8.5
Examples of Applying KFD Transformations

It is much more straightforward to analyze the switching rate Γ than the distribution function P. The switching rate is predicted by a few different models. The simplest models are the Arrhenius model for TAPS, (6.4), and the Giordano model for QPS, which is exactly like (6.4), except that the thermal energy $k_B T$ is replaced by the characteristic energy of zero-point fluctuations, $k_B T_q$. Here, T_q is the quantum temperature which defines the scale of quantum fluctuations and is typically used as a fitting parameter. For short nanowires attached to macroscopic electrodes, one can make a rough estimate of the quantum temperature as $k_B T_q \sim \hbar \omega_p$. The plasma frequency of the device is $\omega_p = 1/\sqrt{L_k C}$, where C is the capacitance between the electrodes and L_k is the kinetic inductance of the condensate in the nanowire, (3.14).

If one combines the TAPS and the QPS into a single expression, the net rate of phase slips can be estimated as

$$\Gamma = \Gamma_{\text{TAPS}} + \Gamma_{\text{QPS}} \tag{8.20}$$

where the rate of thermally activated phase slips is

$$\Gamma_{\text{TAPS}} = \Omega_{\text{TAPS}}(I, T) \exp\left[-\frac{\Delta G(I, T)}{k_B T}\right] \tag{8.21}$$

The barrier is $\Delta G(I, T) = \Delta F(T)(1 - I/I_c)^{5/4}$ (see Figure 8.7), where the barrier at zero current is $\Delta F(T) = \sqrt{6}\hbar I_c(T)/2e$ (see [19]) and the temperature-dependent critical current is given by the Bardeen formula $I_c(T) = I_c(0)(1 - T^2/T_c^2)^{3/2}$ [20]. The attempt frequency for TAPS is known accurately only near T_c (7.11) and equals $\Omega_{\text{TAPS}} \approx (k_B T_c/\hbar)[L/\xi(T)]\sqrt{\Delta G(I, T)/k_B T}$.

According to approximate heuristic Giordano arguments, which have been confirmed in, for example, [108], the expression for the QPS rate should be similar, except that $k_B T$ has to be replaced with $k_B T_q$ everywhere except in the form of the barrier height. Thus,

$$\Gamma_{\text{QPS}} = \Omega_{\text{QPS}} \exp\left[-\frac{\Delta G(I, T)}{k_B T_q}\right] \tag{8.22}$$

Here, $\Delta G(I, T)$ is the same as for the TAPS, and

$$\Omega_{\text{QPS}} \approx \left(\frac{k_B T_c}{\hbar}\right)\left[\frac{L}{\xi(T)}\right]\sqrt{\frac{\Delta G(I, T)}{k_B T_q}}$$

Now, we are ready to extract the switching rate and to fit it using the model outlined above. The input data are the distributions of Figure 8.4. The switching rate can be computed for each distribution using the discrete KFD transformation, (8.19). The results are shown in Figure 8.5. Figure 8.5a,b essentially show the same data, or, to be more precise, the same Γ obtained through (8.19), except that Figure 8.5(b) presents more curves, including those corresponding to the lowest temperatures, which are not shown in Figure 8.5(a).

First, note that the results are plotted in the log-linear format. The fact that the plots look quasilinear implies that the rates (shown by various colored symbols)

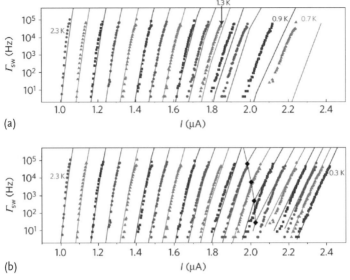

Figure 8.5 Switching rates plotted versus bias current for different temperatures for the sample S1 [131]. The rates are computed from the data of Figure 8.4 using the KFD transformation (8.19). (a) Switching rates for the temperatures between 2.3 K (left-most) and 0.7 K (right-most), for Sample 1. The rates are shown for all temperatures between 2.3 and 1.1 K, with a step of 0.1 K, and, also, for $T = 0.9$ K and $T = 0.7$ K. The symbols are the experimentally obtained rates and the solid curves (using the same colors) are the fits to the overheating model that only includes TAPS-initiated switching events and assumes that the QPS rate is zero. The fits agree well with the data down to $T \approx 1.3$ K. The data and the curve corresponding to $T = 1.3$ K are indicated by an arrow. At low temperatures, such a TAPS-only model does not agree with the data since it predicts switching rates much lower than the rates obtained experimentally. For example, for $T = 0.7$ K, the TAPS-only predicted rate is about 10 000 times lower than the measured one. All the fitting parameters, namely, $T_c = 3.872$ K, $\xi(0) = 5.038$ nm, $I_c = 2917$ nA, are kept the same for all temperatures. (b) The same switching rates for S1 as in (a), extended down to lower temperatures. The rates are plotted for the temperatures between 2.3 K (left-most) and 0.3 K (right-most), the step being 0.1 K. The symbols are the experimentally obtained rates and the solid curves (using matching colors) are the fits to the overheating model that incorporates Γ_{TAPS} as well as Γ_{QPS}, as in (8.20). The single-phase-slip switching occurs to the right of the black curve indicated by the black diamonds. These best fits are obtained assuming that the effective quantum temperature is $T_q = 0.726$ K $+ \Delta T_q$, where the temperature dependent correction is $\Delta T_q = 0.4T$ and T is the temperature. The physical reason for the correction is not well understood yet. It could be either due to the temperature dependence of the superconducting energy gap or due to a survival of multiphase-slip switching even at low temperatures. At $T < 0.7$ K the QPS rate is much larger than the TAPS rate. The QPS based model agrees with the data very well, thus providing a solid evidence that the switching is caused by QPS at $T < 0.7$ K. Please find a color version of this figure in the color plates.

exhibit a general trend, namely, they decrease roughly exponentially via the deceasing of the bias current. This is to be expected since the effect of the current is to reduce the LPS barrier. Therefore, as the current is made weaker, the barrier height becomes larger. On the other hand, the rate of thermal activation over the barrier and the rate of quantum tunneling through the barrier both decrease exponentially with increasing the barrier height. This simple argument provides a qualitative understanding for the curves (symbols) of Figure 8.5. We also find that the rate vs. current curves shift to higher currents as the temperature is reduced. Again, this is expected since the critical current gets larger with cooling and also the rate of thermal activation becomes smaller with cooling, according to (8.21). As the critical current increases, the barrier for phase slips becomes larger. Thus, the rate of QPS also decreases with cooling, according to (8.22), and in agreement with Figure 8.5.

The models used to fit the data in Figure 8.5a,b are different. Namely, the fits in Figure 8.5a are done assuming that the switching events are caused only by TAPS, while the fits in Figure 8.5b are made assuming that both TAPS and QPS contribute the to net phase slip rate, according to (8.20). In both cases, the modeling was done using numerical simulations which took into account the probability of switching caused by a coincidence of the switching events [131, 178]. In other words, the switching events were due to multiple phase slips. Thus, the rate of switching events were not the same as the rates of phase slips, given by (8.21) and (8.22). Thus, to compute the fitting curves, a rather involved numerical simulation was done. The story is simplified though by the fact that the switching events to the right of the curve marked by black diamonds do have a one-to-one correspondence with individual phase slips. This means that each phase slip causes the switching with a probability that is equal to unity or very close to unity. Thus, the switching rates measured at low temperatures ($T < 0.9$ K in the example considered) are expected to match the rates of single phase slips, that is, either (8.21) or (8.22). As shown by the diamond-marked curve of Figure 8.5b, temperatures less than approximately 0.9 K are low enough so single-phase-slip switching is realized.

The fits in Figure 8.5a, done assuming $T_q = 0$, match the data very well at a relatively high temperature. At $T < 1.3$ K, a significant disagreement starts to occur which becomes huge as the temperature is further reduced. Thus, the TAPS-only model does not account for the observed large Γ. The fits shown in Figure 8.5b match the experimental switching rate very well, down to the lowest temperature tested, which is 0.3 K. These fits are done using the full rate of (8.20), which includes the QPS of (8.22). In fact, below $T = 1$ K, the QPS rate is dominant. To see this, consider the example of $T = 0.9$ K. The predicted TAPS-only rate of switching is about two orders of magnitude lower than the rate found experimentally, as Figure 8.5a shows. Yet, as the QPS events are included into the model, the model matches the results very well, as is shown in Figure 8.5b. The conclusion is that already at 0.9 K, the rate of quantum phase slips is roughly a hundred times larger than the rate of thermal phase slips.

Such domination of the QPS over TAPS at high bias current could be understood on the qualitative level as follows. As the bias current is increased, the barrier for phase slips decreases. This fact in itself is not enough to understand why QPS should eventually win the competition. However, it is also known that the rate of quantum tunneling depends exponentially not only on the height, but also on the width of the barrier. This is not true for thermal activation though because the thermal activation does not significantly depend on the barrier width. Thus, as the bias current is increased, the QPS is helped by two factors: the decrease of the barrier height and the decrease of the barrier width, while the TAPS process is only accelerated by the fact that the barrier height is reduced. So, at sufficiently low temperature, where the switching happens close to the depairing current of the wire, the QPS becomes dominant.

The advantage of this method compared to the linear resistance measurements is that the switching is always observed, even at very low temperatures, while the linear resistance might become so low with cooling that the available equipment would not be able to measure it. And this could happen before the QPS becomes dominant. Thus, by measuring the linear resistance, one is not guaranteed to observe the TAPS → QPS crossover. Yet, by measuring the statistics of the switching currents, the crossover is much easier to observe experimentally.

Another important remark to be made here is to emphasize that the crossover from TAPS to QPS is observed close to $T = 0.9$ K. The crossover temperature is supposed to be approximately equal to the quantum temperature, thus an estimate for the quantum temperature is $T_q \approx 0.9$ K. This is in good agreement with the best-fit requirement which leads to the expression $T_q = 0.726 K + 0.4 T$. Thus, the analysis of QPS appears to be self-consistent.

Note also that the fits shown in Figure 8.5a are made using the MH attempt frequency which is less precise that the GZ attempt frequency. Yet, this disagreement is not important since the exact knowledge of the attempt frequency is fairly unimportant compared to the exponential fact and the coefficients involved in it.

The interpretation of the results of Figure 8.5 is complicated by the fact that multiple-phase-slip switching events do occur at higher temperatures. Thus, the Kurkijärvi temperature law for σ versus T dependence, namely, that $\sigma \propto T^{2/3}$ (to be discussed below), was not observed on such samples. The meaning of multiple-phase-slip events is that more than one LPS occurs before the switching happens. Single-phase-slip switching is such a scenario in which each phase slip causes a switch of the wire to the normal state. Single-phase-slip switching is dominant in thicker wires, having larger critical currents and, correspondingly, larger critical temperatures. Such wires will be discussed in the Section 9 which is dedicated to macroscopic quantum tunneling.

8.6
Inverse KFD Transformation

In some cases, one might need to convert the switching rate Γ into the corresponding distribution $P(I)$. To derive the corresponding formula, let us use the master equation (8.15) and combine it with the definition of the number of wires which are not switched by the time the current reaches value I, given in (8.14). Thus, we get

$$N_1 \Gamma(I) = N\Gamma(I) \exp\left[-\int_0^t \Gamma(I(t'))dt'\right] = NP(I)v_I \quad (8.23)$$

Here, the differential dt' can be expressed through the current differential using the function defining the change of the current with time, $I = I(t)$. The differential then can be related as $dt = dI/v_I$ where $v_I(I) \equiv dI/dt$ is the sweep seed. The primed variables are then related as $dI' = dt'/v_I$. Therefore, the integration variable in the equation above can be changed from t' to I' as follows

$$P(I) = \frac{\Gamma(I)}{v_I} \exp\left[-\int_0^I \Gamma(I')v_I^{-1}(I')dI'\right] \quad (8.24)$$

This expression will be referred to as the inverse KFD transformation. In some cases, sweep speed can be taken as a constant. Such is the case when the current is proportional to time. Another situation when v_I can be considered constant is if the distribution width is much smaller than the current amplitude. If the sweeping speed is constant, the expression simplifies to

$$P(I) = \frac{\Gamma(I)}{v_I} \exp\left[-\frac{1}{v_I}\int_0^I \Gamma(I')dI'\right] \quad (8.25)$$

Let us consider an example now. The switching rate is usually given by the Arrhenius law $\Gamma(I) = \Omega(I)\exp[-\Delta F(I)/k_B T]$. The attempt frequency might be approximated as a constant, say $\Omega = 10^{12}$ s^{-1}. The barrier height typically follows the 3/2 power law, at least for short wires, that is, $\Delta F(I) = \Delta F(0)(1 - I/I_c)^{3/2}$. Assume also $\Delta F = 10$ K and $T = 1$ K, $I_c = 10^{-6}$ A. The distribution of the switching currents, obtained by integration (8.24) numerically, is shown in Figure 8.6 for a few different values of the bias current sweep speed v_I. The sweep rate is assumed independent of the bias current in this example. The example illustrates the general fact that the distribution shifts to higher currents as v_I is made larger because the wire does not have time to switch early if the sweeping of the bias is done rapidly.

8.7
Universal 3/2 Power Law for Phase Slip Barrier

Consider a superconducting device composed of a thin wire that serves as a bridge between two macroscopic superconducting electrodes (Figure 1.1). The phase difference between the electrodes is ϕ, which is frequently called "the phase," for brevity. The results provided here are applicable to the cases when the wire is so short that its state is fully defined by the value of ϕ. The results are also applicable to SIS and SNS superconducting junctions because for them the only variable that defines the state of the device is ϕ also.

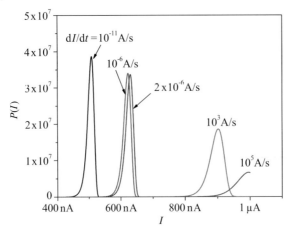

Figure 8.6 Probability distribution of the switching current calculated using the Kurkijärvi theory [179]. Here, P is defined such that $P(I)dI$ is the probability that the switching event occurs in the interval from I to $I+dI$. The critical current in this example is $1\,\mu A$. The curves are given for various sweep speeds $v_I \equiv dI/dt$, which are indicated near corresponding curves. The sweep speed for the blue curve is twice as large than for the red one, illustrating that doubling the sweep speed only causes a minor change of the distribution curve. In practice, the sweep speed is of the order of 10^{-6} A/s. It is clear that the mean switching current is smaller than the critical current by many widths of the distribution curve. This is because the attempt frequency is usually very high, such as $\Omega = 10^{12}$ Hz in this example. Please find a color version of this figure in the color plates.

Here, it is not assumed that Helmholtz energy $F(\phi)$ is necessarily proportional to $\cos\phi$, although it could be if the wire is extremely short and the temperature is close to T_c [18]. The supercurrent through the wire is given by (2.38), which reads $I_s(\phi) = (2e/\hbar)(dF/d\phi)$. The assumption we make here is that the $I_s(\phi)$ has a maximum, called, as usual, the critical current I_c, and we also assume that $I_s(\phi)$ is smooth, in the mathematical sense, near this maximum. According to the Taylor theorem, a smooth function can be approximated near its maximum as

$$I_s = I_c + \frac{Y_2}{2}(\phi - \phi_c)^2 \tag{8.26}$$

where ϕ_c is the "critical phase" defined as $I_s(\phi_c) = I_c$, and $Y_2 = (d^2 I_s/d\phi^2)|_{\phi=\phi_c} < 0$. Note that $Y_2 < 0$ since the supercurrent has a maximum, not a minimum, at $\phi = \phi_c$, which means that the functions $I_s(\phi)$ has a negative curvature at the point $\phi = \phi_c$. Another equivalent condition for ϕ_c is $(dI_s/d\phi)|_{\phi=\phi_c} = 0$. Therefore, the reason we include the quadratic term in the Taylor series, but not the linear term, is that the linear term is zero at the maximum.

Equation 2.38 can be written in the differential form as $dF = (\hbar/2e)I_s(\phi)d\phi$ and then it can be integrated, taking (8.26) into account, to obtain the Taylor series of Helmholtz energy near the critical phase. The result can be presented as

$$\frac{2e}{\hbar}F = \int I_s(\phi)d\phi = Y_0 + I_c(\phi - \phi_c) + \frac{Y_2}{6}(\phi - \phi_c)^3 \tag{8.27}$$

where Y_0 is the constant of integration which can be chosen arbitrarily since the zero energy level can be chosen anywhere. One can directly verify that the equation above, if differentiated, leads to (8.26), as it should.

We now move on to calculate the barrier for phase slips. The usual experimental situation is the condition of a fixed or a controlled bias current I. The externally supplied current I is also a supercurrent provided that the leads connected to the nanowire are superconducting. Note that I might not be equal I_s. In general, $I = I_s + C(dV/dt) + I_n$. Here, C is the electric capacitance between the electrodes connected to the wire and V is the voltage on the electrodes. Here, I_n is the normal current. It can occur if the wire is shunted with a resistor or if bogoliubons are present in the wire. It represents dissipation or loss of the usable energy into heat. Since the barrier calculation does not depend on the presence or absence of the dissipation, we assume, without undermining the generality of the analysis, that $I_n = 0$, that is, that there is no loss of energy in the system. Thus, in the model considered in this section, it is assumed that the injected supercurrent I is generally divided into two parts, namely, the supercurrent going in the wire, I_s, and the current charging the electrodes, namely, CdV/dt. In the stationary case, when $d\phi/dt = 0$ and so $V = 0$, we have the simplest situation, namely, $I_s = I$. Any fluctuations would violate this simple equality because during thermal fluctuations, ϕ is changing and V is changing.

Since the assumed experimental condition is that the current is fixed, the work done on the system by thermal fluctuations is measured as a change of Gibbs, not Helmholtz energy. Therefore, for the barrier calculation, we have to use Gibbs energy $G = F - (\hbar/2e)I\phi$, which can be presented as

$$\frac{2e}{\hbar}G = \frac{2e}{\hbar}F - I\phi = Y_0 - I_c\phi_c + (I_c - I)\phi + \frac{Y_2}{6}(\phi - \phi_c)^3 \tag{8.28}$$

which can also be written in a more convenient form

$$\frac{2e}{\hbar}G = Y_3 + (I_c - I)(\phi - \phi_c) + \frac{1}{6}Y_2(\phi - \phi_c)^3 \tag{8.29}$$

where Y_3 is some constant since in this problem, $I =$ const, that is, the bias current is presumed fixed by an external apparatus. The maxima and minima of Gibbs energy are found through the condition $dG/d\phi = dG/d(\phi - \phi_c) = 0$, which leads to

$$(I_c - I) + \frac{Y_2}{2}(\phi - \phi_c)^2 = 0 \tag{8.30}$$

This simple equation has two solutions, denoted below as ϕ_+, which corresponds to the maximum of Gibbs energy potential, and ϕ_-, corresponding to the minimum. Note that $\phi_- < \phi_c < \phi_+$. The phase particle is trapped in the minimum at ϕ_-. It can escape from the minimum if it somehow reaches the maximum at ϕ_+. The particle can reach the maximum if thermal fluctuations do the work $G(\phi_+) - G(\phi_-)$ on the wire.

These solutions for the equation above are

$$\phi_\pm = \phi_c \pm \frac{\sqrt{2}(I_c - I)^{1/2}}{\sqrt{(-Y_2)}} \tag{8.31}$$

which also gives the distance in the space of the phase of the device between the maximum of Gibbs potential and the minimum of Gibbs potential as

$$\phi_+ - \phi_- = 2\sqrt{2}(I_c - I)^{1/2}|Y_2|^{-1/2} \tag{8.32}$$

Now the phase-slip barrier is found as $\Delta G = G(\phi_+) - G(\phi_-)$, which leads explicitly to

$$\begin{aligned}\frac{2e}{\hbar}\Delta G &= (I_c - I)(\phi_+ - \phi_-) + \frac{Y_2}{6}(\phi_+ - \phi_c)^3 - \frac{Y_2}{6}(\phi_- - \phi_c)^3\\ &= 2\sqrt{2}(I_c - I)^{3/2}|Y_2|^{-1/2} - \frac{2|Y_2|}{6}\left[(I_c - I)^{1/2}(|Y_2|)^{-1/2}\sqrt{2}\right]^3\\ &= \frac{4\sqrt{2}}{3}I_c^{3/2}|Y_2|^{-1/2}\left(\frac{I_c - I}{I_c}\right)^{3/2}\end{aligned} \tag{8.33}$$

Note that the barrier is zero at $I = I_c$, that is, the phase slips can proliferate in a wire biased with a current equal to the critical current. That is why the wire cannot have zero resistance at currents larger than the critical current.

The barrier can also be written as

$$\Delta G = \frac{(2\sqrt{2}\hbar/3e)I_c^{3/2}(1 - I/I_c)^{3/2}}{\sqrt{-(d^2 I_s/d\phi^2)|_{\phi=\phi_c}}} \tag{8.34}$$

For further discussion, it is convenient to formally define a zero-bias-current barrier ΔG_0 as

$$\Delta G_0 = \Delta G_0(T) = \frac{2\sqrt{2}\hbar}{3e}I_c^{3/2}\left(-\frac{d^2 I_s}{d\phi^2}\right)^{-1/2}\Big|_{\phi=\phi_c} \tag{8.35}$$

which is a function of temperature but not the bias current. Of course, if the bias current is zero, the energy barrier to create a phase slip would not be exactly equal ΔG_0. With the new notation, the barrier height becomes

$$\Delta G = \Delta G_0(T)\left(1 - \frac{I_s}{I_c}\right)^{3/2} \tag{8.36}$$

The conclusion we reach here is that the barrier is proportional to $(1 - I_s/I_c)^{3/2}$, independently of the CPR function $I_s(\phi)$, as was first argued by Vakaryuk ([195], V. Vakaryuk, private communication). Note that the power of 5/4 (instead of 3/2) in the GZ theory and McCumber theory of TAPS is obtained assuming that $I_s = I = \text{const}$, while here we do not make such assumption. We only assume that $I = \text{const}$, but I_s can change as the fluctuation occurs.

Let us apply this general equation, (8.35), to a simple example of a JJ, for which $I_s = I_c \sin\phi$. In such cases, $\phi_c = \pi/2$ since the maximum of the $\sin\phi$ function is at $\pi/2$. Then, $\sqrt{-(d^2 I_s/d\phi^2)}|_{\phi=\phi_c} = I_c^{1/2}$. Then, $\Delta G = (2\sqrt{2}\hbar/3e)I_c(1 - I_s/I_c)^{3/2} = (4\sqrt{2}E_J/3)(1 - I_s/I_c)^{3/2}$, where $E_J = (\hbar/2e)I_c$ is a frequently used notation called Josephson energy. The expressions above are valid in SI units. Note also that for a JJ, the exact barrier at $I = 0$ is $2E_J$. Thus, the formulas above are exact only if $1 - I/I_c \ll 1$.

The expression for the barrier can also be expressed using the flux quantum $\Phi_0 = h/2e = \pi\hbar/e$. For example, if the CPR is sinusoidal, then $\Delta G = (2\sqrt{2}/3\pi)\Phi_0 I_c(1 - I_s/I_c)^{3/2}$. Note that our convention is always such that $1 - I_s/I_c \equiv 1 - (I_s/I_c)$.

8.8
Rate of Thermally Activated Phase Slips at High Currents

Now, as we discussed how the rate of switching can be obtained experimentally and we argued that at low temperatures each phase slip causes a switch, it remains to be seen how the rate of phase slips can be obtained theoretically. There are two main factors influencing the rate of phase slips, the temperature and the bias current. The temperature controls the energy of the fluctuations and also suppresses the barrier by suppressing the amplitude of the order parameter. The bias current suppresses the order parameter amplitude, pushes the phase slip core across the wire through the Lorentz force, and thus causes a reduction of the barrier. In a typical experiment, the temperature is fixed and the bias current I_s increases slowly, thus causing a slow decrease of the phase slip barrier ΔG according to the approximate (8.36), therefore making the Kurkijärvi theory applicable. The rate can then be found using the Kramers theory [21] as

$$\Gamma(T,I) = \Omega(T,I)\exp\left(-\frac{\Delta G}{k_B T}\right) = \Omega(T,I)\exp\left[-\frac{\Delta G_0(1 - I_s/I_c)^{3/2}}{k_B T}\right] \quad (8.37)$$

where we use the current dependence of the barrier derived for high currents near the critical current (8.36).

To make a meaningful comparison with the experiment, one needs to know the temperature dependence of the barrier height and, to a lesser extent, of the attempt frequency. Let us start the discussion with the barrier height scale G_0 defined in (8.35). In that equation, the temperature dependence of the critical current is known by virtue of the Bardeen formula, while the second derivative of the CPR is more difficult to write down. Let us make a simplifying assumption that the temperature influences the height of the CPR function $I_s(\phi)$, while the change of its shape is negligible, at least in the range of the measured dispersion of the switching current, the analysis of which we keep in mind here. With this assumption, the CPR can be written as $I_s(\phi) = I_c(T)i_s(\phi)$, where $i_s(\phi) = I_s(T,\phi)/I_c(T)$

is assumed to be temperature independent. Such forms of the CPR can be made plausible by considering the example of a JJ for which $i_s(\phi) = \sin\phi$.

With this assumption, $d^2 I_s/d\phi^2 = I_c(T)i_s''$, where $i_s'' = d^2 i_s/d\phi^2$. Thus, the barrier scale has the same temperature dependence as the critical current of the wire as follows

$$\Delta G_0 = \frac{I_c(T)(2\sqrt{2}\hbar/3e)}{\sqrt{-i_s''|_{\phi=\phi_c}}} \tag{8.38}$$

To find the final expression for the barrier height, we use the Bardeen formula (11.1) $I_c = I_c(T) = I_c(0)(1 - T^2/T_c^2)^{3/2}$, which is valid at all temperatures. Thus, the barrier for phase slips at high currents and various temperatures is

$$\Delta G = \Delta G_{00} \left(\frac{T_c^2 - T^2}{T_c^2}\right)^{3/2} \left(1 - \frac{I_s}{I_c}\right)^{3/2} \tag{8.39}$$

where the barrier corresponding to $T=0$ and $I=0$ is

$$\Delta G_{00} = \frac{4\sqrt{2}}{3} \frac{\hbar I_c(0)/2e}{\sqrt{-i_s''|_{\phi=\phi_c}}}.$$

The solution presented above is approximate and is only good for a short wire in which the fluctuations of the phase difference between the ends are dominant, while the fluctuations of the phase along the wire are not significant. For longer wires, the current exponent 3/2 needs to be replaced by 5/4.

For long wires, the CPR is multivalued and thus it experiences jumps from one branch to another at currents near the critical current [18]. The exact solution for the phase slip barrier height in very long wires under the current-bias condition was derived in the classic paper of McCumber [64] (see Eq. (19) in this reference). It is exact only at temperatures near the critical temperature. The McCumber barrier, ΔG_M, is

$$\Delta G_{M1} = \frac{8\pi \Delta G_M}{H_c^2 \xi A_{cs}} = \frac{8\sqrt{2}}{3}\sqrt{1-3k^2} - 8k(1-k^2)\arctan\left(\sqrt{\frac{1-3k^2}{2k^2}}\right) \tag{8.40}$$

where ΔG_{M1} is the normalized (unitless) barrier height, A_{cs} is the constant cross-sectional area of the wire, H_c is the thermodynamic critical field, $k = \phi\xi(T)/L$ is the normalized phase difference, ϕ is the actual phase difference between the ends of the wire, $\xi(T)$ is the temperature-dependent coherence length, and L is the wire length. The normalized supercurrent density J is related to the normalized phase difference as

$$J(k) = k(1 - k^2) \tag{8.41}$$

The critical current is defined by the maximum of this function. The maximum is achieved at $k_c = 1/\sqrt{3}$. The corresponding normalized critical current density is $J_c = 2/(3\sqrt{3})$. The relation to the actual supercurrent in the McCumber

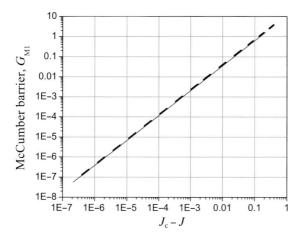

Figure 8.7 Normalized barrier height for thermally activated phase slips in a current-biased thin wire, computed by McCumber [64]. The barrier is plotted, as the dashed black curve, versus $J - J_c$, where J is the normalized supercurrent density and $J_c = 2/3\sqrt{3}$ is the maximum (i.e., the critical) value of the said density. The continuous line, which is simply a straight line, represents the dependence $G_{M1} \propto (J_c - J)^{5/4}$. Apparently, this power low describes the McCumber phase slip barrier with a very high precision.

computation is $I_s = JA_{cs}H_c^2\xi/\Phi_0$, where $\Phi_0 = h/2e$ is the superconducting flux quantum. Thus, the critical current of the wire, with the accepted notations, is $I_c = (2/3\sqrt{3})A_{cs}H_c^2\xi/\Phi_0$. Therefore, if H_c is not known, the critical current and ξ can be used to compute the numerical value for the barrier height. Note also that the factor $H_c^2\xi$, which appears in the normalization factors for the energy barrier and the supercurrent depends on the temperature as $(T_c - T)^{3/2}$, since $H_c = H_c(T) \propto (T_c - T)^2$ and $\xi = \xi(T) \propto (T_c - T)^{-1/2}$.

A graphical analysis of the Mccumber's barrier $\Delta G_{M1}(J)$, which is given by (8.40), is shown in Figure 8.7 where the function is shown as the dashed curve and is plotted in the log–log format. Surprisingly, this rather complicated function appears as a straight line in the plot, indicating that it is indistinguishable from power-law dependence. The straight line has the slope of 5/4. Visibly, it coincides with the plot of the function $\Delta G_{M1}(J)$, in a wide range of J. Thus, we conclude that a good approximation to the McCumber barrier is the power law $\Delta G_M \propto (I_c - I)^{5/4}$. Note that the power is different from the previously found power of 3/2, which describes the barrier versus bias current dependence in short wires, in which case the supercurrent is subject to fluctuations, while the bias current is fixed.

The attempt frequency can also be estimated in the limit of short and long wires. Again, the estimate we can do for short wires will be applicable to all temperatures while the formula for long wire will be correct only near the critical temperature. If the wire is short, it can be considered as an inductive element [154] forming a closed circuit with the capacitor formed by the electrodes. The presence of some effective capacitance should be clear from the illustration of a typical device given in Figure 1.1. Although the electrodes, marked E1 and E2, in this nanowire device do not form a parallel-plate capacitor as in the case of an SIS junction, the capacitance between them is finite and was estimated for the case of electrodes of width about 10 μm to be of the order of $C = 10\,\text{fF}$ [136]. In the section on the kinetic inductance, we have estimated that if the critical current of a nanowire is 1 μA the kinetic inductance, associated with the inertia of the condensate in it is $L_k = 0.3\,\text{nH}$. Thus, one can estimate the resonance frequency of the structure shown in Figure 1.1 as $\omega_p = 1/\sqrt{L_k C} = 0.6\,\text{THz}$. This is a crude estimate of course since the exact shape of the electrodes needs to be taken into account. However, if the resonance frequency is estimated, then the attempt frequency equals it, according to the Kramers theory [21], that is, $\Omega = 1/\sqrt{L_k C}$. This estimate for attempt frequency is not limited to high temperatures and is applicable to the case when the wire is so short that the electrodes, which are always connected to the ends of the wire, are able to dominate the fluctuations dynamics in the wire.

Another limiting case where the analysis is possible is the case in which the length of the wire L is so large that the effect of the electrodes is negligible. Usually, this means that its internal characteristics frequency of fluctuations is much larger than the inverse of the signal travel time, that is, larger than c_{MS}/L, where c_{MS} is the Mooij–Schön velocity [155] in the 1D superconducting wire. The attempt frequency of internal fluctuations of a long wire has been estimated by Langer and Ambegaokar (LA) in [143]. Further theoretical analysis by McCumber and Halperin (MH) [142] has proven that the initial LA estimate is wrong by many orders of magnitude. Interestingly, this fact did not play any significant role in the comparisons of the LA theory to experimental results [23], because, in general, the experimental data are usually very sensitive to the exponential (Arrhenius) factor in the phase slip rate, while the factor placed in front of the exponent (i.e., the attempt frequency) has a weak effect on the fitting curves, as compared to the effect of the exponential factor.

8.9
Kurkijärvi Dispersion Power Laws of 2/3 and 1/3

As was stated above, the switching is a truly stochastic process. Thus, it can be characterized by a mean value $\langle I_{sw} \rangle$ and by the standard deviation σ, even though the distribution is not Gaussian. If the arrays of switching currents is $I_{sw,k}$, which k is the measurement number, then the mean value is $\langle I_{sw} \rangle = N^{-1} \sum_{k=1}^{N} I_{sw,k}$, where N is the total number of measurements. The variance is defined as

$$\text{Var}(I_{sw}) = N^{-1} \sum_{k=1}^{N} (I_{sw,k} - \langle I_{sw} \rangle)^2 \tag{8.42}$$

and the standard deviation is

$$\sigma = \sqrt{\text{Var}(I_{sw})} = \left[N^{-1} \sum_{k=1}^{N} (I_{sw,k} - \langle I_{sw} \rangle)^2 \right]^{1/2} \tag{8.43}$$

Kurkijärvi was able to derive important relations characterizing the dependence of the dispersion on temperature and on the critical current of the device, under the presumption that the bias current starts at zero and increases monotonously, with a constant speed, up to the critical current. The Kurkijärvi result is

$$\sigma(T, I_c) \propto T^{2/3} I_c^{1/3}$$

The requirement of a constant speed is not hard to satisfy. Even if the current is increased as $I \propto \sin(2\pi f t)$, which is quite typical in practice, the sweep speed v_I can be considered approximately constant. This is because the switching events concentrate in a narrow region of currents, so within this region v_I does not change significantly.[3] We will see below that v_I occurs under the logarithm, so even if it changes slightly within the distribution width, the effect of this change is usually weak.

The expression above can be presented as two separate rules: (a) the Kurkijärvi temperature law $\sigma \propto T^{2/3}$ (KTL), and (b) the Kurkijärvi critical current law $\sigma \propto I_c^{1/3}$ (KCCL). The former was confirmed in multiple experiments on various systems [35, 39, 80]. It is applicable to the cases when the temperature is varied while any possible variations of the critical current are negligible, for example, in the case of a JJ measured at temperatures much lower than the critical temperature of the electrodes. The latter is valid for a fixed temperature, when the critical current is varied. A wide range variation of the critical current is possible in, for example, graphene superconducting proximity junctions in which I_c is controlled by the gate voltage.

The KCCL and KTL can be derived rigorously [179, 181] if one assumes the switching rate is of the form given by (8.36). Here, we only provide a simplified semiquantitative derivation which allows the reader to understand the physical origin of these scaling laws. First, we should realize the simple fact that σ approximately equals the width of the distribution of the type shown in Figure 8.6. The time the current is swept through the distribution of width σ is σ/v_I. This time must be sufficient for the system to switch. This means $\Gamma\sigma/v_I \approx 1$. Therefore, using (8.37) for the rate, we get $(\sigma\Omega/v_I)\exp(-\Delta G/k_B T) \approx 1$. Taking the natural logarithm of the left and the right parts, we get $k_B T \ln(\sigma\Omega/v_I) = \Delta G$. Now, we use (8.36) for the barrier to get $(k_B T/\Delta G_0)\ln(\sigma\Omega/v_I) = (1 - I/I_c)^{3/2}$. Finally, realizing that the latter equations is the condition for the switching to occur with certainty, we can replace I with $\langle I_{sw}\rangle$ and get the following dependence of the switching current on the temperature and the specifics of the device

$$\langle I_{sw}\rangle = I_c - I_c\left(\frac{k_B T}{\Delta G_0}\right)^{2/3}\ln^{2/3}\left(\frac{\sigma\Omega}{v_I}\right) \tag{8.44}$$

where Ω is the attempt frequency and σ/v_I approximates the time over which the bias current sweeps within the dense part of the distribution $P(I_{sw})$. Since the

---

3) Of course, if the mean switching current shifts to a different position due to a temperature change for example, then the corresponding v_I changes also.

logarithm depends weakly on its argument, one can approximate the experimental data pretty well without knowing this time precisely.

It is more difficult to derive σ. A rigorous derivation is given [179, 181]. Here, we make a simplified estimate. First, one must introduce some notations for simplicity: $i_1 = 1 - \langle I_{sw} \rangle / I_c$, $i_2 = 1 - (\langle I_{sw} \rangle - \sigma)/I_c = i_1 + \sigma/I_c$, and $M = \ln(\sigma \Omega / v_I)$. Therefore, (8.44) can be written as

$$\left(\frac{k_B T}{\Delta G_0}\right)^{2/3} M^{2/3} = i_1 \tag{8.45}$$

Now, assume that the distribution is approximately normal or Gaussian. This means that the probability of switching in the interval shifted by one standard deviation from the center is roughly 10%. More precisely, this means that the probability of switching in the interval between $\langle I_{sw} \rangle - 3\sigma/2$ and $\langle I_{sw} \rangle - \sigma/2$ is roughly 0.1. Thus, we can estimate, assuming that in this interval the rate of switching is constant and equals G_2, that $\Gamma_2 \sigma / v_I = 0.1$ where $\Gamma_2 = \Omega \exp[-\Delta G(i_2)/k_B T]$. From the last two equations, it follows that $\Delta G(i_2) = k_B T[M - \ln(0.1)]$. Also, by definition, $\Delta G(i_2) = \Delta G_0 i_2^{3/2}$. Thus, $i_2^{3/2} = (k_B T/\Delta G_0)[M - \ln(0.1)]$. Therefore, $i_2 = i_1 + \sigma/I_c = (k_B T/\Delta G_0)^{2/3}[M - \ln(0.1)]^{2/3}$.

We are now close to obtaining the final result for the σ. To find the solution, the last equation needs to be combined with (8.45) to get

$$(k_B T/\Delta G_0)^{2/3} M^{2/3} + \frac{\sigma}{I_c} = \left(\frac{k_B T}{\Delta G_0}\right)^{2/3} [M - \ln(0.1)]^{2/3}$$

This simplifies to $\sigma/I_c = (k_B T/\Delta G_0)^{2/3} M^{2/3} \{[1 + \ln(10)/M]^{2/3} - 1\}$. Since the attempt frequency is usually very large, in the range of GHz or even THz, and the sweep speed is of the order of Hz, one expects that $M/\ln(10) \ll 1$. Thus, according to the Taylor theorem, we can write $[1 + \ln(10)/M]^{2/3} - 1 \approx (2/3)[\ln(10)/M]$. Therefore, we arrive at the final expression for the standard deviation of the switching current

$$\sigma = I_c \left(\frac{k_B T}{\Delta G_0}\right)^{2/3} M_2 \tag{8.46}$$

where $M_2 = M^{-1/3}[2\ln(10)/3]$ depends very weakly (as a logarithm) on the temperature, so it can be assumed approximately constant.

To obtain the Kurkijärvi power law in its usual form, remember that the zero-bias-current barrier for phase slips, ΔG_0, is proportional to the critical current. For example, for a nanowire, there is an estimate [19] $\Delta G_0 = (\sqrt{6}\hbar/2e) I_c$. Finally, we get

$$\sigma \approx I_c^{1/3} \left(\frac{k_B T}{\Phi_0}\right)^{2/3} M_3 \tag{8.47}$$

where $M_3 = [2\ln(10)/3](2\pi/\sqrt{6})^{2/3} \ln^{-1/3}(\sigma \Omega / v_I)$ and $\Phi_0 = h/2e$ is the magnetic flux quantum (in SI units). So, clearly, if $I_c = $ const, then $\sigma \propto T^{2/3}$, which is the KTL. If $T = $ const and I_c can be varied, then the prediction is $\sigma \propto I_c^{1/3}$, which is the KCCL.

9
Macroscopic Quantum Tunneling in Thin Wires

The quantum mechanics theory [173] predicts that a particle can penetrate into energetically forbidden regions of space. Thus, a quantum particle can jump over a potential barrier, even if, speaking classically, its energy E_p is less than the maximum height of the potential energy barrier E_b. Such quantum overcoming of barriers is called quantum tunneling or simply tunneling. It is a phenomenon unique to quantum physics and has no analogy in classical physics. Thus, its observation indicates that the system under investigation behaves quantum-mechanically. Such a conclusion could be of importance since then one can use other laws of the quantum theory as guidance and reshape the system into a useful quantum device, for example, a qubit.

Qualitatively speaking, the phenomenon of tunneling is possible due to the Heisenberg uncertainty principle which states that the uncertainty in energy ΔE and the uncertainty in time Δt are related as $\Delta t \approx h/\Delta E$, where $h = 2\pi\hbar$ is the Planck's constant. The meaning of the principle is that the particle's energy is a subject to quantum fluctuations. The particle can "borrow" additional energy ΔE from so-called "zero-point" fluctuations for a time Δt. In the example above, the missing energy needed to overcome the barrier is $\Delta E = E_b - E_p$. Thus, that amount can be taken from zero-point fluctuations (or thermal fluctuations, if the temperature is finite) for a time as large as $\Delta t \approx h/(E_b - E_p)$. This time might be sufficient to cross the barrier. Then, the particle would tunnel over the barrier easily. If the time needed to cross the barrier, t_b is larger than the zero-point fluctuation time, then tunneling is still possible, but the probability is exponentially low, $p_t \propto \exp(-t_b/\Delta t) = \exp[-t_b(E_b - E_p)/h]$.

The tunneling phenomenon is one of the causes of relaxation of the supercurrent in a thin wire. The relaxation means that the supercurrent slows down with time, and its kinetic energy eventually goes into heat, possibly through some intermediate energy forms. The relaxation is slowed very strongly by the topological effect, namely, by the fact that for the supercurrent to diminish, the Little's helix, representing the winding order parameter (Figure 6.3), must cross the x-axis. In other words, there is a barrier to the supercurrent relaxation which is defined by the free energy needed to suppress the order parameter to zero at some spot along the wire. Then, one can think of tunneling over the barrier in Hilbert space of all configurations the superconducting wavefunction. This is important in order to realize that

Superconductivity in Nanowires, First Edition. Alexey Bezryadin.
© 2013 WILEY-VCH Verlag GmbH & Co. KGaA. Published 2013 by WILEY-VCH Verlag GmbH & Co. KGaA.

tunneling can not only happen in real space, but also in Hilbert space of quantum states. Such tunneling of the superconducting order parameter in a thin wire is called the quantum phase slip (QPS). It is different from "classical" thermally activated phase slips (TAPS) because in TAPS, the energy to cross the barrier comes from thermal fluctuations. However, in QPS, the energy comes from zero-point or quantum fluctuations. Since QPS is a tunneling of the entire BCS condensate, which involves a macroscopic number of electrons acting collectively, it is classified as macroscopic quantum tunneling (MQT) [174].

The quest for quantum behavior exhibited by macroscopic degrees of freedom represents one of the most exciting and promising branches of modern physics. Some important examples of macroscopic quantum behavior include delocalization of the center of mass of C_{60} molecules [175], tunneling of the magnetization of a small magnetic particle [35], or tunneling of core-less vortices [39]. This field of research is promising regarding the development of novel information processing devices based on qubits and, in general, leads to a better understanding of the crossover between classical and quantum world views [24, 25, 27, 28].

Superconducting nanowires represent one of many small-scale systems in which macroscopic quantum phenomena have been observed. The most common approach to observe QPS is through linear transport measurements [53, 108, 184–186]. The signature of QPS is a "tail" on the $\ln(R)$ versus T plots, signifying that the resistance drops slower with cooling than the exponential decline predicted by the Arrhenius activation law. As the temperature of the sample is reduced, such an excess resistance tail begins at T_q, that is, at the temperature at which the QPS rate becomes larger than the TAPS rate. Yet, this technique failed to provide evidence for QPS in some of the shortest wires having low normal state resistance [183, 187]. This fact might indicate that QPS is either completely suppressed in wires having a low normal resistance or in wires having a small length due to the strong coupling to the electrodes. Another possibility is that T_q was too low, and thus the total resistance at T_q was lower than the noise floor of the measurement setup. If a sample has $R(T_q)$ lower than the resolution of the measurement apparatus, it is then not possible to observe the crossover from TAPS to QPS, even if it does exist in reality.

A more straightforward approach to observe QPS is through a statistical analysis of the switching current events occurring at high bias currents. As the bias current (I) is increased, the wire remains superconducting and shows zero voltage until the switching current I_{sw} is reached. Usually, $I_{sw} \sim I_c$ and, always, $I_{sw} < I_c$. At $I = I_{sw}$, the wire suddenly switches to the normal state and the voltage increases sharply from zero to $\approx R_n I_{sw}$. Such switching events can be easily detected. In a wide range of parameters, the destruction of superconductivity occurring at I_{sw} is triggered by a single microscopic event, either a TAPS or QPS. Thus, the probability of switching as a function of current equals the probability of phase slips, which can be predicted theoretically. So, to distinguish between TAPS and QPS, one simply needs to measure the dispersion of the switching current as a function of temperature and compare it to the predictions of the TAPS and QPS theories. This method of the QPS detection was pioneered by Kurkijärvi [179] and has been successfully applied to the detection of MQT in Josephson junctions by Kurkijärvi and collaborators [80]. Afterwards, there results were reproduced and improved on SIS junctions [34, 81], magnetic nanoparticles [35] and homogeneous nanowires [131]. Thus, the phenomenon of macroscopic quantum tunneling should be considered well-proven experimentally at the time of this writing, which is 2012.

It turns out that such a method is especially efficient when applied to wires which are far from SIT, having the SIT driving parameter $\iota_{\text{SIT}} = (R_q - R_n)/R_n$ in the range $2.5 < \iota_{\text{SIT}} < 14$ [114–116]. The switching statistics on thinner wires, which are near SIT, namely, in the range $0.57 < \iota_{\text{SIT}} < 3.51$, have also been studied [131]. It was found that the wires which are relatively close to SIT have a tendency to multiple-phase-slip switching events. Due to this the interpretation of switching probability functions requires numerical analysis. Such an analysis indicates that at low temperatures, the switching is caused by single phase slips and that QPS is the dominant switching mechanism. These results and conclusions have been reproduced and confirmed by an independent research group [59]. This fact is significant since most other results related to QPS in nanowires have not be confirmed by more than one group as of the time of this writing. For example, so-called antiproximity effect in nanowires [60], which was explained by the QPS physics [61], has not been reproduced by an independent research group yet.

9.1
Giordano Model of Quantum Phase Slips (QPS) in Thin Wires

Nicholas Giordano was the first to obtain evidence for the existence of QPS in thin superconducting wires [53–55]. Although initially there was some skepticism with respect to his conclusions [156], recent numerous publications on similar systems make it quite certain that Giordano's conclusions about the reality of QPS in thin wires are conceptually correct [108, 170], at least for sufficiently long wires. Short wires, empirically defined as ~ 200 nm or shorter, might require a more advanced model that would take various damping effects into account, which can exponentially slow down the tunneling rate [30], according to Caldeira and Leggett MQT theory.

In addition to publishing ground-breaking experimental results, Giordano developed a phenomenological model that predicts the rate of quantum phase slips. Giordano's model is a triumph of physical intuition. It is based on a simple analogy between thin nanowires and SIS junctions, and the notion of the Little's phase slips. Here, we outline the Giordano model following [56, 108].

First, the model assumes that the external magnetic field is zero. Second, it is also assumed that the wire is very thin, and thus the magnetic field generated by the wire is negligible everywhere. Thus, the vector-potential is assumed to be zero. Such neglect of the magnetic field is justifiable on the following grounds. Assume the critical current density of the material from which the wire is made is j_c. The wire radius is r. Hence, the maximum supercurrent in the wire is $I_{\max} = \pi r^2 j_c$. Accordingly, the maximum magnetic field on the surface of a long cylindrical wire, in SI units, is $B_{\max} = \mu_0 I_{\max}/2\pi r = \mu_0 r j_c/2$. Thus, the maximum field approaches zero if the radius approaches zero. Let us take reasonable estimates for the parameters involved, namely, $j_c \sim 10^7 \, \text{A/m}^2$ and $r = 5 \, \text{nm}$. Thus, we estimate $B_{max} \approx 3 \times 10^{-8} \, \text{T}$. Such a weak magnetic field is many orders of magnitude lower than the critical fields of a typical superconductor. For example, the thermodynamic critical field for optimized MoGe is $\approx 5.5 \times 10^{-2} \, \text{T}$. Thus, the effect of the self-induced magnetic field on the behavior of the nanowire is negligible.

To make an analogy with the nanowire, Giordano used a TDGL equation, which, according to [57, 142], can be written as

$$\frac{\partial \psi}{\partial t} = -\frac{4\pi}{A_{cs}\xi H_c^2 \tau_r} \frac{\delta F}{\delta \psi^*}$$

According to the argument above, we neglect the magnetic field and assume $A = 0$ everywhere. The TDGL equation describes the diffusive relaxation of the superconducting order parameter ψ if it is initially out of equilibrium. The first order, diffusion-type, TDGL equation does not predict any oscillatory phenomenon, such as waves. The equation has a limited validity; it is approximately correct only if the superconducting energy gap is zero, $\Delta = 0$, or if $\Delta \ll k_B T$. The characteristic relaxation time in the equation is

$$\tau_r = \frac{h}{16 k_B (T_c - T)} \tag{9.1}$$

By an extrapolation, we can roughly estimate that at $T = 0$, the timescale is $\tau_r = \pi \hbar / 8 k_B T_c \approx 0.4 \hbar / k_B T_c$. Such a timescale is in agreement with the Heisenberg uncertainty principle, provided that the energy uncertainty is $k_B T_c$.

The Giordano model is based on the analogy with the macroscopic quantum tunneling in Josephson junctions, which is well understood [34]. The general formula for the MQT rate, (3.42), is equally applicable to JJs and nanowires, or to any tunneling over a barrier, when damping is negligible. Little's phase slips (see Figures 6.3 and 6.4) produce a region having a suppressed order parameter. Such a region acts as a temporary superconducting weak link on the wire. In that respect, a nanowire is similar to a JJ, in which the weak link is permanently created by the presence of the thin oxide barrier. Such similarity is the basis for the assumption that the MQT phenomenon in thin wires should be described by equations similar to those used to describe MQT in JJs.

To understand the physical idea of the analogy notice that in the formula $S/\hbar \propto \Delta U / \hbar \omega_p$ derived for SIS junctions (3.44) the quantity $\hbar \omega_p$ defines the scale of zero-point energy in the device. Thus the action is proportional to the barrier height divided by $E_{ZPE} = \hbar \omega_p / 2$, which is the zero-point energy of the junction. Thus the action can be presented in a more general form as $S/\hbar \propto \Delta U / E_{ZPE}$. To make a transition to superconducting wire one needs to estimate its zero-point energy. The only available timescale in a long nanowire is the relaxation time τ_r. This timescale, according to Heisenberg Uncertainty, defines the strength of the zero-point energy fluctuations as $E_{ZPE} = \hbar / \tau_r$. Thus, to obtain reasonably good estimates for the MQT rate in a thin superconducting wire one can replace $\hbar \omega_p$ with $\hbar / \tau_r$ in formulas describing TAPS, to get expressions describing QPS.

So, the action for quantum phase slips, in analogy to (3.44), can be written as

$$\frac{S_G}{\hbar} = a_G \frac{\Delta F}{\hbar \tau_r^{-1}} \tag{9.2}$$

Since such an estimate is not exact, the constant a_G is added as an unknown fitting parameter which is expected to be of order unity. Its value for MoGe nanowires is $a_G \approx 1.3$, according to [108]. For Al nanowires, a value of $a_G \approx 1.2$ was reported in [58]. Note that the free energy barrier ΔU, which is correct for JJs, is replaced by the corresponding free energy barrier ΔF, which is the minimum work needed to create a phase slip in a nanowire.

The expression for the action, (9.2), combined with (3.42), leads to the Giordano formula for the QPS rate in thin wires

$$\Gamma_{QPS} = \Omega_{QPS} \exp\left(-\frac{S_G}{\hbar}\right) = \Omega_{QPS} \exp\left(-\frac{a_G \Delta F \tau_r}{\hbar}\right) \tag{9.3}$$

Note how it is similar in the general form to the rate of TAPS

$$\Gamma_{TAPS} = \Omega_{TAPS} \exp\left(-\frac{\Delta F}{k_B T}\right) \tag{9.4}$$

By now, we have learned that phase slips can occur either by thermal activation or by quantum tunneling. The crossover temperature T^* is the temperature at which these two sources of fluctuations, that is, thermal and quantum, contribute equally to the total frequency of phase slips. Thus, T^* is defined through $\Gamma_{QPS}(T^*) = \Gamma_{TAPS}(T^*)$. Let us estimate T^*. This can be achieved, assuming that the attempt frequency is constant and the same both for TAPS and QPS. Then, the quantities under the exponents must be equal, which is equivalent to $\hbar = a_G k_B T^* \tau_r(T^*)$. Now, using (9.1), we arrive at $8(T_c - T^*) = a_G \pi T^*$, which leads to

$$\frac{T_c}{T^*} = 1 + \frac{\pi a_G}{8} \tag{9.5}$$

If one takes a generic value $a_G = 1$, then $T^* \approx 0.72 T_c$. If $a_G = 1.3$, as was found experimentally in [108], then $T^* \approx 0.66$. The experiments reported in [108] indicate that $T^* \sim 0.5 T_{c,film}$. Taking into account the fact that the critical temperature of the film electrodes $T_{c,film} > T_c$, one estimates $0.5 < T^*/T_c < 1$. An examination of multiple MoGe nanowire samples in [109] shows that $T^*/T_c = 0.7 \pm 0.15$, which agrees with the model estimate above.

The significance of T^* is that it gives the temperature scale below which QPS can be observed through low-bias, linear, resistance measurements. If the wire is thick, its resistance drops quickly with cooling. By the time the wire is cooled down to $T^* \sim 0.7 T_c$, the resistance is immeasurably low. Such is the case in early experiments on tin whisker [23] in which the resistance dropped by six orders of magnitude within an interval of just one or two milliKelvin, that is, in a narrow interval near T_c, such as $(T_c - T)/T_c \sim 10^{-3}$. Consequently, no QPS was detected since the QPS rate should become comparable to the TAPS rate only at temperatures as low as $T^* \approx 0.7 T_c$. Thus, to observe QPS in linear transport measurements, one needs very thin wires, so thin that they remain resistive, in the experimental sense, down to $T \sim 0.7 T_c$ or lower.

According to the LAMH model, the attempt frequency of thermally activated phase slips is

$$\Omega_{\text{TAPS}} = \Omega_{\text{MH}} = \frac{\sqrt{3}}{2\pi^{3/2}} \frac{1}{\tau_r} \frac{L}{\xi} \sqrt{\frac{\Delta F}{k_B T}}$$

Since $\tau_r \propto (T_c - T)^{-1}$, therefore $\Omega_{\text{MH}} \propto (T_c - T)^{9/4}$. This frequency defines the number of times the system "attempts" to overcome the barrier. Not all of these attempts are successful. The fraction of successful attempts is given by the Arrhenius exponential factor.

Giordano made a plausible suggestion that at low temperatures, when the thermal energy is negligible, the phase slips are energized by zero-point fluctuations. Thus, the thermal energy $k_B T$ needs to be replaced by the estimate for the quantum zero-point energy, E_{ZPE}. Such an estimate is based on the only available timescale, τ_r, and the general principle of Heisenberg uncertainty, which is $E_{\text{ZPE}} \approx \hbar/\tau_r$. The attempt frequency can then be obtained from the LAMH expression by replacing $k_B T$ with E_{ZPE}. The result is

$$\Omega_{\text{QPS}} = \frac{\sqrt{3}}{2\pi^{3/2}} \frac{L}{\xi(T)} \frac{1}{\tau_r} \sqrt{\frac{\Delta F \tau_r}{\hbar}} = \frac{\sqrt{3}}{2\pi^{3/2}} \frac{L}{\xi(T)} \sqrt{\frac{\Delta F}{\hbar \tau_r}}$$

Since $\Delta F \propto (T_c - T)^{3/2}$ and $\tau_r \propto (T_c - T)^{-1}$, one can easily see that $\Omega_{\text{QPS}} \propto (T_c - T)^{7/4}$. Until now, the accuracy of experiments was not sufficient to confirm or reject the exact form of the attempt frequency or even to confirm the exponent 7/4. However, the Giordano model is frequently used to fit data with satisfactory precision because the attempt frequency plays a negligible role, compared to the rapidly changing exponential factor by which the attempt frequency is always multiplied.

Using (9.3), one obtains the QPS rate as

$$\Gamma_{\text{QPS}} = \Gamma_G = \frac{\sqrt{3}}{2\pi^{3/2}} \frac{L}{\xi} \frac{1}{\tau_r} \left(\frac{\Delta F \tau_r}{\hbar}\right)^{1/2} \exp\left(-a_G \frac{\Delta F \tau_r}{\hbar}\right) \tag{9.6}$$

To obtain a prediction for the linear resistance as a function of temperature, $R(T)$, we need the relationship between the rate of phase slips, Γ, and the corresponding resistance at low bias. Such a relationship is given in (6.29) for thermally activated phase slips

$$R_{\text{TAPS}} = R_q \frac{\hbar \Gamma_{\text{TAPS}}}{k_B T}$$

Following Giordano's prescription, we replace the thermal energy $k_B T$ with the quantum or zero-point-fluctuations energy $\hbar/\tau_r$. Thus, the resistance due to QPS can be written as

$$R_{\text{QPS}} = \sqrt{\frac{\pi}{3}} B_G R_q \frac{\hbar \Gamma_{\text{QPS}}}{\hbar/\tau_r} \tag{9.7}$$

Here, B_G is an unknown factor which is introduced because the exact time dependence of the phase slippage events is not known. Using the expression for Γ_{QPS} given in (9.6), we arrive at the following more explicit form

$$R_{QPS} = B_G R_q \frac{L}{\xi} \left(\frac{\Delta F \tau_r}{\hbar}\right)^{1/2} \exp\left(-a_G \frac{\Delta F \tau_r}{\hbar}\right) \tag{9.8}$$

Measurements of the low-bias resistance [108] led to an estimate of $B_G = 7.2$.

To calculate the nanowire resistance, one needs to take into account that at a finite temperature, both QPS and TAPS can contribute to the supercurrent decay. Thus, R_{QPS} needs to be added to R_{TAPS}. In addition to this, a conductance of bogoliubons in the superconductor could provide a conductance of the order of R_n^{-1}. Such normal current contribution is only present near T_c. Its temperature dependence is not well known. Thus, we simply add a constant conductance R_n^{-1} to the total wire conductance. At low temperatures, this contribution of the normal conductance automatically becomes negligible since the conductance of the condensate diverges. Thus, the best estimate of the resistance, according to the Giordano model, is

$$R_G(T) = \left[R_n^{-1} + (R_{TAPS} + R_{QPS})^{-1}\right]^{-1} \tag{9.9}$$

The energy barrier for QPS is always chosen the same as for TAPS, at least within the Giordano model. A justification for the assumption that the barrier height for QPS equals that one for TAPS comes from the analogy with the tunneling and the thermal activation of single electrons or any other elementary particles. In such typical situations, the barrier that defines the tunneling is the same one as the barrier controlling the tunneling rate.

Thus, the barrier is the same as in the LAMH model and/or in the Little's fit, namely,

$$\Delta F = \Delta F(T) = \frac{8\sqrt{2}}{3} \frac{H_c^2}{8\pi} A_{cs} \xi = \frac{8\sqrt{2}}{3} \frac{\hbar A_{cs}}{2\beta \sqrt{4m}} |\alpha|^{3/2} \tag{9.10}$$

The expression is given in Gaussian units. The thermodynamic critical field H_c and the coherence length ξ are a function of the temperature (see (6.17)). This expression is equivalent to Eq. (3.23) of [143], assuming that the bias current is negligible. According to the usual definitions of the GL theory, the condensation energy density is $H_c^2/8\pi = \alpha^2/2\beta$ and the coherence length is $\xi = \hbar/\sqrt{4m|\alpha|}$. If one also takes into account that $\alpha \propto T - T_c$, one concludes that the energy barrier for a phase slip in a thin wire is proportional to $(T_c - T)^{3/2}$.

The expression for the phase-slip barrier can be presented in various ways, some of which we will discuss below. For example, one can express the barrier through the critical current [19]. This can be done using standard results of the GL theory. According to GL theory, the critical field is

$$H_c = \frac{\Phi_0}{2\pi \sqrt{2} \lambda \xi} \tag{9.11}$$

where λ is the magnetic field penetration depth. Therefore, the LPS barrier is

$$\Delta F = \frac{8\sqrt{2}}{3}\frac{H_c^2}{8\pi}A_{cs}\xi = \frac{\Phi_0 H_c}{6\pi^2 \lambda}A_{cs}$$

The GL expression for the critical current of a thin wire is

$$I_c = A_{cs}\frac{cH_c}{3\sqrt{6}\pi\lambda} \tag{9.12}$$

Combining with the expression above, we get a simple relation between the critical current and the energy barrier

$$\Delta F = \frac{\sqrt{6}I_c\Phi_0}{2c\pi} = \frac{\sqrt{6}\hbar I_c}{2e} \tag{9.13}$$

Note that this is very similar to the expression for JJs, which are SIS tunnel junctions. In the latter case, the barrier is $\Delta U = 2E_J = \hbar I_c/e$. Since the critical current of a thin wire is well described by the Bardeen formula, (11.1), which provides accurate predictions at all temperatures, we can now write an expression for the barrier as

$$\Delta F(T) = \frac{\sqrt{6}I_c\Phi_0}{2c\pi} = \frac{\sqrt{6}\hbar I_c(0)}{2e}\left(1 - \frac{T^2}{T_c^2}\right)^{3/2} \tag{9.14}$$

Such an expression is expected to be valid at any temperature, at least approximately.

Let us now transform the expression for the barrier further and express it through the normal state resistance of the wire, R_n. To achieve this, we need to express the depairing current of the wire through R_n. Equations (9.11) and (9.12) lead to

$$I_c(T) = \frac{c\Phi_0 A_{cs}}{6\sqrt{12}\pi^2 \lambda^2(T)\xi(T)}$$

The above equality is derived in the GL framework, which assumes that T is near T_c. We can now make a plausible assumption that the above equation is also valid at zero temperature, at least approximately. Such an assumption is based on the following argument. The above equation is valid at temperatures near T_c, in which case the quantities involved increase rapidly with cooling. As the temperature is reduced, the variation of the involved quantities, $I_c(T)$, $\lambda(T)$, and $\xi(T)$ slows down. They almost stop changing at $T \approx T_c/2$. The GL theory is still approximately correct at this temperature. At lower temperatures, the variations of the quantities involved are insignificant, and thus the mutual relationship between them, given by the formula above, "freezes" at $T_c/2$ and remains approximately the same down to $T = 0$. Thus, we can write an approximate expression

$$I_c(0) = \frac{c\Phi_0 A_{cs}}{6\sqrt{12}\pi^2 \lambda^2(0)\xi(0)}$$

Now, it is possible to include the result of the BCS theory for zero temperature, namely, the magnetic penetration depth expression for dirty superconductors [2] $\lambda^2 = (\xi_0/l_e)\lambda_L^2$, where $\lambda_L^2 = mc^2/4\pi ne^2$ and $\xi_0 = \hbar v_F/\pi\Delta$. Also, according to Drude model, $\rho_n = mv_F/ne^2 l_e$. Therefore, the mean free path is $l_e = mv_F/ne^2\rho_n$, where ρ_n is the normal resistivity related to the normal resistance as $R_n = \rho_n L/A_{cs}$ or $\rho_n = R_n A_{cs}/L$. One more well-known zero-temperature expression is $\xi^2(0) = l_e \xi_0$, which is valid for a dirty superconductor, that is, if $l_e \ll \xi_0$ [2]. We will also use the BCS relation $\Delta = 1.76 k_B T_c$. After some algebra, one gets

$$I_c(0) = \frac{\pi L \Delta}{3\sqrt{3} e R_n \xi(0)} \tag{9.15}$$

This expression does not involve the speed of light, c. This fact gives us a hint that it is valid both in SI units as well as in Gaussian units, which is in fact true. The next step is to apply (9.13), which we will also assume to be approximately correct for $T = 0$, and to get

$$C_T \equiv \frac{\Delta F(0)}{k_B T_c} = 0.83 \frac{R_q}{R_n} \frac{L}{\xi(0)} = 0.83 \frac{R_q}{R_\xi}$$

The barrier-height constant C_T depends only on the length of the wire L, the normal resistance of the wire R_n and on the coherence length at zero temperature in the wire $\xi(0)$. The resistance of a segment of length $\xi(0)$ of the wire is defined as $R_\xi = \xi(0) R_n/L$. If we treat (9.13) as valid for all temperatures and use the Bardeen formula for the critical current, $I_c(T) = I_c(0)(1 - T^2/T_c^2)^{3/2}$, which is valid at all temperatures, then the barrier for phase slips at all temperatures can be written as

$$\Delta F(T) = C_T k_B T_c \left(1 - \frac{T^2}{T_c^2}\right)^{3/2}$$

The constant C_T defines the barrier at $T = 0$ as

$$\Delta F(0) = C_T k_B T_c$$

Now, it is possible to write the QPS rate for $T = 0$ by using $\tau_r(0) \approx \pi\hbar/8k_B T_c$. From (9.6), the rate at $T = 0$ is

$$\Gamma_{QPS}(0) = \Gamma_G(0) = \Omega_{QPS}(0) \exp\left(-\frac{\pi a_G C_T}{8}\right) \tag{9.16}$$

where the attempt frequency at zero temperature is

$$\Omega_{QPS}(0) = \frac{\sqrt{6 C_T}}{\pi^2} \frac{L}{\xi} \frac{k_B T_c}{\hbar}$$

Let us take an example of a nanowire having $L = 100\,\text{nm}$, $\xi(0) = 10\,\text{nm}$, $C_T = 10$, $T_c = 3\,\text{K}$. Then, the attempt frequency is $\Omega_{QPS}(0) = 21\,\text{THz}$. Suppose we want to make a QPS qubit [45, 50, 51] having energy $\Gamma_{QPS}(0) = 7\,\text{GHz}$. Then, the exponential factor has to be equal, that is, $\Gamma_{\text{QPS}}(0)/\Omega_{\text{QPS}}(0) \approx 1/3000$. Therefore, from (9.16), it follows that

$$\pi a_G C_T = 8 \ln\left[\frac{\Omega_{\text{QPS}}(0)}{\Gamma_{\text{QPS}}(0)}\right]$$

In the example, we consider $\pi a_G C_T \approx 64$. Then, assuming $a_G = 1.3$, we estimate $C_T \approx 16$. This is sufficiently close to the starting "seed" value of $C_T = 10$. Thus, we accept the value $C_T = 16$. Since $C_T = 0.83(R_q/R_n)[L/\xi(0)]$, it follows that

$$\frac{R_q}{R_n} \frac{L}{\xi(0)} = \frac{3.07}{a_G} \ln\left[\frac{\Omega_{\text{QPS}}(0)}{\Gamma_{\text{QPS}}(0)}\right]$$

This formula determines, within the Giordano approximate, what should be R_n of the wire to obtain the desired rate of QPS. Using $C_T = 16$, we get an estimate $R_n = 3.6\,\text{k}\Omega$, which is not difficult to achieve in practice, with MoGe nanowires. Yet, in experiments on the short wires, that is, shorter than 200 nm, the QPS was not observed at low bias currents, and it was concluded that $a_G > 4$ [187]. For such a value of a_G, one gets $C_T = 8$ and therefore one would need $R_n = 11\,\text{k}\Omega$ in order to make a working qubit. Such highly resistive wires would be in the insulating phase already, as is discussed in the section about SIT (Chapter 10). This is because the insulating part of the phased diagram is defined as $R_n > R_q \approx 6.5\,\text{k}\Omega$, at least for relatively short wires [130]. Thus, we see that different estimates produce results, placing the wire on both sides of the SIT. Therefore, a necessary requirement for making a QPS qubit is that the QPS-active nanowire should be near the SIT critical point. Wires which are deeply in the superconducting phase or deeply in the insulating phase should not be suitable for making QPS qubits.

The constant C_T appears under the exponent. Thus, its impact on the QPS rate is very strong. Let us consider another example. Suppose one needs a wire that is always phase coherent. Absolute phase coherence is impossible within the Giordano model. However, following [19], we can set a practical limit of less than one phase slip per second in a 1 cm long MoGe wire. Such low rate of QPS can be achieved if $C_T \approx 100$ or higher. According to [19], the wire should be larger than 15 nm in diameter to achieve such a level of phase stability.

Note that in principle, there is another option of how to choose the timescale, which defines the attempt frequency as well as the energy scale of zero-point fluctuations. If the wire is short and can be considered as a lumped inductive element, the device formed by the wire and the electrodes resembles an LC-circuit, where L_K is the kinetic inductance of the wire and C is the capacitance between the superconducting electrodes to which the wire is connected (Figure 1.1). In such cases, the characteristic frequency is $\omega_p = 1/\sqrt{L_k C}$, which is the plasma frequency. Here, L_k is the wire's kinetic inductance. In such cases, the analogy with the MQT in JJs is much more direct. Then, the QPS rate might be estimated using the general tunneling rate (3.42), combined with the quantum action (3.48) and the attempt frequency (3.47). The barrier for MQT in JJs, called ΔU, needs to be replaced by the QPS barrier, denoted above as ΔF. Such an option was not considered by Giordano

9.1 Giordano Model of Quantum Phase Slips (QPS) in Thin Wires

since his interest was to analyze long wires in which the quantum fluctuations are completely intrinsic and independent of the electrodes attached to the wire.

So far, we have analyzed the case of zero or very low bias currents. In the case of a strong bias current, the Helmholtz energy barrier ΔF needs to be replaced in all formulas by Gibbs free energy barrier. Both the GZ theory and the McCumber theory predict the same dependence on I, namely, the Gibbs barrier is proportional to $(1 - I/I_c)^{5/4}$. In our notations, Gibbs barrier is

$$\Delta G(T) = 0.83 k_B T_c \frac{R_q L}{R_n \xi(0)} \left(1 - \frac{T^2}{T_c^2}\right)^{3/2} \left(1 - \frac{I}{I_c}\right)^{5/4} \tag{9.17}$$

To get the rate of QPS at high bias currents, one ought to replace ΔF in (9.6) by ΔG. Thus, the equation for the QPS at $I_c - I \ll I_c$ is

$$\Gamma_{\text{QPS}} = \Gamma_G = \frac{\sqrt{3}}{2\pi^{3/2}} \frac{L}{\xi} \frac{1}{\tau_r} \left(\frac{\Delta G \tau_r}{\hbar}\right)^{1/2} \exp\left(-a'_G \frac{\Delta G \tau_r}{\hbar}\right) \tag{9.18}$$

where the adjustment constants a'_G can be different from the zero-bias case.

Now, we can repeat the arguments leading to (9.5) and derive, for high bias currents, the following estimate for the crossover temperature, that is,

$$\frac{T_c}{T^*_{\text{hb}}} = 1 + \frac{\pi a'_G}{8} \tag{9.19}$$

Here, T^*_{hb} is the temperature at which the rates of QPS and TAPS are equal, if measured at high bias currents, that is, in the switching experiments, which are discussed in Section 8. Such experiments have been done recently in [116] (note that in this reference, T^*_{hb} is denoted as T_q). The conclusion reached is that $T_c/T^*_{\text{hb}} \approx 6.3$ for optimized MoGe nanowires. From an independent study of Al nanowires, a value $T_c/T^*_{\text{hb}} \approx 3.5$ can be inferred [59]. Thus, the average value is $T_c/T^*_{\text{hb}} \approx 5$. Therefore, according to (9.19), $a'_G \approx 10$. This illustrates a trend that, according to published experimental results, $a'_G \gg a_G$. A possible explanation to this is that most of the high bias experiments are done on wires having $R_n < R_q$, in which QPS is probably suppressed by the Caldeira–Leggett mechanism, while most of the low bias experiments are done on wires having $R_n > R_q$. Such wires are expected to show a significantly higher rate of QPS since they belong to the insulating or not superconducting region of the SIT phase diagram, in which QPS proliferate. The SIT is further discussed in Chapter 10. Interestingly, there is one reported low bias experiment on wires having $R_n < R_q$. This experiment [187] gave a much higher than usual a_G, estimated as $a_G > 4$.

Sometimes, zero-point energy is measured in Kelvins and is called quantum temperature, T_q. In such language, the QPS action takes the form $S/\hbar \propto \Delta G/k_B T_q$. This form is obtained if the thermal energy $k_B T$ is replaced by the quantum zero-point energy $k_B T_q$. Thus, the QPS rate, at high bias, for example, can be written as

$$\Gamma_{\text{QPS}} = \Omega_{\text{QPS}}(T, I) \exp\left(-\frac{\Delta G(T, I)}{k_B T_q}\right) \tag{9.20}$$

Note that the temperature of the wire, T, has a role to play also. It defines the density of the condensate and through this, defines the barrier $\Delta G(T, I)$. This form can also be formally applied to low or zero-bias current cases. For this, one needs to remember that at $I = 0$, we have $\Delta G(T, 0) = \Delta F(T)$, that is, Gibbs energy taken at zero-bias equals Helmholtz free energy.

Such presentation of the quantum rate, involving T_q, is a variation of the Giordano model. In this version, the exact origin of T_q is not specified and it is used as a fitting parameter. Now, let us compare the exponential factors in the two versions of the model, (9.20) and (9.18). A formal comparison leads to a conclusion that $k_B T_q = \hbar(\tau_r a'_G)^{-1} \propto (T_c - T)$. Thus, T_q might be expected to increase linearly with cooling. When applying this model to fit the switching rates, measured at high bias currents and low temperatures, it was found that T_q is either constant [116, 131] or decreases linearly with cooling [131], unlike the theoretical expectation stated above. This fact shows that extrapolating the linear temperature dependence of τ_r^{-1} down to zero temperature is incorrect, at least in the applications of the Giordano model. Also, the extrapolation to zero temperature leads to a conclusion that, at $T = 0$, $T_q = 2.54 T_c / a'_G$. Experiments [116] indicate that $T_q = 0.16 T_c$. This requires $a'_G \approx 16$, which is not of order unity already. Such large values of a'_G might also be due to the fact that the linear dependence of the relaxation time is only valid near T_c and cannot be extended to zero temperature by simple linear extrapolation. In order to obtain a theoretical prediction for the QPS rate as a function of temperature in the region $T \ll T_c$, one needs to use the Golubev–Zaikin QPS theory, which will be discussed in Section 9.3.

Below, we briefly discuss various topics of relevance to the QPS process. The power of 5/4 in the Gibbs barrier given by (9.17) was calculated for long wires, in which the energy of a phase slip is independent on the distance to the electrodes (and $I_s = $ const). For short wires, on the other hand, one could expect $\Delta G \propto (1 - I/I_c)^{3/2}$, as was argued in Section 8.7. Such a conclusion can also be expected on general grounds because a short wire is similar to a Josephson junction. As always, the barrier height found for TAPS is also valid for the computation of the QPS rate.

In experiments involving low temperatures, compared to T_c, an important question arises on how to extend the Giordano model to lower temperatures. This can be done by utilizing phenomenological polynomial expressions for the critical field and the coherence length. One such wide-temperature-range formula for $H_c(T)$ is [165]

$$\frac{H_c(T)}{H_c(0)} = 1.732 t_2 - 0.40087 t_2^2 - 0.33844 t_2^3 + 0.00722 t_2^4$$

where $t_2 \equiv 1 - T/T_c$ is the normalized deviation from the critical temperature. Such an expression for the critical field can be introduced, for example, into (9.10) in order to find an expression for the LPS barrier applicable at all temperatures.

If a simpler formula is desirable, another quite accurate empirical approximation is

$$\frac{H_c(T)}{H_c(0)} = 1 - \frac{T^2}{T_c^2}$$

This expression is sometimes called the "parabolic law" for the thermodynamic critical field.

Sheahen [166] found an approximate formula for the energy gap, which is also valid at any $T < T_c$

$$\frac{\Delta(T)}{\Delta(0)} = \cos\left(\frac{\pi T^2}{2T_c^2}\right) = \sin\left[\frac{\pi}{2}\left(1 - \frac{T^2}{T_c^2}\right)\right]$$

The coherence length can be approximated, again at all temperatures, as

$$\frac{\xi(T)}{\xi(0)} = \left(1 - t_1^4\right)^{1/2}\left(1 - t_1^2\right)^{-1}$$

where $t_1 \equiv T/T_c$ is the reduced temperature.

Even with all of these improvements, the model cannot be made exact at low temperatures since the relaxation time τ_r is not accurately known at low T. The expression (9.1) can only be used to get a rough estimate at $T \approx 0$. Thus, the only reasonable approach, it seems, is to assume τ_r is a constant at low T, or to use T_q as a fitting parameter.

9.2
Experimental Tests of the Giordano Model

Building an *exact* analogy between plasma oscillations in a JJs and the relaxation of the order parameter in nanowires is not possible. This follows from the fact that a JJ is described by a second-order differential equation, (3.19), while TDGL is a first-order differential equation. Thus, the Giordano model is not exact. Its justification comes from its good agreement with experiments, such as in [53, 108, 131]. Examples of such favorable comparisons are given in this section. The model provides rather good predictions for the QPS rate in the cases when damping is not significant. If damping is strong, meaning that the effective shunting resistance for the superconducting device is lower than R_q, then the quantum Schmid–Bulgadaev (SB) transition occurs [46, 93, 130]. Thus, the wire enters a state in which QPS is suppressed entirely, at least at zero temperature. In such cases, the Giordano model overestimates the rate of quantum phase slips because it does not take into account the dissipative suppression of the QPS rate of the type discussed in the Caldeira and Leggett theory [30].

The QPS can be observed experimentally as a change in the sign of the curvature of log R vs. T plots. Namely, in the TAPS-dominated region, that is, at $T > T^*$, such plots exhibit a negative curvature while in the QPS-dominated region, that is, at $T < T^*$, the curvature becomes positive [108], as is shown in Figure 9.1. The low-temperature region, having a positive curvature, is sometimes called a resistive tail of the $R(T)$ curve. The fitting curves in Figure 9.1 are generated using the Giordano model, namely, (9.9). There, the resistive contribution due to QPS is given by (9.8). The contribution of TAPS is defined in (6.30), and R_n is the resistance of the wire in the normal state. Taking into account that there are only two adjustable parameters for the entire family of curves, namely, $a_G = 1.3$ and $B_G = 7.2$, we conclude that the agreement between the model and the data is good.

It is also clear from Figure 9.1, that C_T determines the strength of QPS. The $R(T)$ curve of the wire having large $C_T = 0.83 R_q L / R_n \xi(0) \approx 48$, namely, Sample #8, agrees well with the Arrhenius activation law and shows no resistive tail, having a positive curvature. Thus, the contribution of QPS to the total resistance is very weak, if any, for such high C_T. As C_T is lowered, the contribution of QPS becomes more and more significant. Another example is Sample #4, which has $C_T \approx 23$ and shows a well pronounced resistive tail, having a positive curvature. Thus, QPS

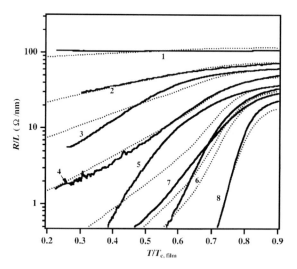

Figure 9.1 Examples of $R(T)$ curves (solid curves) measured on different MoGe nanowires [108]. The dashed curves are the fits to the Giordano model (9.9). The fits are generated under the assumption that the film has the same critical temperature as the wire, that is, $T_{c,film} = T_c$. The plot is log-linear, meaning that the y-axis is in the logarithmic format and the x-axis is in the linear format. The curvature of each curve is negative at higher temperatures and positive at low temperatures. Such change reflects a crossover from the TAPS dominated regime to the QPS dominated regime. The inflection point is called T^*. It represents the temperature at which the rate of thermal activation of LPS equals the rate of quantum tunneling of LPS. For example, for the Sample #4, the inflection point is $T^* \approx 0.6 T_{c,film} \approx 3$ K. The normal state resistances of the nanowires, R_n, and their lengths, L, are 1: 14.8 kΩ, 135 nm; 2: 10.7 kΩ, 135 nm; 3: 47 kΩ, 745 nm; 4: 17.3 kΩ, 310 nm; 5: 32 kΩ, 730 nm; 6: 40 kΩ, 1050 nm; 7: 10 kΩ, 310 nm; and 8: 4.5 kΩ, 165 nm. Thus, for all samples except one (Sample #8), the condition $R_n > R_q$ is true. Assuming $\xi(0) = 5$ nm, the sample specific constant $C_T = 10, 14, 17, 23, 25, 34, 40,$ and 48 for samples 1 through 8 respectively.

contribute strongly to the resistance of this sample. In wires having even lower C_T, the QPS is completely dominant. Sample #1, for example, has $C_T \approx 10$ and shows no sign of superconductivity at all. A possible explanation to this observation is that the quantum fluctuations are so strong that they have destroyed superconductivity completely. This is possible because, unlike in JJs, each phase slip in a thin wire brings a normal core with it. So if QPS occur frequently enough, the mean value of the order parameter can be reduced to zero or close to zero.

The implicit assumption of the Giordano model is that all samples should exhibit such a crossover from TAPS to QPS at sufficiently low temperatures. Thus truly superconducting samples, that is, such samples that have exactly zero resistance at zero temperature, do not exist according to the Giordano model. So, the model does not allow quantum phase transitions, for example, superconductor-insulator or superconductor-normal quantum transitions, simply because S-type wires do not exist according to the model.

Even if the wire remains resistive down to temperatures considerably lower than T_c, it is not always possible to observe QPS. For example, in [187] and [130], the log R vs. T curves only showed negative curvature and a good agreement with TAPS model, that is, showed nothing other than Arrhenius activation behavior. Examples of such data, contradicting the Giordano model, are shown in Figure 9.2. In this case, Nb wires were studied. The exponent-control-parameter C_T is listed in the figure caption. According to the findings of [108], illustrated in Figure 9.1, the QPS resistance tail appears clearly if $C_T < 45$. The collection of Nb wires of Figure 9.2 includes wires having $C_T = 32$, 8.7, and 3.9, corresponding to samples Nb3, Nb5, and Nb6. They are expected to show resistive tails having positive curvature. Such tails are not observed.

It was previously discussed that a condition of observing QPS is $T < T^* \approx 0.7 T_c$. For some samples, it is satisfied. For example, for the sample Nb5, $T_c \approx 2.5$ K and, therefore, $T^* \approx 1.75$. The measurement of the resistance for this sample is traced to as low as 0.9 K and yet no transition to a QPS regime is detected. Giordano model fits, shown by dashed and dotted curves for samples Nb3, Nb5, and Nb6, exceed the experimental resistance very strongly. Note the y-axis is in the log scale. Thus, it is clear that this group of nanowires contradicts the Giordano model of QPS.

Our explanation for the fact that QPS is not observed where it is expected is that most of these samples satisfy the $R_n < R_q$ condition. Thus, they are on the superconducting side of the superconductor-insulator SB phase diagram. Therefore, QPS is expected to have zero frequency in them due to the damping of tunneling provided by the electrodes. Such an explanation, which will be discussed in more detail below, is based on two quite well-established facts, namely, that (1) a superconducting nanowire or a JJ, does not have QPS and behaves as a coherent superconductor if shunted by an impedance of less than $R_q = h/4e^2 \approx 6.5$ kΩ [93], and (2) the electrodes or transport measurement leads connected to a nanowire always have an AC impedance close the impedance of free space, $Z_0 = 0.377$ kΩ. In a simplified view, since $Z_0 \ll R_q$, each wire connected to any macroscopic lead is heavily shunted and thus has to act as a true superconductor, that is, has to have

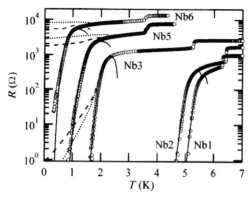

Figure 9.2 Resistance versus temperature measurements for five different Nb nanowires [187]. As usual, the resistance is measured at a low bias current, such that further reduction of the bias current does not cause any change in the measured resistance. Solid lines show fits to the LAMH model of TAPS, R_{TAPS}. The samples Nb1, Nb2, Nb3, Nb5, and Nb6 required the following fitting parameters: The critical temperatures are $T_c = 5.8, 5.6, 2.7, 2.5$, and 1.9 K, correspondingly. The normal resistances and the lengths of the wires are Nb1: 0.47 kΩ, 137 nm; Nb2: 0.65 kΩ, 120 nm; Nb3: 1.61 kΩ, 172 nm; Nb5: 4.25 kΩ, 110 nm; Nb6: 9.5 kΩ, 113 nm; Best-fit coherence lengths, extrapolated to zero temperature, are $\xi(0) = 8.5, 8.1, 18, 16$, and 16.5 nm, respectively. The corresponding exponent-control parameter is $C_T = 185, 123, 32, 8.7, 3.9$. The dashed lines represent the Giordano model, $R_G(T)$, (9.9), with generic constants $a_G = 1$ and $B_G = 1$. The dotted lines also represent $R_G(T)$, computed using $a_G = 1.3$ and $B_G = 7.2$. For these samples, the Giordano model overestimates the resistance. All but one sample satisfy $R_n < R_q$. The QPS might be suppressed by the Schmid–Bulgadaev quantum transition, which sets the rate of QPS to zero due to coupling to the environment [92].

zero rate of QPS at zero temperature. In reality, the situation is somewhat more complicated than this though since the wire itself has a resistance of the order R_q. The wire resistance needs to be considered as though it is included in series with the external shunt. If one considered a small segment of the wire, then for such segment, the effective shunt approximately equals $R_n + Z_0$. Since $R_n \gg Z_0$, one can approximately take R_n as the only shunt. According to the Schmid–Bulgadaev quantum transition diagram [83, 84, 90, 91], generalized to thin wires by Büchler et al. [194], Khlebnikov and Pryadko [46], and Meidan et al. [144], if the shunt is lower than R_q, then the QPS rate is zero at $T = 0$ and the shunted nanowire behaves as a true coherent superconductor. (Note that QPS, if present, would break the coherence of the condensate and would lift the resistance of the wire above zero.) Thus, we arrive at a conclusion that if $R_n < R_q$, then each element of the wire is shunted sufficiently strongly, and thus the QPS rate is zero in it (at $T = 0$). Qualitatively speaking, such a point of view explains why the Nb wires of Figure 9.2 do not have QPS resistive tails and do not exhibit a positive curvature, as observed in Figure 9.1. The reason is simple: In Figure 9.2, all samples except one satisfy the $R_n < R_q$ condition, so they belong to the type of true superconductors and, thus, do not exhibit the usual QPS signatures. This comparison of the two sets of data (Figures 9.2 and

9.1) can be regarded, in fact, as experimental evidence in favor of the SB quantum phase transition in nanowires.

The sample Nb6 satisfies $R_n > R_q$. According to the picture outlined above, it should be on the resistive side of the Schmid–Bulgadaev phase diagram. In agreement with this expectation, the sample shows some deviation from the usual negative-curvature Arrhenius-like activation dependence of the resistance on temperature. This is manifested by the fact that its resistance exceeds the LAMH fit at low temperature, roughly below 0.5 K. Such deviation can be a sign that this sample has some nonzero rate of QPS. This is exactly what is expected based on the SB transition picture.

It is not only Nb nanowires that show deviations from the Giordano model. An example of data obtained on $Mo_{79}Ge_{21}$ nanowires is shown in Figure 9.3. The figure shows three samples, A, B, and C, having the exponent control value $C_T = 29$, 31, and 16, correspondingly. Again, for such low values of C_T, according to the Giordano model, one expects to see a pronounced QPS contribution to the wire resistance. Yet, the $R(T)$ curves of the three samples agree perfectly well with the LAMH model, which assume only TAPS contribution and no QPS contribution. The corresponding LAMH fits are shown as solid curves.

The critical temperatures of these samples, obtained as adjustable parameters in the LAMH best-matching fits, are $T_c = 4.8$, 4.37, and 3.86. Therefore the crossover

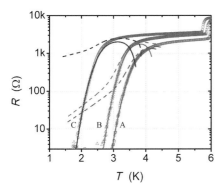

Figure 9.3 Resistance versus temperature curves of three representative $Mo_{79}Ge_{21}$ nanowires [132]. The data are shown by symbols. The continuous curves are the best fits to the LAMH model, generated using $C_T = 29$, 31, and 16 for the samples A, B, and C, correspondingly. The dashed curves represent fits to the Giordano model which are generated using the same parameter plus two additional Giordano parameters $a_G = 1.3$ and $B_G = 7.2$. Apparently, the model predicts a strong contribution of QPS, which is in disagreement with experimental curves. An explanation for the unobservable low rate of QPS is the low normal resistance of all three wires, which satisfy $R_n < R_q$, as is evident from the plots. According to Khlebnikov and Pryadko [46], coupling a nanowire to an impedance that is lower than $R_q = h/4e^2 \approx 6.5\,k\Omega$ makes the wire a true superconductor, characterized by zero rate of QPS. Note that the Giordano model predicts a finite QPS rate at $T = 0$, that is, according to the Giordano model the wires should be classified as normal or resistive, not as superconducting. Please find a color version of this figure in the color plates.

temperatures are expected to be $T^* = 3.36$, 3.06, and 2.7, for samples A, B, and C, correspondingly. The plots illustrate that each sample shows some measurable resistance in a finite region of temperatures below the corresponding T^*. However, none of the samples shows any deviation from the Arrhenius resistance drop with cooling.

Finally, the Giordano model fits are generated using the same fitting parameters as in the best-matching LAMH fits. The resulting expectations for the $R_G(T)$ are shown in Figure 9.3 (the dashed curves). Again, it is clear that the QPS contribution is expected to push the resistance to a much higher value than the resistance measured experimentally. The explanation that seems most probable is, related to the SB quantum transition, as with Nb nanowires. The Giordano model neglects the Caldeira–Leggett damping effects [30]. Thus, it only involves the suppression of the phase slip tunneling by the barrier height. The damping, that is, the interaction with the environment, interacts with the tunneling events. The environment can take away energy from a tunneling event and thus can prohibit the tunneling event. According to the Schmid–Bulgadaev phase diagram, the tunneling is suppressed to zero at $T = 0$ by this sort of interaction with the environment in the cases where there is an effective normal shunt of value less than R_q. All three samples, A, B, and C, are of such quality, assuming that R_n can be identified as a normal shunt.

The reasoning presented above is in accordance with the SB theory, which was clearly summarized by Werner and Troyer [92]. The theory predicts a quantum dissipative transition. It was confirmed experimentally by Penttilä et al. [93]. The essence of this transition is as follows: The MQT rate is zero, in the limit of zero temperature, and the sample acts as a true superconductor if $R_n < R_q$, where R_n is the normal shunt. If, on the other hand, $R_n > R_q$, then the phase "delocalizes," that is, it becomes not a good quantum number. A well-defined phase having quantum fluctuations much smaller than the mean value is needed to define the supercurrent. If the phase is not well-defined due to strong quantum fluctuations, then the wire acts as an insulator or as a normal metal, depending on the fugacity of QPS [194]. The fugacity is the total number of quantum phase slips occurring per second, without taking the damping into account or the pairing [168, 169] of phase slips to antiphase-slips into account. Thus, we distinguish here and in general the rate of QPS, which is the fugacity, and the net rate of phase slippage, which is the difference between the rates of phase slips (PS) and antiphase-slips (APS). In a superconducting wire, the fugacity can be large, but due to the pairing of PS and APS, the net rate of phase slippage can be zero and, correspondingly, the resistance can be zero. This is exactly what happens if the wire is shunted with a normal conductor having resistance $R_n < R_q$ [46, 194]. Apparently, the wires in [108] had relatively low fugacity. However, the PS-APS pairs were broken since $R_n > R_q$ for most of the samples. Because of the low fugacity, the wires of [108] do not appear insulating. However, they are not truly superconducting either: They agree with the Giordano theory, which predicts, if extrapolated, a finite resistance at zero temperature. Therefore, all wires in Figure 9.1, except #8, should be considered normal and not superconducting.

It is important to remember though that in the cited experiments, the nanowires did not have any external shunts, like a JJ has within the SB transition theory [92]. Thus, the applicability of this model to nanowires is still under debate. The reason for such doubts is that it is not completely clear yet if the wire populated with QPS can act as a normal shunt for itself or it cannot. If the answer is yes, then the existing data can be naturally explained by the Schmid–Bulgadaev transition [130].

A possible approach to resolve this somewhat controversial topic is to assert that the electrodes connected to the wire provide a shunt of the order of 100 Ω. Such impedance is typical for any coaxial cable or any two macroscopic signal wires in general due to the vacuum impedance being in the same range. Thus, the QPS in the wire are shunted with $R_n \sim 100\,\Omega$. Such a situation was analyzed theoretically by Khlebnikov and Pryadko [46]. They concluded that a SIT occurs when the wave impedance of the environment, for example, the impedance of the electrodes, equals the quantum resistance $R_q \approx 6.5\,k\Omega$. The superconducting phase of the SIT is such that the QPS rate is zero. Thus, based on the Khlebnikov and Pryadko theory, one can conclude that QPS will always be suppressed in nanowires since the electrodes would always provide a shunting impedance much less than R_q.

Yet, there is one technical correction to be made. Namely, the resistance of the wire itself is connected in a series with the shunting electrodes. Thus, the shunting impedance experienced by each individual QPS in the wire is $R_n + Z_e$, where Z_e is the high-frequency impedance of the electrodes [130]. Since usually $Z_e \ll R_n$, the value of R_n is the parameter which determines whether the wire is in S-phase or I-phase (to learn more about SIT, see Chapter 10) [130].

A more detailed analysis of such an approach involves a renormalization group technique [144] which allows one to include the effect of all other phase slips on one particular QPS in a self-consistent manner. It takes into consideration the fact that the resistance of the wire at low temperatures can be R_n only if the rate of QPS is sufficiently high to suppress superconductivity in the entire wire. In other cases, the resistance might be less than R_n. If R is low, then the shunting influence of the electrodes is stronger. This is because each QPS is better coupled to the leads if the wire has low impedance. If the coupling of QPS to the leads is strong, this causes further suppression of the QPS. Thus, a low QPS rate becomes even lower, causing a rapid drop of resistance with cooling. A strong dichotomy is present in the Meidan et al. model, which matches experimental results. The general conclusion is the same: the wires having $R_n < R_q$ are superconducting, while the wires having $R_n > R_q$ are insulating. The renormalization group model predicts a much sharper drop of the resistance with cooling compared to the traditional SB transition model [92].

Now, we turn our attention to the high-bias limit. To detect QPS at high bias, one needs to measure $V(I)$ curves. In most cases involving nanowires, the $V(I)$ curve is hysteretic. At sufficiently low temperatures, each phase slip causes an easily detectable switching from superconducting to the normal state. By measuring the dispersion of such switching events, it is possible to obtain evidence for QPS, as explained in detail in Section 8.

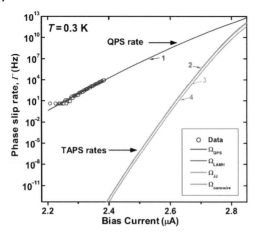

Figure 9.4 A comparison of various models predicting the rate of Little's phase slips to the experimentally obtained rate (circles) plotted versus bias current. This is a high bias current result which appeared in [131]. The blue curve is the fit to the Giordano model of QPS, which agrees with the data well. The other color curves represent TAPS models computed using different choices of the attempt frequency. It is illustrated that the choice of the attempted frequency has a very minor impact on the fitting curve. Predicted TAPS rates are many orders lower than the experimental switching rates. Thus QPS is confirmed. Please find a color version of this figure in the color plates.

The expression for the LPS rate, (9.18), which involves a specific LPS barrier dependence on the bias current, (9.17), was compared to experimental data in [116, 131]. An example is reproduced in Figure 9.4. The quantum phase-slip theoretical rate, predicted by the Giordano model, is shown by the blue curve. It is in good agreement with the data. The fit is made using T_q as a fitting parameter, which turns out of the order of 1 K. The other curves represent the TAPS rate. They deviate from the data (circles) by many orders of magnitude.

It should be noted that in most cases, the role of the attempt frequency, which is the preexponential factor, is not significant. This also means that it is not easy to determine B_G experimentally. Its value might also depend on the value of the bias current at which the phenomenon of QPS is observed. To illustrate the fact that the attempt frequency has a negligible effect on the rate of phase slips, consider three curves in Figure 9.4, which are generated by the LAMH TAPS model in which the attempt frequency was chosen either as Ω_{MH} or as $\Omega = 1/\sqrt{L_k C}$, or a combination of the two. All three theoretical curves are very similar, which illustrates that the attempt frequency effect is usually negligible compared to the influence of the exponential factor. A good first approximation to any modeling of the phase slip rate is to choose a current-independent attempt frequency.

9.3
Golubev and Zaikin QPS Theory

Golubev and Zaikin (GZ) [172] use the microscopic effective action formalism developed in [168, 182] to compute the rate of phase slips, also called QPS fugacity, in thin wires. An important result of their theory is the action of a QPS, which takes the form

$$\frac{S_{GZ}}{\hbar} = A_{GZ} \frac{R_q}{R_n} \frac{L}{\xi(T)} \tag{9.21}$$

where A_{GZ} is an unknown constant of order unity.

The mean number of QPS occurring per second, that is, the QPS rate or the fugacity, is

$$\Gamma_{QPS} = \Omega_{GZ} \exp\left(-\frac{S_{GZ}}{\hbar}\right)$$

where Ω_{GZ} is the preexponential factor which plays the role of the attempt frequency.

The attempt frequency according to GZ is

$$\Omega_{GZ} = B_1 \frac{L}{x_{QPS}} \frac{S_{GZ}}{\hbar} \frac{1}{t_{QPS}} = B_1 \frac{S_{GZ} \Delta}{\hbar^2} \frac{L}{\xi} = B_2 \frac{\Delta}{\hbar} \frac{R_q}{R_n} \frac{L^2}{\xi^2}$$

where B_1 and B_2 are unknown unitless constants of order unity, $\xi = \xi(T)$ is the temperature-dependent coherence length in the nanowire and $\Delta = \Delta(T)$ is the temperature-dependent energy gap. The new notations are t_{QPS}, which is the typical duration of a quantum phase slip, and x_{QPS}, which is the typical size of the QPS. It is estimated that $t_{QPS} \approx \hbar/\Delta$ and $x_{QPS} \approx \xi$.

Similar to (9.7), the QPS rate can be converted into the resistance as

$$R_{GZ} = B_3 R_q \frac{h \Gamma_{QPS}}{\hbar/\tau_{QPS}} = B_4 R_q \frac{S_{GZ}}{\hbar} \frac{L}{\xi} \exp\left(-\frac{S_{GZ}}{\hbar}\right)$$

As an example, consider a wire such that $L = \xi$ and $R_n = R_q$. Then, $S_{GZ} \approx \hbar$, meaning that $R \approx R_q$.

It is also possible to estimate the crossover temperature T^* between the TAPS dominated regime and the QPS dominated regime. As usual, it is approximately defined by the condition that the expressions inside the corresponding exponents are equal, that is, $\Delta F/k_B T^* = A_{GZ}(R_q/R_n)[L/\xi(T^*)] = (A_{GZ} C_T/0.83)(1 - T^*/T_c)^{1/2}$, where $C_T = 0.83(R_q/R_n)[L/\xi(0)]$. It was shown previously that $\Delta F = C_T k_B T_c (1 - T/T_c)^{3/2}$. Thus, we get $T_c/T^* = 1 + A_{GZ}/0.83$. Experimentally, a rough estimate for the crossover temperature, at low current bias, is $T^* \sim 0.7 T_c$ (see, e.g., [108]). This implies that $T_c/T^* \approx 1.4$, leading to an estimate $A_{GZ} \approx 0.35$.

Note that $\xi R_n/L$ is the resistance of each ξ-long segment of the wire. Therefore, $L/\xi R_n$ is the conductance of a wire section of length ξ. Since $1/R_q$ is the conductance quantum for electronic pairs, it follows that the normalized conductance of any wire segment of length ξ equals $g_\xi = R_q L / \xi R_n$. Thus, the action can be written simply as

$$\frac{S_{GZ}}{\hbar} = g_\xi A_{GZ}$$

A proliferation of quantum phase slips, which eliminates superconductivity in the wire, is expected at $S_{GZ} \sim \hbar$, which translates into $g_\xi \sim 1$.

According to the analysis of Golubev and Zaikin, (9.21) is valid only if the wire is sufficiently short, that is, if the capacitive effects are negligible. The formula for this condition of shortness is

$$2 N_0 \xi A_{cs} \frac{e^2}{2 C_w} \gg 1 \tag{9.22}$$

where C_w is the total electric capacitance of the wire. The wire is usually suspended in vacuum during measurements. For a rough estimation of the capacitance of the wire, one can use the expression for the capacitance of a cylinder, which, in SI units, is $C_w = 2\pi\epsilon_0 L / \ln(r_g/r_w) \approx 10^{-18}$ F. Here, r_w is the radius of the wire and r_g is the typical distance to the ground. Consider an example: $r_d = 5$ nm, $r_g = 1$ μm, and $L = 100$ nm. Note that the choice of r_g is not critical because even if we put $r_g = 1$ cm, the estimated capacitance would only be three times lower than for the accepted generic value $r_g = 1$ μm. The charging energy is $e^2/2C_w \approx 10^{-20}$ J. The order of magnitude for the phase slip core volume is $\xi A_{cs} \approx 4 \times 10^{-25}$ m^3. Finally, using $v_F \approx 10^6$ m/s for MoGe, we can estimate the density of states as $N_0 = m^2 v_F / 2\pi^2 \hbar^3 \approx 4 \times 10^{46}$ J^{-1} m^{-3}. Therefore, the quantity that determines the shortness of the wire, according to the formula above, is $\xi A_{cs} N_0 (e^2/2C_w) = 320 \gg 1$. Thus, the condition for the sufficient shortness, set by (9.22), is satisfied. Even if the wire would be ten microns long, (9.22) would still be satisfied. Thus, the MoGe nanowires produced by molecular templating as well as many other types of nanowires discussed in literature are sufficiently short for GZ theory to apply.

The expression for S_{GZ} given in (9.21) takes into account the energy required to produce a normal core of a QPS as well as the contribution to the quantum action of the dissipative normal currents occurring in the core. Thus, the theory seems to be in compliance with the Caldeira–Leggett principle of dissipative suppression of tunneling. Yet, there is one important limitation of the GZ theory: The theory does not take into account the shunting effect of the electrodes, which usually have impedance of the order of the vacuum impedance $Z_0 = \mu_0 c \approx 377$ Ω. Thus, the results of the GZ theory are only applicable to sufficiently long wires, such that the shunting effects of the electrodes, which are typically connected to the ends of the wire for the purpose of transport measurements, can be neglected.

Qualitatively speaking, one can speculate that the GZ theory can only be possibly valid in two cases: (1) If the electrodes are highly resistive themselves and compact, and thus they act as lumped elements. (2) If the resistance of the wires is $R_n > R_q$. Since the impedance of the wire, as seen by individual phase slips, is connected in a series with the effective shunt provided by the electrodes, the total shunt may be large and thus negligible. The latter scenario is discussed in detail in Section 10.1.

In the opposite case, $R_n < R_q$, the damping might cause a much lower rate of QPS due to the localization of the phase variable ϕ, as predicted by Chakravarty [82] and Schmid [83, 84].

The case of an extra short wire that is strongly influenced by the electrodes was considered in [46, 47, 144]. Generally, in all situations, the role of the electrodes is to add damping and suppress the QPS rate. Thus, if the electrodes have a strong impact, the GZ theory might underestimate the QPS rate, for the obvious reason that it is not designed to take into account the effect of the electrodes.

9.4 Khlebnikov Theory

One can consider the opposite limit in which the action of the QPS core is negligible and the leading contribution to the action is due to gapless plasma modes in the electrodes. The electrodes are assumed to have zero QPS and/or TAPS rate but can have phase gradients. Such a case was analyzed by Khlebnikov [47]. His theory computes the rate of thermally assisted tunneling, as opposed to over-barrier phase slips, which are described by various TAPS models. Yet, his theory predicts an exponential decline of the resistance with cooling, which is usually considered to be a TAPS signature.

In the Khlebnikov theory (K-theory), the core of the phase slip is considered point-like, while the quantum action is computed for the long-range disturbance that a phase slip produces. A characteristic length scale can be defined as $l_K = \hbar c_{MS}/k_B T$. A nanowire is considered short if $L \ll l_K$. If the Mooij–Schön velocity [155] is estimated as $c_{MS} \sim 10^5$ m/s and the temperature is 1 K, for example, then the length scale is $l_K \sim 700$ nm. The reason to consider short wires is that many existing QPS theories predict that the resistance should change with temperature as a power law [46, 168]. Some experiments confirm the expected power law [185, 186], yet others [130] showed that the dependence of the resistance on temperature is like the Arrhenius exponential law, not a power law. It can intuitively be expected that for the short wires ($L \ll l_K$), the effect of the electrodes would be to stabilize superconductivity and make phase slips less frequent.

The K-theory is formulated in terms of the 2D condensate "stiffness" K_s, which has dimensions of energy. To define it, suppose E_{grad} is the energy related to phase gradients present in a thin superconducting film, that is, in 2D geometry. Then, we can write $E_{grad} = (1/2) \int K_s (\nabla \phi(x, y))^2 dx dy$, where $dx dy$ is a surface infinitesimal element of the considered thin film. This definition leads, according to the GL theory, to $K_s = \hbar^2 n_s/2m$, where m is the mass of one electron and n_s is the density of pairs of electrons participating in the BCS quantum state. In the dirty limit (i.e., if $l_e \ll \xi$, which is the case for MoGe wires), the 2D superfluid stiffness is [157]

$$K_s = \frac{\pi \hbar \Delta}{4e^2 \rho_{2D}} \tanh \frac{\Delta}{2k_B T} = \frac{R_q \Delta}{2\rho_{2D}} \tanh \frac{\Delta}{2k_B T} = \frac{\Delta R_q L}{2 R_N w} \tanh \frac{\Delta}{2k_B T} \quad (9.23)$$

where $\rho_{2D} = \rho/d$ is the 2D sheet resistance and d is the superconducting film thickness. For example, for a film of thickness $d = 5$ nm and the generic resistivity of the optimized MoGe $\rho = 200\ \mu\Omega$ cm, one estimates $\rho_{2D} = 400\ \Omega$. Note that according to this notation of the 2D sheet resistance, the normal resistance of the wire can be expressed as $R_N = L\rho_{2D}/w$, where d is the wire thickness and w is the wire width.

In terms of K_s, the K-theory predicts that the nanowire ohmic resistance as a function of temperature is

$$R_{\text{Khl}} \approx R_n \exp\left(-\frac{\pi^2 w K_s}{2 L k_B T}\right)$$

The prefactor is chosen here to have $R \to R_n$ of $T = T_c$ since at $T = T_c$, the stiffness is $K_s = 0$.

Using the expression for K_s of (9.23), the resistance can be written as

$$R_{\text{Khl}} \approx R_n \exp\left(-\frac{\pi^2 R_q \Delta}{4 R_N k_B T} \tanh \frac{\Delta}{2 k_B T}\right)$$

In the limit of zero temperature, the tanh term becomes unity. Then, the QPS action, that is, the term under the exponent becomes similar to GZ action, except that the constant factor L/ξ is now replaced with the divergent term Δ/T. This factor makes a key difference since $R_{\text{GZ}} \to$ const while $R_{\text{Khl}} \to 0$ as $T \to 0$. (The term "resistance" is always used in the sense of zero bias resistance.) Thus, the K-theory predicts a superconducting wire while the GZ theory of QPS predicts a resistive or normal wire. Such a difference is reasonable since the GZ theory is intended for the cases when the electrodes are very resistive or absent, so their effect is not noticeable. The K-theory, on the other hand, is developed to describe experimental situations when the wire is coupled strongly to superconducting electrodes. The electrodes provide a stabilizing effect that suppresses quantum fluctuations of the order parameter in the wire and convert a resistive GZ wire into a superconducting one. This stabilizing effect is present, provided that the electrodes are hard superconductors in which fluctuations are negligible. Experimentally, this is usually true.

The exponential suppression of the wire resistance predicted by the K-theory is caused by two factors: one is the population of the initial state (the entry point of tunneling) and the other is the tunneling exponential. The first factor is more important, that is, it gives exponentially stronger suppression of the phase-slip tunneling at low temperatures. The population of the initial state has, as expected, the form $\exp(-E_i/k_B T)$, where E_i is the energy of the initial state from which the system prefers to tunnel. The initial energy E_i is the energy arising from the phase gradients; it is accumulated over the entire wire, including the regions outside the phase slip core. The boundary conditions used in the calculation are that the phase at the ends of the wire does not fluctuate during tunneling. In other words, the ends are assumed to be attached to macroscopic superconductors. The conclusion of the K-theory is that all wires having strongly superconducting electrodes attached to them are themselves superconducting at sufficiently low temperature. Yet, the wires might appear resistive if the temperature is not sufficiently low. The theory was further developed by including interactions among the phase slips. Such a theory of [48], considers the same problem, but in terms of the variable that is dual to the phase, that is, the electric dipole moment of the wire. The electrical dipole moments in superconducting nanowires have been analyzed in detail in [49].

9.5
Spheres of Influence of QPS and TAPS Regimes

Little's phase slips, as any barrier-crossing event, can occur either by thermal activation or by tunneling. The thermally activated phase slips (TAPS) regime is usually observed in two cases. First, TAPS is observed at sufficiently high temperatures, that is, when the thermal fluctuations are strong and much more likely to generate an LPS than quantum fluctuations. Second, if the normal resistance of the wire is less than the quantum resistance, that is, if $R_n < R_q = h/4e^2$. The normal resistance defines the critical point of the Schmid–Bulgadaev transition, which can appear as superconductor-insulator or as superconductor-normal transition. In both cases, the QPS is suppressed completely for low-resistance wires having $R_n < R_q$, and the QPS is allowed for highly resistive wires having $R_n > R_q$. Thus, the wires having $R_n < R_q$ are in the superconducting phase and they exhibit TAPS behavior as temperature is increased. The wires having $R_n > R_q$ are either normal or insulating, depending on their diameter and the corresponding QPS fugacity.

There is a quantum phase transition between superconducting (S) and nonsuperconducting (NS) wires, and there is, probably, only a crossover between normal and insulating wires. Thus, the term SIT (superconductor-to-insulator transition) could be justifiably replaced with SRT (superconductor-resistor transition). The meaning of the SRT is that the wires change in the limit of zero temperature and zero-bias current from a true superconducting regime, having zero QPS rate, to a resistive regime, having a nonzero QPS rate. The control parameter which causes such a change is the normal-state resistance of the wire. The resistive regime can be highly resistive, characterized by a high rate of QPS, sufficient to destroy all measurable traces of superconductivity. It is usually called insulating. The crossover from the normal to the insulating behavior should be controlled by the wire diameter. Such a crossover was probably observed in the series of samples reported in [108].

One efficient method to identify TAPS and QPS is to measure the dispersion, σ, of the switching current. The standard behavior of σ versus temperature was previously observed by Jackel et al. [80] and Voss and Webb [81] in experiments on Josephson junctions. It consists of a decrease of the standard deviation according to the Kurkijärvi power law $\sigma \propto T^{2/3}$, followed, as the temperature is reduced further, by a saturation, attributed to QPS. Such standard behavior of σ is also found on nanowires if the critical current is high enough. An example is shown in Figure 9.5. The nanowires having relatively higher critical currents (A–D), in which the energy of zero-point fluctuations is expected to be larger, exhibit the standard

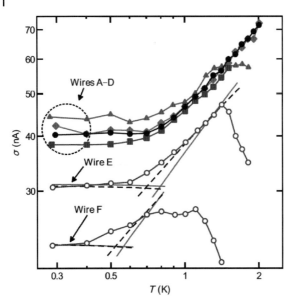

Figure 9.5 Standard deviation of the switching current for six different nanowires [115, 116]. The nanowires having relatively higher critical currents (A–D) exhibit the standard behavior, that is, a decrease according to the Kurkijärvi law, $\sigma \propto T^{2/3}$ (dashed lines), at higher temperatures and a saturation at low temperatures. The crossover or quantum temperature, T_q, is typically between 0.5 and 1 K. The samples E and F, which have relatively lower critical temperatures, showed a multiphase-slip switching process and the corresponding decline of σ with heating at the higher temperature end. The solid lines show the dependence $\sigma \propto T^{4/5}$, for comparison. The axes are in the log–log format.

behavior, that is, a decrease according to the Kurkijärvi law, $\sigma \propto T^{2/3}$, at higher temperatures and a saturation at lower temperatures. The crossover, T_q, is typically between 0.5 and 1 K. The samples E and F, which have relatively lower critical temperatures, showed a multiphase-slip switching process and the corresponding decline of σ with heating at the higher temperature end.

The multiphase-slip switching represents the situation when detected switching events are caused by a near coincidence of two or more phase slips which are needed to overheat the wire and switch it to the normal state [131]. If such a coincidence does not happen, then the phase slip remains undetected since it cannot, on its own, switch the wire. As the temperature increases, a coincidence of a larger number of phase slips is needed to produce a switch. Thus, more phase slips remain undetected. The switching becomes a collective effect. In thermodynamics, fluctuations related to collective phenomena, involving many particles or many degrees of freedom, exhibit lesser fluctuations compared to single-particle phenomena. This fact provides a qualitative hint of why σ becomes smaller as the number of phase slips needed to cause a switch increases.

The QPS becomes dominant at a crossover temperature, called T^*. For JJs, the crossover from thermal to quantum fluctuations occurs at $T^* \approx \hbar\omega_p/2\pi$ [1]. In the case of nanowires, the crossover temperature T^* probably depends on the strength of the bias current. In high bias current measurements, it was observed that T^* scales linearly with the critical temperature of the wire. The experimentally found relation, observed on short MoGe nanowires, is $T^* \approx 0.16 T_c$ (see [116]).

9.6 Kurkijärvi–Garg Model

The Kurkijärvi and Garg (KG) model provides a unifying description of the effects of thermal and quantum escape from a metastable (local) energy minimum [179, 181]. Applications of the KG model to superconducting nanowires were advanced in [115, 116], providing the basis for the present section. The KG model is applicable at $I_c - I \ll I_c$ and describes the cases when the bias current is slowly ramped from zero to some high value, I_{max}, such that $I_{max} > I_c$. In such experiments, the wire always switches from the S-state to N-state at some current, which is called the switching current, I_{sw}. (If such switch does not occur for any reason then KG model does not apply.) Since some fluctuations, either thermal or quantum, or both, are always present, a premature switching is commonly the case. Thus, $I_{sw} < I_c$. Here, as usual, I_c is the fluctuation-free critical current of the junction or a depairing current of the nanowire.

In a great variety of situations involving TAPS or QPS, the escape rate can be written as

$$\Gamma = \Omega \exp\left[-\frac{U_c}{k_B T_{esc}}\left(1 - \frac{I}{I_c}\right)^b\right] \tag{9.24}$$

where I and I_c is the bias and critical currents respectively, $\Omega = \Omega_0(1 - I/I_c)^a$ is the attempt frequency, which depends on the current as a power law having the exponent a. The energy barrier for phase slips is $U_c(T)(1 - I/I_c)^b$. It is a power law of power b. The value $U_c = U_c(T)$ is the temperature-dependent barrier extrapolated to zero current. Such a dependence of the barrier is found, for example, in SIS junctions, as is reviewed in Section 3, and, in particular, in Sections 3.2.3 and 3.3.1.

The parameter T_{esc} is the effective escape temperature. In the case of thermal escape, $T_{esc} = T$, according to the Arrhenius activation law. In the QPS regime, T_{esc} represents a new intrinsic scale, T_q, which is set by zero-point energy of the system.

The main focus of the analysis pioneered by Kurkijärvi, and later extended by Garg, is to find the width of the distribution of the switching current, σ. The switching current is assumed to have a one-to-one correspondence with the phase slip events. In other words, each phase slip is presumed to cause a switching in the nanowire, from the superconducting to the normal state. The switching is stochastic for the very reason that a phase slip can occur at a random moment. The width, σ, is also called the standard deviation and is formally defined in (8.43). It is the square root of the mean square deviation of the switching current from the mean value of the switching current, $\langle I_{sw} \rangle$.

The KG analysis, which we do not reproduce here, leads to the following prediction for the width of the distribution

$$\sigma = \frac{\pi I_c}{\sqrt{6b\kappa}} \left(\frac{U_c}{k_B T_{esc}} \right)^{-1/b} \kappa^{1/b} \qquad (9.25)$$

Here, $U_c = U_c(T)$ is the temperature-dependent barrier for a phase slip, $\kappa = \ln(\Omega t_o)$, and $t_o = \sigma/v_I$ is the time spent sweeping through the width of the switching distribution. Note that the product Ωt_o equals the number of attempts to overcome the barrier, which the system makes as it moves through the interval σ, in which the switching typically occurs. The sweep speed is defined as usual as $v_I = dI/dt$.

Equation (9.25) can be solved approximately by substituting some approximate value for Ω and σ into the logarithm. For example, κ can be found in the first approximation as $\kappa = \ln(\Omega_0 t_o)$. If, for instance, $\Omega_0 = 100\,\text{GHz}$, $\Omega = 30\,\text{GHz}$, and $t_o = 0.01\,\text{s}$ then $\ln(\Omega_0 t_o) = 20.7$ and $\ln(\Omega t_o) = 19.5$, that is, reducing the value of the attempt frequency by factor 3 leads to a reduction of κ only by $\sim 6\%$. In many practical cases, such uncertainty is acceptable, especially because the main effect of interest is the temperature dependence of σ, which originates from the fact that at high temperatures $T_{esc} = T$, while in the quantum regimes, $T_{esc} = T_q \approx \text{const}$. The barrier U_c may be temperature-independent (e.g., SIS junctions at low temperatures) or may be temperature-dependent (e.g., SNS graphene junctions [195] or nanowires at temperatures not too low).

Equation (9.25) is valid both for TAPS and QPS, but separately. Fortunately the temperature interval in which both TAPS and QPS have comparable rates is very narrow due to the exponential dependence of the TAPS rate on temperature. If QPS is the dominant type of fluctuations, then T_{esc} has to be replaced by some characteristic quantum temperature T_q which depends on the strength of zero-point fluctuations in the device considered. For example, for an SIS Josephson junction having negligible damping, it is well-known (Section 3.3.1) that $b = 3/2$, $k_B T_q = (5/36)\hbar\omega_p$, and $U_c = 2\sqrt{2}\hbar I_c/3e$.

If QPS is dominant, the tunneling rate is defined by the action. The KG model works only if the action near the critical current is a power law, that is, $S(I) = S_0(1 - I/I_c)^b$. Then, the switching rate is $\Gamma = \Omega \exp[-S_0(1 - I/I_c)^b/\hbar]$. The dispersion then is $\sigma = (\pi I_c/\sqrt{6b\kappa})(S_0/\hbar)^{-1/b}\kappa^{1/b}$. In what follows, we will assume that the action can be presented as the barrier height divided by some characteristic quantum temperature, that is, $S_0/\hbar = U_c/k_B T_q$, following the analogy with SIS junctions, in which case $k_B T_q = (5/36)\hbar\omega_p$ (Section 3.3.1). Take the simplest possible example and assume $b \approx 1$. Then, $\sigma/I_c \approx 1.3(S_0/\hbar)^{-1}$. Thus, the dispersion of the switching current can be used to determine the normalized action of quantum phase slips. A strong quantum behavior is such that $S_0/\hbar \sim 1$.

If TAPS dominates the escape process, one can use the KG general model, outlined above, to get the Kurkijärvi power law for the TAPS induced dispersion of the switching current. Indeed, in the case of TAPS, the effective temperature is just the bath temperature, $T_{esc} = T$. Assume that a junction is considered such that $b = 3/2$. If the temperature is sufficiently low, then I_c can be assumed constant because the energy gap in typical superconductors does not change much below $T_c/2$. Therefore, U_c is also temperature-independent, at least approximately, at temperatures $T < T_c/2$, because $U_c \sim \hbar I_c/2e$. Thus, according to the general (9.25), we get $\sigma \propto (U_c/k_B T)^{-2/3} \propto T^{2/3}$, which is the Kurkijärvi power law of 2/3. If, on the other hand, the critical current is temperature dependent, then one can use the general rule for the barrier, namely, that $U_c(T) \propto I_c(T)$, and get $\sigma \propto I_c(I_c/T)^{-2/3} \propto T^{2/3} I_c^{1/3}$, which is the general Kurkijärvi law. This law was confirmed experimentally on tunable graphene proximity junctions [195]. The derivations above are based on reasonable assumptions that the temperature dependence of the logarithm κ can be neglected.

The KG analysis leads to a prediction about the mean value of the switching current, that is,

$$\langle I_{sw} \rangle = I_c \left[1 - \left(\frac{U_c}{k_B T_{esc}} \right)^{-1/b} \kappa^{1/b} \right] \tag{9.26}$$

If it is combined with the above expression for σ, then one can obtain a formula that connects the dispersion and the mean value of the switching current, namely,

$$\sigma = \frac{\pi I_c}{\sqrt{6b\kappa}} \left[1 - \frac{\langle I_{sw} \rangle}{I_c} \right] \quad \text{or} \quad I_c = \langle I_{sw} \rangle + \frac{\sqrt{6b\kappa}}{\pi} \sigma \tag{9.27}$$

An experimental verification of (9.24) can be done as follows. First, statistical distributions of the switching current are measured at various temperatures [115, 116]. Examples of such data are shown in Figure 9.6a. Then, the discrete KFD transformation, (8.19), is applied to the experimentally obtained distributions in order to obtain corresponding switching rates. Examples of such switching rates, plotted versus the bias current, are shown in Figure 9.6b. Then, (9.24) is used to generate the fits which are shown as continuous curves in Figure 9.6b. Obviously, the agreement is quite good. Then, one can apply the inverse KFD transformation (8.24) and obtain the fitting curves to the distribution functions. Such are the continuous curves in Figure 9.6a.

In order to generate the theory fits of Figure 9.6a, the barrier is chosen as $U_c = D_x \sqrt{6}\hbar I_c/2e$, where $D_x = 1.095$ is some adjustable parameter which is needed because the wire is not usually an ideal cylinder and not infinite long. The critical or depairing current I_c is obtained from (9.27), which is possible since σ and $\langle I_{sw} \rangle$ are known from the experiment and κ can also be estimated. One of the fitting parameters needed to generate the fits of Figure 9.6a is the attempt frequency Ω. It can be approximated as a constant since the I_{sw} statistical distributions are usually much narrower than the critical current and shifted considerably from the critical current [179]. Therefore, the attempt frequency cannot change significantly within the interval of bias currents in which switching events are observed. The corresponding best-fit values for Ω are shown in Figure 9.7b. The values are close to the model estimate, represented by the horizontal green line.

In order to estimate Ω theoretically, the simplest approach is to use the analogy with Josephson junctions, for which the attempt frequency equals the plasma

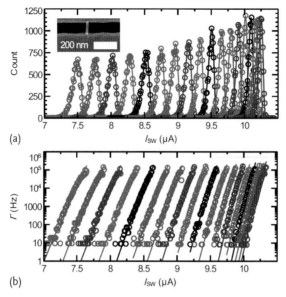

Figure 9.6 Measurements and analysis of switching current fluctuations [115, 116] in a MoGe nanowire. (a) Examples of switching current distributions (circles) measured at various temperatures ranging from 2 K (the left-most curve) to 0.3 K (the right-most distribution curve); the step is 0.1 K. The fits, shown as solid curves of the same color, are generated using the inverse KFD transformation (8.24) applied to the switching rate fits shown in (b). Insert: SEM image of a nanowire which was annealed in vacuum using high current pulses [114] and thus was crystallized (see the discussion about such Joule-heat-treated wires in the text). (b) Switching rates obtained from the distribution plotted in (a) using the discrete KFD transformation (8.19). The experimental switching rate is shown by circles. The solid curves of the same color are the fits to the KD model, namely, to (9.24). The exponent is $b = 3/2$ for all curves. The fits are done using two fitting parameters, namely, the attempt frequency Ω and the escape temperature T_{esc}. Both of these parameters are analyzed in the next figure. Please find a color version of this figure in the color plates.

frequency, which is $\Omega = 1/2\pi\sqrt{L_k C}$, where C is the capacitance and L_k is the kinetic inductance. For a nanowire connected to 10 μm wide electrodes, the capacitance is of the order of 10 fF, according to [136]. The kinetic inductance is defined as $L_k = (2/3\sqrt{3})(L/\xi)(\hbar/2eI_c)$. If $\xi \approx 5$ nm and $I_c \approx 11$ μA, which is the case for sample A, taken in this section as an example [115, 116], then $L_k \sim 0.2$ nH. Thus, the expected attempt frequency roughly is $\Omega \sim 100$ GHz.

Another essential fitting parameter in (9.24) is the effective escape temperature, T_{esc}. The corresponding best-fit values of T_{esc} are shown in Figure 9.7a. At higher temperatures, $T_{esc} = T$, as is expected for TAPS. At low temperatures, $T_{esc} = $ const. By our definition, the saturation temperature is defined as quantum fluctuations temperature T_q. In other words, at low temperatures, $T_{esc} = T_q$. For the sample considered, $T_q \approx 0.8$ K. The graph allows us to determine the crossover temperature T^*. It is the temperature at which the rates of the TAPS and the QPS are

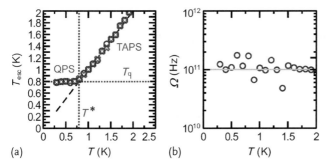

Figure 9.7 Plots of the fitting parameters used to generate fits in Figure 9.6b (see [115, 116]). (a) The fitting parameter T_{esc}, which defines the escape rate in (9.24), is plotted versus temperature (blue circles). The crossover temperature is marked T^* and the quantum fluctuations temperature is T_q. They are approximately equal, as expected from the KG model. (b) The attempt frequency Ω, which is another adjustable parameter in (9.24), is plotted as a function of temperature. The horizontal green line represents a rough theoretical estimation which is explained in the text. Please find a color version of this figure in the color plates.

equal. The example considered gives $T^* \approx 0.8\,\text{K}$. Thus, these two temperature scales coincide, $T^* = T_q$.

Such a conclusion is in agreement with the phase slip rate formula (9.24). Indeed, if we assume the same attempt frequency for TAPS and QPS, then the crossover from one type of fluctuation to another should happen at $T = T^*$, defined as $\Gamma_{TAPS}(T^*) = \Gamma_{QPS}(T^*)$. Therefore,

$$\Omega \exp\left[-\frac{U_c}{k_B T^*}\left(1-\frac{I}{I_c}\right)^b\right] = \Omega \exp\left[-\frac{U_c}{k_B T_q}\left(1-\frac{I}{I_c}\right)^b\right]$$

Such equality is achieved at $T = T^* \approx T_q$, where T_q is defined by the device parameters and its zero-point fluctuations.

Another way to estimate T^* is to use (9.26) and require that at $T = T^*$, the mean value of the switching current computed for QPS is equal to $\langle I_{sw}\rangle$ computed for TAPS. In the QPS case, we are supposed to put $T_{esc} = T_q$ and in the TAPS case, $T_{esc} = T$, where T is the temperature of the sample. The condition for the crossover temperature is then

$$I_c\left[1-\left(\frac{U_c}{k_B T_q}\right)^{-1/b}\kappa^{1/b}\right] = I_c\left[1-\left(\frac{U_c}{k_B T^*}\right)^{-1/b}\kappa^{1/b}\right]$$

From this, it follows again that, according to the KG model, $T^* \approx T_q$.

To complete the test of the KG model, we need to check whether σ and I_{sw} change with the sample temperature T according to the model. The results of such tests are presented in Figure 9.8. The fits are shown by dashed curves ($b = 3/2$) and by the solid curves ($b = 5/4$).

Note that the sample C is high-voltage-pulsed and crystallized. A typical TEM image of the wire after such high-current crystallization is shown in Figure 9.6(in-

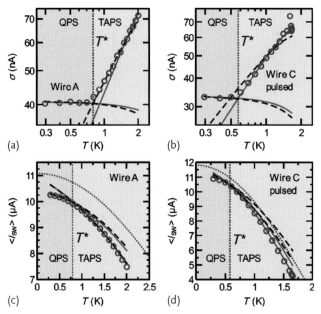

Figure 9.8 Mean switching current (c, d) and its dispersion (a, b) are plotted versus temperature [115, 116]. The plots (a) and (b) are in the log–log format. (a) and (c) – Sample A, unpulsed. (b) and (d) – Sample C, voltage-pulsed and crystallized before I_{sw} and σ were measured. In panels (a) and (b), the fits are generated by (9.25) in which the temperature dependence of the critical current, I_c, is computed according to the Bardeen formula. The fits shown by dashed black curves correspond to $b = 3/2$, while the red line fits represent the case $b = 5/4$. The two almost horizontal curves (solid and dashed) in (a) and (b), fitting well the low-temperature part, correspond to the QPS-dominated regime, meaning $T_{esc} = T_q$. They are computed assuming $T_q = 0.8$ K for sample A, and $T_q = 0.6$ K for sample C. The two other curves (solid and dashed), which fit well the high-temperature part of the data, represent TAPS according to (9.25), in which $T_{esc} = T$. In (c) unpulsed and (d) pulsed, the mean switching current is plotted, having the fits generated by (9.26), according to the same rules as in (a) and (b). The crossover temperature T^* is shown by the vertical dotted lines. The T^* is defined as the temperature below which $\sigma = $ const. The green dotted curves are $I_c(T)$ theoretical curves generated by the Bardeen formula (Section 11). The most important observation here is that σ does saturate below T^*, while $\langle I_{sw} \rangle$ keeps growing with cooling below T^*. This is exactly what the KG model predicts, judging by the fits. Yet, this would not be the case if the saturation of sigma would be due to noise-generated heating of the electrons in the nanowire. Please find a color version of this figure in the color plates.

sert). The experiments show that the crystallization does not change the behavior of σ versus T qualitatively. Thus, weak links cannot be responsible for the saturation of σ observed at low T. Such a conclusion can be reached on the following grounds: If some sort of weak links, and not MQT in a homogeneous wire, is responsible for the observed saturation of the σ, then the crystallized wires would show a qualitatively different behavior of σ. This is not observed. The crystallized MoGe wires (e.g., wire C-pulsed) exhibit the same qualitative trends in the dependence of σ on T. Thus, weak links cannot be the origin of the observed QPS saturation behavior, generally speaking. The fits with $b = 3/2$ are in good agreement with the data corresponding to amorphous wires, while $b = 5/4$ is a better match for the crystallized (pulsed) wires.

9.7
Theorem: Inverse Relationship between Dispersion and the Slope of the Switching Rate Curve

In this section, we derive a relation $\sigma \sim 1/z$ where $z \equiv d(\ln \Gamma)/dI$. The derivation is done within the KG model. The derivation goes as follows. The rate of phase slips is $\Gamma = \Omega \exp[-(U_c/k_B T_{esc})(1 - I/I_c)^b]$. Take the logarithm and get $\ln \Gamma = \ln \Omega - (U_c/k_B T_{esc})(1 - I/I_c)^b$. Define $z = d(\ln \Gamma)/dI = d(\ln \Omega)/dI - (U_c/k_B T_{esc})b(1 - I/I_c)^{b-1}(-1/I_c)$.

Assume that $d(\ln \Omega)/dI = 0$, meaning that the dependence of the attempt frequency on the bias current is neglected. It can be shown, using experimental data, that $|d(\ln \Omega)/dI| \ll |(U_c/k_B T_{esc})b(1 - I/I_c)^{b-1}(-1/I_c)|$. Thus, we can write $z = -(U_c/k_B T_{esc})b(1 - I/I_c)^{b-1}(-1/I_c)$. Now, remember that for thin wires, $U_c = \sqrt{6}\hbar I_c/2e$, thus $z = -(\sqrt{6}\hbar I_c/2ek_B T_{esc})b(1 - I/I_c)^{b-1}(-1/I_c)$. After dividing the nominator and the denominator by I_c, we get $z = (\sqrt{6}\hbar/2ek_B T_{esc})b(1 - I/I_c)^{b-1}$. It is convenient to introduce a normalized current as $i \equiv (1 - I/I_c)$. Then, $z = (\sqrt{6}\hbar b/2ek_B T_{esc})i^{b-1}$. For the discussion which follows, we need the quantity $i^{1-b} = 1/i^{b-1}$. From the above, it follows that $(1/z) = (2ek_B T_{esc}/\sqrt{6}b\hbar)i^{1-b}$. Therefore,

$$i^{1-b} = \frac{1}{z} \frac{\sqrt{6}b\hbar}{2ek_B T_{esc}} \tag{9.28}$$

Now, let us examine (9.26), which can be written as $\langle I_{sw} \rangle = I_c - DI_c$, where the coefficient D is $D = (\kappa k_B T_{esc}/U_c)^{1/b}$ or $D^b = \kappa k_B T_{esc}/U_c$. It is known [19] that the barrier for phase slips in a thin wire is $U_c = \sqrt{6}\hbar I_c/2e$. Therefore,

$$D^{-b} = \frac{\sqrt{6}\hbar I_c}{2e\kappa k_B T_{esc}} \tag{9.29}$$

This can be also presented as

$$\frac{\kappa}{I_c} = \frac{\sqrt{6}\hbar D^b}{2ek_B T_{esc}} \tag{9.30}$$

Remember that $\kappa = \ln(\Omega t_\sigma)$.

Equation (9.26) also leads to $I_c - \langle I_{sw} \rangle = DI_c$. Using the normalized current i, the coefficient D can be transformed to

$$D = \frac{I_c - \langle I_{sw} \rangle}{I_c} = i \tag{9.31}$$

Here, we assume $I_{sw} = I$. This is reasonable since at the moment of a switching event, it is indeed true that $I = I_{sw}$. On the other hand, if σ is small, one can also take $\langle I_{sw} \rangle = I_{sw}$, which is done here.

Thus, using $D = i$, (9.29) becomes

$$i^{-b} = \frac{\sqrt{6}\hbar I_c}{2ek_B T_{esc}} K \qquad (9.32)$$

The next step in the derivation is to use the expression for the dispersion, (9.25). It can be written as $\sigma = (\pi/\sqrt{6}b)(I_c/\kappa)D$. This can be further transformed using (9.30) in $\sigma = (\pi/6b)(2ek_B T_{esc}/\hbar)D^{1-b}$. Since we have seen (9.31) that $D = i$, we can also write $\sigma = (\pi/6b)(2eT_{esc}/\hbar)i^{1-b}$. This can be further transformed, using (9.28), into

$$\sigma = \frac{\pi}{6b}\frac{2eT_{esc}}{\hbar}\frac{\sqrt{6}b\hbar}{2eT_{esc}}\frac{1}{z}$$

Now, it is a matter of simple algebra to get the final result that, at $I = I_{sw}$,

$$\sigma = \frac{\pi}{\sqrt{6}}\frac{1}{z} = 1.28 \left[\frac{d(\ln \Gamma)}{dI}\right]^{-1} \qquad (9.33)$$

Of course, the switching rate Γ depends on the bias current I. For the purpose of applying this theorem, it is assumed that the average slope should be taken near the point $I = \langle I_{sw} \rangle$. Such an assumption is valid judging by the typical experimental results for the switching rate presented in Figure 9.6b. To confirm the theorem derived in this section, we plot the dispersion versus $1/z$ in Figure 9.9. Obviously, the agreement with the data is excellent. This graph confirms the validity of the KG model in the quantum regime.

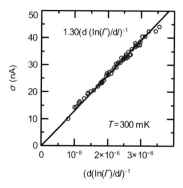

Figure 9.9 Standard deviation versus the inverse $d\ln\Gamma/dI$ function at base temperature $T = 0.3$ K, which is already deep into the quantum regime [115, 116]. The data points are obtained by making linear fits to the plots of $\ln\Gamma$ versus I, as in Figure 9.6b. Here, $\Gamma(I)$ is, as usual, the phase slip rate as a function of the bias current. The data points shown here represent many samples; some are as-fabricated and amorphous and some are crystallized by the high-current pulsing method [114]. The slope of the best linear fit is 1.3, which agrees well with the theoretically expected slope of 1.28.

10
Superconductor–Insulator Transition (SIT) in Thin and Short Wires

The information technology development strongly depends on miniaturization and integration of its electronic circuits. Superconducting nanowires can play a role in this miniaturization drive since they, in the ideal case, dissipate no heat. Thus, it is important to know *how thin* a nanowire made of a superconducting material can be made and still retain its superconducting qualifications, such as, the ability to carry an electric current under zero voltage applied. It is well-known, due to the theory of Abrikosov and Gor'kov, that by changing the composition of a superconducting material, for example, by adding magnetic atoms, one can weaken and even destroy superconductivity. Yet, it is less know that simply by changing the dimensions of nanowires, one can also modify their superconducting characteristics and even render them nonsuperconducting. Quite surprisingly, the existing experiments indicate that for thin and *short* wires, the total normal resistance has a more pronounced effect on superconductivity of the wire than its diameter [100, 104, 130, 136].

The destruction of superconductivity in superconducting nanowires poses a fundamental physics question: It remains to be fully understood which mechanisms are responsible for the destruction of superconductivity in nanometer-thin wires. Some existing evidence suggests that the SIT in nanowires can be related to the phenomenon of macroscopic quantum tunneling, which is one of the most interesting and puzzling topics of modern condensed matter physics, that is, the classical versus quantum world description. Here, we discuss some of the most recent advances in the understanding of this problem, although the final solution still appears missing. One sign demonstrating that the problem is not fully resolved yet is the fact that no standard textbook on superconductivity (among those known to the present author) includes a chapter on superconductor-insulator transition in thin wires or thin superconducting films. Therefore, we will try to fill this gap to the extent possible.

In this chapter, we focus on thin and short wires. The reason is that longer wires tend to show a crossover rather than a quantum phase transition, as their diameters are reduced [108]. Although there are experiments indicating that long superconducting wires can be driven into an insulating regime, though they either show no universal behavior [184] or qualitatively show the same behavior as in thin films, in which the resistance per square controls the quantum transition [189]. On the con-

Superconductivity in Nanowires, First Edition. Alexey Bezryadin.
© 2013 WILEY-VCH Verlag GmbH & Co. KGaA. Published 2013 by WILEY-VCH Verlag GmbH & Co. KGaA.

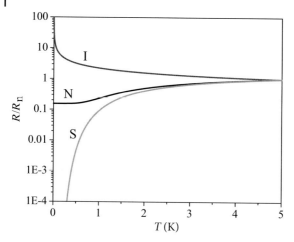

Figure 10.1 Suggested qualitative classification of nanowires. The classification includes truly superconducting wires, "S," insulating wires, "I," and normal wires "N." The S-wire is such that its resistance is only due to TAPS. Thus, as the temperature is lowered, TAPS freezes out, and the resistance rapidly approaches zero. The insulating wire is characterized by the resistance that continuously increases with cooling. The normal wires exhibit a saturation of the resistance. The resistance is understood as taken in the limit of zero bias current.

trary, short wires exhibit a universal behavior which is qualitatively distinct from thin films. Such behavior is the subject of this chapter.

It is convenient to classify the wires into three main classes: superconducting (S), insulating (I), and normal (N). Using the illustration of Figure 10.1, we can make this classification quantitative. Let us define the three classes as follows. (1) The "S-"wires, or "truly superconducting" wires, are such that $dR/dT > 0$ and $d^2(\ln R)/dT^2 < 0$ in the limit of zero temperatures. The resistance of such wires must approach zero as the temperature approaches zero.[1] The S-type wires exhibit a rapid decline of the resistance with cooling and their $\ln R$ vs. T curves appear concave in the log–linear plots (Figure 10.1). (2) Nanowires will be classified as insulating (I-type) if their resistance increases with cooling indefinitely ($dR/dT < 0$ as $T \to 0$) and if the $R(T)$ curves are convex (i.e., $d^2 R/dT^2 > 0$). The condition that $d^2 R/dT^2 > 0$ as $T \to 0$ is equivalent to the requirement that the $R(T)$ curve has a positive curvature at temperatures near zero, which means that the resistance increases with ever faster pace as the temperature is reduced.

(3) The resistance of normal wires, "N" (Figure 10.1), approaches a nonzero constant level with cooling, that is, $R(T) \to R(0) > 0$ as $T \to 0$. Since one cannot achieve zero temperature, the normal wires can be defined as such that either $dR/dT > 0$ and $d^2 R/dT^2 > 0$ (as is the case for the N curve of Figure 10.1) or, in addition, by the following two conditions: $dR/dT < 0$ and $d^2 R/dT^2 < 0$. The case

---

1) Note that due to the occurrence of Little's phase slips, no wire has zero resistance at a nonzero temperature. Note also that the discussed resistance always corresponds to the limit of zero bias current.

($dR/dT > 0$, $d^2R/dT^2 > 0$) is normal and strongly conducting, while the case ($dR/dT < 0$, $d^2R/dT^2 < 0$), which is not shown on the graph, would be a wire which is normal and weakly conducting.

Consider an example: The theory of QPS by Zaikin, Golubev, van Otterlo, and Zimányi (ZGOZ) [168] predicts that the nanowire resistance depends on temperature according to a power law, that is, as $R = C_r T^{\alpha_r}$, where the constant C_r and the temperature-independent power α_r depend on various characteristics of the wire. According to ZGOZ theory, the wires having $\alpha_r > 0$ are superconducting while the wires $\alpha_r < 0$ are insulating. Let us see if this is in agreement with our own classification defined above. For the power law, the derivative is $dR/dT = \alpha_r C_r T^{\alpha_r - 1}$ and the log of the resistance is $\ln R = \ln C_r + \alpha_r \ln T$. Thus, $d(\ln R)/dT = \alpha_r/T$ and therefore the second derivative, which, generally speaking, defines the curvature, $d^2(\ln R)/dT^2 = -\alpha_r/T^2 < 0$. Thus, it is clear that if $\alpha_r > 0$, then $dR/dT > 0$ and $d^2(\ln R)/dT^2 < 0$, so that the wire is superconducting according to our classifications, as well as ZGOZ classifications. If $\alpha_r < 0$, then one finds that $dR/dT < 0$. Also, in this case, the second derivative is $d^2R/dT^2 = C_r \alpha_r (\alpha_r - 1) T^{\alpha_r - 2} > 0$. Obviously, the resistance increases indefinitely with cooling if the power of the power law is negative. Thus, the wire is insulating according to our classification as well as ZGOZ.

By analyzing the illustrative $R(T)$ curves of Figure 10.1, one concludes that at any finite temperature, the difference between the three classes of wires, S, N, and I, is only quantitative if only their resistance is compared. Namely, all classes have a finite resistance, although some may be more resistive than others. Yet, at zero temperature, the three classes become qualitatively distinct. Consider a thought experiment, that is, imaging a setup of the sort shown in Figure 1.1, in which the wire links two macroscopic superconducting electrodes. Assume the temperature is zero and the initial condition is such that the capacitor formed by the electrodes, C_e, has a nonzero charge on it. With such a starting point, the device having an S-wire would undergo oscillations of the voltage and current in which the charging energy of the capacitor converts to the kinetic energy of the condensate in the wire and then converts back to the charging energy, periodically. Such oscillations would last indefinitely at zero temperature if the wire is a true superconductor. The period of such oscillations is $2\pi\sqrt{L_k C_e}$, where L_k is the kinetic inductance of the wire. If the wire is N-type, then the voltage would decay to zero exponentially over the characteristic time $R(0)C_e$, where $R(0)$ is the wire resistance at zero temperature. Finally, if the wire is insulating, then the voltage on the capacitor would remain unchanged. This thought experiment defines the three main classes of wires or weak links in general.

If the class of the wire is defined by some parameter, say the wire resistance or the wire diameter, then, at least in theory, one can consider another thought experiment in which such parameter is continuously varied and the wire experiences transitions from one class of behavior to another. Such transitions would happen at critical values of the control parameter and would be called quantum critical points. The existence of the critical points is only possible if there are classes or phases which have qualitatively different ground states. For example, if the BCS condensate has a nonzero amplitude in wires having diameters larger than some critical value d_c, while in wires thinner than this value, the condensate is completely destroyed, then, d_c is a critical point.

Since it is impossible to reach zero temperature, we can learn about such quantum critical points only indirectly, by studying the crossover from one type of behavior to another at finite temperatures when the control parameter is changed. The number of quantum phase transitions equals the number of distinct phase minus one. For example, if the wire can be in the S, N, and I states, then one expects to find two quantum transitions, namely, the superconductor-normal transition and the normal-to-insulator transition. In practice, most of the experiments on nanowires suggest that there are only two states, the true superconducting state and the insulating (or a resistive) state. Thus, there is only one quantum phase transition, namely, the SIT (or an SRT) [100, 130].

A conclusion about the presence of SIT can be reached by analyzing a large number of $R(T)$ curves for an ensemble of superconducting samples. If the ensemble is dense enough so that it contains samples characterized by a broad distribution of the control parameter, then one can try, by measuring various characteristics of the samples, to establish how many distinct classes of samples are present in the ensemble. If there is more than one class, then a quantum transition might be present.

In the case of nanowires, the control parameter is not easy to identify. A common expectation is that the diameter is the one. Such expectation is based on the understanding that the electron–electron Coulomb repulsion is inversely proportional to the electric capacitance of the wire, which is proportional to its diameter. That as the diameter is reduced, the Coulomb or charging energy should increase, eventually leading to destruction of superconductivity. On the other hand, Anderson localization of electrons can also have an effect on superconductivity. Such localization is known to become strong as the total normal resistance of the wire exceeds the quantum resistance. Thus, the total resistance could also be a control parameter. Below, we consider related experimental results.

Resistance versus temperature curves for a large ensemble of nanowires are shown in Figure 10.2. All of these wires are identical in composition and similar in dimensions. They are made using the same method of molecular templating. There are only two aspects in which the wires differ. First, they are made on trenches of variable widths, so they differ by length, varying in the range from 29 to 490 nm. Second, the samples differ by the nominal thickness of the deposited film, which varies in the range from 4 to 8.5 nm. It should be also noted that the width of wires produced by molecular templating is always roughly equal to the thickness of the deposited film. The variation of the geometric characteristics of the wires generates a significant variation of the normal resistance, in the range between 1.17 up to 32.46 kΩ.

The $R(T)$ measurements allow the wires to be sorted into two qualitatively distinct classes. Those whose resistance drops with cooling are placed on Figure 10.2a. Apparently, they are "true superconductors" or S-type, according to the classification proposed above. What that means is that they are expected to approach zero

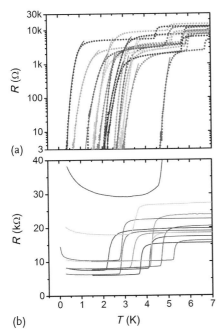

Figure 10.2 Resistance versus temperature curves for a large ensemble of nanowires [130]. All wires are made using the same fabrication process, namely, molecular templating. There are two geometric parameters in which the wires differ. First, the length is different and varies between 29 and 490 nm. Second, the nominal thickness of the wires is also different and varies from 5 to 10 nm. The width of the templating molecule changes between the samples also. The wires have been sorted into two qualitatively distinct classes. Those whose resistance drops with cooling are placed on the top panel (a), and those having the resistance increasing with cooling are collected at the bottom panel (b). The top-panel samples are S-type. All of the curves are in excellent agreement with the Little's fit $R(T) = R_N \exp(-\Delta F/k_B T)$ (solid curves). Each of the bottom-panel wires shows all the characteristics with the I-type ($dR/dT < 0$, $d^2 R/dT^2 > 0$). Normal wires are not observed among short (less than $\approx$ 500 nm) and homogeneous wires, at least in the experiments in which the wires are made of the superconducting amorphous alloy of $Mo_{79}Ge_{21}$. Please find a color version of this figure on the color plates.

resistance as the temperature approaches zero. This is evidenced by the fact that all of these samples exhibit an excellent agreement with the Little's fit $R(T) = R_N \exp(-\Delta F/k_B T)$ (6.6), shown by solid lines. The Little's fit represents an activation type of behavior. The resistance is due to TAPS, which "freeze out" according to the Arrhenius activation law as the temperature is reduced. The fact that the $R(T)$ curves follow the activation law $R \propto \exp(-\Delta F/k_B T)$ strongly suggests that the resistance is due to a thermal activation process (TAPS) and not due to the quantum tunneling phenomenon (QPS).

Note that our conclusion that the $R(T)$ curves of S-samples can be modeled with a TAPS theory was contested in at least two theoretical papers [144, 194]. These references imply that the entire $R(T)$ curve can be correctly modeled by only including quantum phase slips into the model, and by neglecting TAPS. On the contrary, the GZ theory of TAPS [171] gives support to the existence of TAPS at higher temperatures. Yet, the GZ theory is only valid near T_c, while the curves shown in Figure 10.2a are extended quite far from T_c. The ultimate argument which gives support to the accepted approach of fitting the $R(T)$ curves using a model based on TAPS (not the QPS) is that the Arrhenius activation law is valid at all temperatures, sufficiently far from the T_c. Thus, the employed Little's fit, which is essentially based on the Arrhenius law, should be valid at all temperatures, including $T \approx 0$, at least approximately. This is confirmed by the fits presented in Figure 10.2a.

The wires which show a resistance increase with cooling are placed in Figure 10.2b. They appear to be I-type, according to our classification. This is because their resistance increases with cooling and all these curves satisfy the positive curvature condition $d^2 R/dT^2 > 0$ (confirmed for all of them). Transport properties of these samples can be understood in terms of weak Coulomb blockade. Here, we emphasize the fact that the I-samples and S-samples are qualitatively different in that the former approach a high, potentially infinite resistance with cooling while the latter rapidly approach zero resistance. Thus, a reasonable expectation is that there is a quantum phase transition from the S-type to the I-type in the limit of zero temperature. Such a transition is called the superconductor-insulator transition (SIT).

Another strong confirmation to the presence of SIT comes from the high bias transport measurements displayed in Figure 10.3. The term "high bias" means that the bias current is sufficiently high so that the Ohm's law breaks down, that is, the $V(I)$ curve exhibits some nonlinearity. The nonlinearity usually begins as soon as the change of the kinetic energy of the electrons by the applied bias becomes comparable or larger than their thermal energy. An important conclusion can be drawn from the sign (positive or negative) of the nonlinearity, or the curvature, $V'' = d^2V/dI^2$. If $V'' > 0$, then one concludes that the current weakens superconductivity and causes the resistance to increase. Just the opposite happens in the I-samples. They have very low conductance if not perturbed. A stronger bias causes more perturbation, weakens the insulating state, and cause the resistance to drop. That is why $V'' < 0$ for the I-samples. In conclusion, the nonlinear measurements, illustrated in Figure 10.3, provide an independent way to establish the type of the wire: S or I. If the type is S, the bias current weakens superconductivity and makes the wire more resistive. If the type is I, then higher bias currents and correspondingly high voltages weaken the insulating property of the wire (possibly associated with some localization of the condensate) and cause its resistance to diminish.

Thus, in the limit of zero-bias and zero temperature, one expects to have two distinct type: superconducting and insulating. One arrives at such conclusions on the basis of the analysis of the $R(T)$ curves and, independently, on the basis of the analysis of the $V(I)$ curves. Let us summarize the logical steps of the analysis here again. By measuring $R(T)$ curves of a large number of samples, it is found that the function either drops rapidly with cooling or increases. If an increase is observed, then it continues as the temperature reduces further, and never reverses. Thus, by extrapolating to zero temperature, one concludes that the sample resistance would

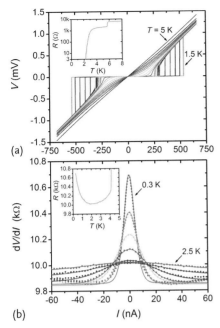

Figure 10.3 The figure shows typical nonlinear transport measurements for a representative (a) superconducting and a representative (b) insulating sample. (a) Voltage–current characteristics of a wire classified as S-type. The wire length is $L = 61$ nm and $T_c = 3.3$ K. The curves are taken at $T = 1.5, 1.6, 1.7, 1.8, 1.9, 2.0, 2.1, 2.2, 2.3, 2.4, 2.6, 2.8, 3.0, 3.25, 3.5, 4.0$, and 5.0 K. The key fact illustrated by these curves is that S-samples always become more resistive (i.e., less superconducting) as the bias current is increased. Mathematically speaking, as the current is increased, the $V(I)$ curves exhibit nonlinearity and the curvature, d^2V/dI^2, is zero or positive at sufficiently low bias. The insert shows the zero-bias resistance measured at low bias, about 10 nA or less. (b) The differential resistance is plotted versus bias current for an I-type sample. The wire length is $L = 140$ nm. The curves are taken at $T = 0.3, 0.5, 0.75, 1.0, 1.5, 2.0$, and 2.5 K. The fits, shown by solid curves, represent the weak Coulomb blockade theory, due to Golubev and Zaikin [190]. The key fact is that the differential resistance, dV/dI, decreases with increasing the bias current I, which means that the second derivative d^2V/dI^2 is negative. The insert shows the linear resistance, which always increases with cooling for I-samples after it reaches its minimum around 2 K. Please find a color version of this figure on the color plates.

be either zero or some very high value. Thus, it is possible to conclude that samples can be either superconducting or insulating.

A similar qualitative analysis is done with respect to the bias. If the temperature is as low as possible and the bias is made weaker and weaker, one finds two distinct types of samples, namely, those which show an increasing differential resistance and those characterized by a decreasing differential resistance. Again, one extrapolating to zero-bias, one would expect that the first type of samples would show a very high or an infinite resistance, while the second type would show a decreasing resistance. Again, one concludes that samples can be separated into two distinct classes.

In summary, since one cannot reach absolute zero temperature, it is not possible to directly measure whether at $T = 0$ the sample resistance is zero (S-type), or a finite value (N-type), or infinity (I-type). Therefore, one can use the following indirect criteria to classify samples. The S-samples satisfy the following conditions at sufficiently low temperatures:

$$\frac{dR}{dT} > 0, \quad \frac{d^2(\ln R)}{dT^2} < 0, \quad \frac{dR_d}{dI} > 0, \quad \text{and} \quad R(T) \to 0 \quad \text{for} \quad T \to 0$$

The I-samples, at sufficiently low temperatures, satisfy the following conditions:

$$\frac{dR}{dT} < 0, \quad \frac{d^2 R}{dT^2} > 0, \quad \text{and} \quad \frac{dR_d}{dI} < 0$$

Here, $R_d = dV/dI$ is the differential resistance. It is assumed that $R_d > 0$ in all samples considered, which is in agreement with the majority, if not all experiments on superconducting nanowires, provided that the bias current I is not too high, namely, that $I < I_c$ and $I < I_{sw}$.

Using these classification conditions, all or almost all homogeneous nanowires can be labeled either as S-type or I-type [130], based on the pronounced qualitative difference between these two groups, illustrated in Figures 10.2 and 10.3. Such classification is summarized in Figure 10.4 for a large group of nanowires made of the superconducting alloy $Mo_{79}Ge_{21}$. In this phase diagram, the superconducting wires are shown as circles and the insulating wires are shown as squares.

Understanding which parameter(s) define whether the wire is superconducting or insulating is a challenging task. The most natural assumption is that the wire

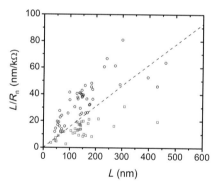

Figure 10.4 A phase diagram based on a large group of nanowires. This quantum transition can be classified either as SIT or SRT. All wires were fabricated following the same method, namely, by using molecular templating [130] with a superconducting alloy $Mo_{79}Ge_{21}$. The only difference between the samples is their length, width and the nominal thickness. According to our classification (see text), it turns out that homogeneous nanowires can be classified into two qualitatively distinct groups, namely, S-wires (circles) and I-wires (squares). The axis are the length of the nanowire L (horizontal) and the ratio $L/R_n = A_{cs}/\rho$ (vertical), where R_n is the normal resistance, A_{cs} is the cross-section area of the wire, and $\rho \approx 200\ \mu\Omega$ cm is the resistivity of the used MoGe alloy. The dashed line is given by equation $R_n = h/4e^2$. Apparently, this line serves as a good phase boundary for the observed SIT, at least for short wires, empirically defined as those having $L < 300$ nm.

diameter d or the cross-section area $A_{cs} = \pi d^2/4$ should define the class of the wire. The diagram of Figure 10.4 clearly shows that this is not the case. The vertical axis in the diagram is $L/R_n = A_{cs}/\rho$, where $\rho \approx 200\,\mu\Omega$ cm is the resistivity of the wire material, which does not depend on the thickness (of studied films) according to Graybeal and Beasley [191]. Thus, the vertical axis can be considered to be, at least approximately, the wire cross-section area in arbitrary units. Thus, an $A_{cs} = $ const phase boundary for this diagram would be a horizontal line. Obviously, it is not possible to choose any horizontal line that would divide the diagram in such a way that all I-samples would be on one side and all S-samples would be on the other. Thus, the phase boundary for the considered quantum phase transition cannot be formulated as the diameter being equal to some critical value. Therefore, the wire diameter is not the control parameter for the SIT.

A much better phase-separation boundary for this diagram is a straight line passing through the origin. Such lines correspond to the condition of constant normal resistance. Such is the dashed line in Figure 10.4, which is defined by the condition $R_n = h/4e^2$. This is equivalent to $L = A_{cs} R_n/\rho$ and $L/R_n = A_{cs}/\rho$. Therefore, $y = x/R_n = 4e^2 x/h$ if the x-axis is $x = L$ and the y-axis is $y = L/R_n$. That is why, on the diagram considered, a graphic representation of the condition that the normal wire resistance equals the resistance quantum is a straight line having the slope that equals the quantum of conductance, $4e^2/h$ (the dashed line). Obviously, it provides an almost perfect border between the two possible classes of wires: insulating and truly superconducting. Thus, the total normal resistance of the wire is a good control parameter for the SIT. The wires having $R_n > h/4e^2 \approx 6.5\,\text{k}\Omega$ are insulating, while wires having a lower resistance are true superconductors, in the sense discussed above. It also appears that longer wires might deviate from this simple rule. Based on the available data, one concludes that the wires shorter than about 300 nm obey the rule stated above. Longer wires can have their normal resistance larger than the quantum resistance and still appear to be superconducting, as the three circles on the diagram, corresponding to wires of length larger than 400 nm.

Within the SM model, a quantum transition, namely, the Schmid–Bulgadaev (SB) transition, occurs at $R_n = R_q = h/4e^2$. The transition is found in the theory of JJs, after quantizing the Stewart–McCumber model and adding to it the Caldeira–Leggett quantum dissipation model. For a more detailed discussion of this transition, see Section 3.4. Since the critical point of the SIT found in superconducting nanowires is also $R_n = R_q$, a natural question arises of whether it is, in fact, possible to explain the observed loss of superconductivity in nanowires in terms of a generalized SB transition [144, 194].

Büchler et al. have applied a renormalization group analysis to compute the phase diagram of a nanowire shunted by a normal resistor, connected to the ends of the nanowire [194]. The results from such a model were, in fact, very similar to the Schmid–Bulgadaev transition in SIS junctions. Namely, if the shunt is less than R_q, then the wire is superconducting, while if the shunt is larger than R_q or if there is no shunt at all, then the quantum phase slips prevent any steady supercurrent and the wire acts as an insulator (for high fugasity) or as a normal metal (for a low QPS fugasity).

In the experiments (Figure 10.2), the wire is obviously shunted by the leads, which always have an impedance of the order of 100 Ω ≪ R_q, that is, close to the vacuum impedance. This external shunting has to be counted in series with the impedance of the wire itself. This is because each phase slip event occurring on a segment of the wire is effectively shunted by other segments of the wire which are connected in series with the low-impedance electrodes, as seen by the phase slip. This is why the wire's normal resistance might play the definitive role in controlling SIT. Since R_n is much larger than the impedance of the leads, R_n effectively acts as the only control parameter for the SIT phase boundary. One objection to this might be that if a phase slip happens in some segment of the wire, the other segments are superconducting and therefore cannot act as normal shunts. Yet, the QPS events occur in all spots of the homogeneous wire with equal amplitudes. Additionally, each QPS has a normal core. Thus, if the QPS process occurs everywhere on the wire, with approximately equal amplitudes, it means that the whole wire is in a superposition of the normal and the superconducting state. Thus, the possibility of some wire segments acting as shunts with respect to other segments appears realistic.

Such a qualitative picture can be made quantitative using the renormalization group approach. In [144], the corresponding $R(T)$ curves were computed and agreed well with the $R(T)$ experimental measurements of [130]. The theory implies that the entire family of curves of Figure 10.2a, including the high-temperature region of each curve, can be explained by QPS, without involving TAPS in the picture. On the other hand, the Little's fits (solid curves in the Figure) generated using the TAPS concept also agree with the data very well. The Little's fit is essentially the Arrhenius thermal activation law which is valid at any temperature. Therefore, a question remains, that is, whether TAPS or the QPS is the dominant dissipation mechanism that explains the shape of the $R(T)$ curves of Figure 10.2a. Note that the Golubev and Zaikin theory [171] of TAPS supports the expectation that at sufficiently high temperatures TAPS dominates QPS.

A possible solution to the puzzle, discussed in detail in the next section, is to take into account the inductive impedance of the wire. Such impedance, simply the product of the kinetic inductance and the characteristic frequency of a QPS, is equal, at least approximately, to R_n. This wire impedance is connected with a very low impedance of the leads. Thus, the total shunting impedance is close to R_n, that is, close to the normal resistance of the wire, even if the condensate is not much suppressed. That, combined with the Büchler et al. theory, can explain why the experimental transition occurs at $R_n = R_q$.

It should be noted that the discussion of SIT presented here is a simplified one. Additional complexity arises, or might arise, because of the presence of localized magnetic moments or spins on the surface of the wire. There is a strong experimental indication that such spins do exist and they can couple to the condensate. Thus, Rogachev et al. [188] have found that the critical current of the wire is somewhat suppressed at zero magnetic field due to such surface spins. If a magnetic field is applied, either parallel or perpendicular to the nanowire, the spins are polarized, and their suppressive effect is diminished. Thus, the critical current of the wire is observed to increase with magnetic field. Influence of such localized spins on any aspect of the superconductor-insulator quantum transition is not well understood.

10.1
Simple Model of SIT in Thin Wires

Consider the case when a nanowire is connected to a pair of measurement leads (Figure 10.5). Such a pair (or a few pairs) is a necessary component of any DC transport measurement setup. Usually, there is more than just one pair of leads, but for our present analysis, the exact number of leads is not of any importance, as long as the number is greater than one. The pair shown in the figure serves to bring the bias current to the sample. However, there is one other, usually unwanted function of the leads – they effectively shunt the nanowire. This is because the leads form a transmission line permitting propagation of photons of any frequency. Effectively, they act as an ohmic resistor, connecting the ends of the wire, similar to the shunt in the SM model (Figure 3.1).

This might seem strange, given the fact that the leads are not directly connected to one another. Yet, at any amplitude of AC voltage V present on the wire, the same voltage is present between the tips of the leads since they are directly connected to the wire. This means that the power P_{dis} dissipated into the leads by the wire is $P_{dis} \approx V^2/Z_l$. On the other hand, if one would connect a normal resistor of the same value Z_l to the ends of the wire and if the voltage on the wire would be V, then the dissipated power would again be V^2/Z_l. No matter the exact shape of the leads, they almost always form a transmission line having the impedance of the order of 100 Ω. Thus, the leads and a normal shunt produce the same dissipative effect on the wire, under the presumption that the wire generates an AC voltage V between its ends due to thermal or quantum fluctuations.

On the other hand, it is well-known that a shunt makes a strong impact on the QPS. It was shown, for example, in [194], that a normal shunt connected to the ends of a nominally superconducting nanowire can induce the Schmid–Bulgadaev quantum phase transition. What this means is that if an ohmic shunt has a value of less than R_q, then phase slips are suppressed and so the wire has zero resistance in the limit of zero temperature. On the other hand, if the shunt is larger than R_q, then the QPS proliferate and the wire is resistive. In other words, an SRT takes place as the shunt matches the quantum resistance.

If applied naively, this model suggests that any practical nanowire would be a true superconductor and would exhibit $R = 0$ at $T = 0$. (The term "resistance" is always used in the sense of zero bias resistance.) This is based on the fact that, as was argued above, the leads always provide an effective shunting resistance $Z_l \sim 100\,\Omega$. Thus, any wire is strongly shunted. This contradicts Figure 10.2 which demonstrate that the wires come in two categories, superconducting and resistive (i.e., slightly

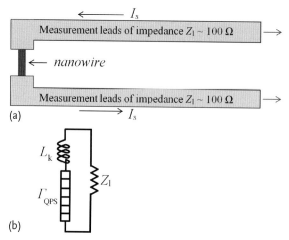

Figure 10.5 (a) A simplified schematic for a DC measurement setup. The nanowire is connected to two macroscopic leads which bring the DC bias current I_s to the wire and take it back. These two parallel leads form a transmission line which permits photons of any frequency to propagate away from the wire, effectively to infinity. The leads are assumed infinitely long, as is stressed by the arrows, so that any photon which leaves the nanowire propagates away and never comes back to the nanowire. If the leads have a large but finite length and also somewhat lossy, the damping effect is the same as having infinite length. Thus, the nanowire can dissipate energy by emitting photons into the leads. In other words, the phase slips in the wire, including QPS, couple to the photon modes of the leads. Such coupling is the essence of the quantum dissipation, which is the key player in the Caldeira–Leggett macroscopic quantum tunneling theory [30]. Since the wire, if it has any AC voltage on it, loses its energy by the emission of photons into the leads, the leads effectively act as a normal shunt. This is because a shunt would also take energy from the nanowire in form of Joule heating. Note that a voltage might occur since phase slips, if present, produce a time variation of the phase difference ϕ on the wire and the voltage is $V \propto d\phi/dt$. The analogy with an ohmic shunt is based on the fact that the dissipated power is proportional to V^2 in both cases. (b) An effective electric diagram of a superconducting nanowire connected to a pair of measurement leads. The pair act as a shunting resistor Z_l for any high frequency fluctuating voltage which develops on the wire as phase slips occur. The nanowire is modeled as two elements, representing two of its most important functions. The first is the wire's kinetic inductance L_k. The kinetic inductance is perfectly coherent since it does not allow any QPS. The QPS ability of the wire is presented schematically by a rectangle crossed by horizontal lines which represent the flow of QPS.

insulating). The experiment strongly suggests that the choice of which group any given wire belongs to depends on its normal state resistance R_n. This fact needs an explanation; we suggest one in what follows.

A model that gives correct predictions can be obtained by taking into account the kinetic inductance of the wire. The role of the kinetic inductance was recently illuminated in [196]. There, it was demonstrated that a quantum device, for example, a low capacitance JJ, can show a large amplitude of MQT if the junction is isolated from the environment by a compact chain of kinetic inductors. In [196], the chain of compact inductive elements was also represented by JJs, but having a relatively large capacitance, so they acted as classical inductors.

An analogous situation probably happens in nanowires, in which case the kinetic inductance L_k of the nanowire itself, and the impedance $Z_k = \omega_{QPS} L_k$, associated with it, serve to isolate the QPS in the nanowire from the shunting effect of the leads connected to the wire (Figure 10.5).

To make this analysis more concrete, we need to choose the characteristic frequency of the phase slip. Following the theory of Golubev and Zaikin [172], we assume that the duration of a phase slip can be estimated as $t_{QPS} = \hbar/\Delta$. During a single phase slip, the order parameter decreases to zero and then increases again to its equilibrium value. Such a process represents only half of the period of the oscillation of the order parameter. Thus, the period should be twice as large as t_{QPS}. Consequently, a good estimate for the angular frequency is $\omega_{QPS} = 2\pi/2t_{QPS} = \pi/t_{QPS} = \pi\Delta/\hbar$. The impedance of an inductor equals the inductance times the angular frequency. Thus, the impedance of the wire is

$$Z_k = L_k \omega_{QPS} = \frac{\pi L_k \Delta}{\hbar}$$

The kinetic inductance of a thin wire is defined in (2.80) as

$$L_k = \frac{2}{3\sqrt{3}} \frac{L}{\xi} \frac{\hbar}{2e I_c} \tag{10.1}$$

Thus, the impedance can be written as

$$Z_k = \frac{\pi \Delta}{3\sqrt{3} e I_c} \frac{L}{\xi}$$

We have seen in (9.15) that at $T \approx 0$, the critical current can be expressed through the normal resistance of the wire according to [19]

$$I_c = \frac{\pi L \Delta}{3\sqrt{3} e R_n \xi} \tag{10.2}$$

By combining this expression with the previous one, we get a remarkably simple expression for the impedance of a thin wire at a characteristic frequency of quantum phase slips

$$Z_k = R_n \tag{10.3}$$

It shows that the impedance of a wire at the QPS frequency is about the normal resistance of the wire. Note that this expression does not assume that the condensate is weakened and there is a normal component in the wire. On the contrary, the expression is valid for a fully superconducting wire. The obtained impedance originates from the kinetic inductance (i.e., the inertia) of the condensate in the wire.

The Schmid–Bulgadaev (SB) transition occurs as the total shunt impedance equals R_q, that is,

$$Z_k + Z_l = R_q \qquad (10.4)$$

To justify this condition, one can consider the wire as being made of quasi-independent segments of size ξ. Each segment acts more or less as a Josephson junction. Such junctions would schematically correspond to the squares in the QPS wire of Figure 10.5. It is clear from the schematic that each square is shunted by the impedance of the leads Z_l, connected in series with the impedance of the other segments in the wire. Since segments are small, the impedance of all segments but one is approximately equal to the total impedance of the wire, which is denoted as Z_k. Thus, the total shunt is $Z_k + Z_l$. On the other hand, it is known that the SB transition takes place in a shunted JJ when the shunt equals R_q. Thus, we justify the condition written above.

Since $Z_l \ll R_q$, the condition for the occurrence of the SB superconductor-insulator quantum transition can be simplified to

$$Z_k = R_q$$

Taking into account the estimate $Z_k = R_q$, derived above (10.3), we arrive at

$$R_n = R_q \qquad (10.5)$$

This equation defines the value of R_n at which phase slips proliferate and the wire becomes resistive as it enters the phase-delocalized region of the Schmid–Bulgadaev phase diagram [82–84, 90, 91]. Of course, the calculation presented above is not exact since it relies on certain extrapolations of the results of the GL theory down to zero temperature.

Note also that the wires having $R_n > R_q$, although located in the insulating part of the SB diagram, might not act insulating when tested with a typical DC measurement setup. The reason is that although the dissipative effects in such a sample are not sufficiently strong to completely suppress QPS and make the phase difference a good quantum number, the rate of phase slips can be low if the wire diameter is large. This means that long wires (empirically, much longer than 200 nm) having the resistance larger than R_q can only appear slightly resistive or even truly superconducting if their large diameter does not allow any measurably high QPS rate or QPS fugacity. The basic principle is that the SB transition is able to localize the phase and put QPS rate to zero, but it is not able to make the fugacity of QPS larger than the value predicted by the Giordano model, for instance, based on the wire diameter. Thus, a macroscopic solenoid made of a wire which is, say, one hundred microns in diameter, will appear perfectly superconducting, even if its normal resistance is much larger than R_q. The reason is that the probability of quantum tunneling of a phase slip or a vortex across such thick wire is unmeasurably low [19].

The phase boundary $R_n = R_q$ is in agreement with the results of [100, 130]. The phase diagram summarizing these results is shown in Figure 10.4. The diagram demonstrates that for the great majority of tested samples, the rule is that the samples having $R_n < R_q$ behave as true superconductors (circles), while the wires having $R_n > R_q$ are insulating (squares). They can also be called resistive, not insulating, since their resistance is not infinite. The boundary between the two groups, defined as $R_n = R_q$, is shown as the dashed line (Figure 10.4).

The same rule applies to the samples of [108]. In this publication (see Figure 9.1), all samples, except Sample 8, have a large normal resistance such that $R_n > R_q$. According to the rule explained above, all such samples should have a delocalized phase, that is, they should be either normal or insulating, depending on the diameter, which limits the maximum possible QPS rate. In fact, all of them can be classified as "resistive". This is exactly what was found experimentally. Most of the samples satisfying the $R_n > R_q$ condition are resistive, not superconducting. Such a conclusion can be reached by observing that most of the samples follow the Giordano, $R_G(T)$, model. This model predicts that the resistance is finite at $T = 0$. Thus, the resistance does not drop according to the Arrhenius law with cooling. Sample 8, on the other hand, shows an Arrhenius type drop because for this sample, $R_n < R_q$ and so QPS is strongly suppressed. One can expect $R \to 0$ if $T \to 0$ for the Sample 8.

Another reference that indicates that the Schmid–Bulgadaev rule applies to thin wires is [184]. In this paper, the wire was etched by argon bombardment, causing a gradual reduction of the diameter and the corresponding gradual increase of the normal resistance. The $R(T)$ measurements were done between the etching steps. As the resistance R_n exceeded R_q, the $R(T)$ curve started to show strong deviations from the Arrhenius activation law, suggesting that the phase becomes delocalized. Further etching made the sample highly resistive and possibly insulating.

However, one more confirmation that $R_n = R_q$ represents a critical point at which true superconductivity is lost comes from [187], in which Nb wires were studied. The results are reproduced in Figure 9.2. Among five samples, only one (Nb6) satisfied the condition $R_n > R_q$. The sample is supposed to be resistive, that is, normal or insulating at $T \to 0$. Indeed, a deviation from the Arrhenius type negative curvature $R(T)$ curve is observed at low temperatures. Such a resistive tail confirms that the phase in not perfectly coherent in the wire.

11
Bardeen Formula for the Temperature Dependence of the Critical Current

All physical quantities describing the effect of superconductivity depend on temperature. Deriving the temperature dependence of such quantities analytically is usually possible using the GL theory, but it is valid only near the critical temperature T_c. The BCS theory is valid at all temperatures, though it is more complicated and usually does not allow derivations of simple analytic expressions. Thus, finding exact analytic formulas valid in a wide temperature interval is usually impossible.

However, there is one exception. Bardeen derived an expression for the critical current of a thin wire which is supposed to be valid at any temperature below T_c. The expression is remarkably simple [20]

$$I_c(T) = I_c(0)\left(1 - \frac{T^2}{T_c^2}\right)^{3/2} \tag{11.1}$$

It defines the depairing current. The term "depairing current" is used everywhere interchangeably with the "critical current." Here, $I_c(0)$ is the depairing (critical) current at $T = 0$.

Testing the depairing current experimentally is generally difficult because the experimental critical current can be lower than the depairing current due to the effects of premature switching, phase slippage, and/or vortex motion across the sample.

Recently, two papers appeared in which results of such testing appeared. It was indeed confirmed that the depairing current is in agreement with the Bardeen formula. One such example is shown in Figure 11.1. There, the switching current is measured on nanowires which were shunted using normal resistors. The test is based on the finding of [197] that the switching current increases and then saturates as the shunting resistor is made smaller and smaller. Such saturation of the switching current, combined with the observation that the fluctuations of the switching current diminish drastically as the shunt is added, allows one to conclude that the switching current approaches the depairing current as a low-value shunt is added. Empirically, it was found that a shunt of the order of 10 Ω is sufficient. Thus, the switching current measured on nanowires connected in parallel with normal resistors is taken as the depairing current. Such results are compared to the Bardeen formula for $I_c(T)$, (11.1). The theory shows an excellent agreement with the data (Figure 11.1).

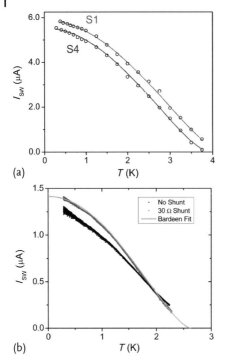

Figure 11.1 Experimental confirmation of Bardeen formula for the depairing current in thin superconducting filaments (nanowires) [197]. (a) The mean value of the switching current is plotted versus temperature for two samples (S1 and S4) which were shunted using 5 and 10 Ω resistors respectively. The shunting is found to suppress the fluctuations of the switching current and push the switching current up to the depairing current. The solid lines are fits to the Bardeen formula for the depairing current, (11.1). The fitting parameters used to generate the fits are $I_c(0) = 5.88$ and 5.53 μA and $T_c = 4.2$ and 3.9 K for S1 and S4, correspondingly. (b) Mean switching current I_{sw} is plotted versus T for sample S5 for the unshunted bare wire (black) and for the same wire being shunted using a 30 Ω normal resistor (the red data points). The Bardeen fit (yellow) is in excellent agreement with the shunted wire date, but not with the unshunted case data. The fitting parameters were $I_c(0) = 1.415$ μA and $T_c = 2.62$ K. Please find a color version of this figure on the color plates.

Appendix A
Superconductivity in MoGe Alloys

The MoGe is the material of choice for the fabrication of superconducting nanoscale experimental structures and devices for which a high degree of homogeneity is essential. The most prominent example where the homogeneity is essential is the fabrication of nanowires. It is well known that if the chosen material is granular, such as the case with most pure metals, including gold and aluminum, then the resulting nanowire is not continuous, at least when thin wires are attempted [106]. The unique property of MoGe to form homogeneous amorphous films, even if the film thickness is only 1.5 or 2 nm, is known from the pioneering work of Graybeal and Beasley [151]. In this section, we summarize important characteristics of the MoGe alloy in its optimized composition, $Mo_{79}Ge_{21}$, that is, the composition characterized by the highest critical temperature.

The $Mo_{1-x}Ge_x$ alloy is known to undergo, at $x = 0.195$ [133, 134], a structural phase transition from a BCC phase to the amorphous phase.

Superconducting critical temperature, bulk, $T_{c,bulk}$	7.36 K
Superconducting energy gap (Δ)	1.1 meV (bulk)
Resistivity	180–220 µΩ (bulk-film-wire)
Elastic mean free path (l_e)	0.3–0.4 nm
Fermi velocity (v_F)	10^6 m/s
Diffusion constant ($D = v_F l_e/3$)	1.2×10^{-4} m^2/s
Density	9.7 g/cm^3
Fermi wavevector	8.63 nm^{-1}
Pippard (intrinsic) coherence length ($\xi_0 = \hbar v_F/\pi\Delta$)	190 nm (bulk)
Thermodynamic critical field (B_c)	55 mT (film)
Dirty-limit coherence length ($0.85\sqrt{l\xi_0}$)	4.4 nm (bulk) 4.9 nm (film)
Debye temperature	266 K
London penetration depth	18.5 nm (film)
Effective magnetic penetration depth	424 nm (film)
Perpendicular magnetic penetration depth	18 µm (for a 10 nm thick film)
Lower critical field (H_{C1})	~ 4 mT (10 nm thick and 10 µm wide film)
Upper critical field (H_{C2})	12.2 T (bulk); 6.7 T (film)
Electron–phonon coupling constant	0.672
Bare density of states	0.775 states/eV-atom
Thickness of the natural oxide	≈ 2.5 nm

Superconductivity in Nanowires, First Edition. Alexey Bezryadin.
© 2013 WILEY-VCH Verlag GmbH & Co. KGaA. Published 2013 by WILEY-VCH Verlag GmbH & Co. KGaA.

Appendix B
Variance and the Variance Estimator

Consider some random or stochastic quantity or variable x. As an example of such a quantity, keep in mind the switching current I_{sw}, which is fundamentally impossible to predict and which differs from one measurement to the next, even if all external parameters are kept unchanged. Such a random quantity can be represented, among other things, by its mean value $\langle x \rangle = \mu$, which is also called the expectation value and is denoted as $E(x) = \mu = \langle x \rangle$. All three notations are equivalent and assume averaging over an infinite set of measurements of x. The definition of the mean value is $\mu = \int x P(x) dx$, where $P(x)$ is the probability density of the random quantity x. The integral is taken over all possible values of x. The definition of the probability density is such that $P(x)dx$ is the probability that a measurement of the random quantity gives a value in the interval between x to $x + dx$. So the probability density must be normalized as $\int P(x) dx = 1$. This is because it is assumed to be certain (i.e., has a unity probability) that each measurement will give one of possible values.

The degree of the random deviation of x from its mean value μ is given by its standard deviation σ or the variance $\mathrm{Var}(x) = \sigma^2$. The term "dispersion" is loosely defined; it can mean either σ or σ^2, or both. The variance is defined as $\mathrm{Var}(x) = E[(x-\mu)^2] = \int P(x)(x-\mu)^2 dx$, and the standard deviation is defined as $\sigma(x) = \sqrt{\int P(x)(x-\mu)^2 dx}$.

We can now derive a few useful relations, assuming that C_1 is some constant:
$$E(C_1 x) = \int C_1 x P(x) dx = C_1 \int x P(x) dx = C_1 E(x) = C_1 \mu$$

$$\mathrm{Var}(C_1 x) = \int (C_1 x - C_1 \mu)^2 P(x) dx = C_1^2 \mathrm{Var}(x) = C_1^2 \sigma^2 \quad \text{(B.1)}$$

The deviation from the mean value is $\Delta x = x - \mu$. Then, $E(\Delta x) = 0$ and $\mathrm{Var}(\Delta x) = \mathrm{Var}(x)$.

If y is another random variable, then the joint probability density for both of them is $P(x, y) = P(x)P(y)$ (which is true only if they are independent). That means $P(x, y) dx dy = P(x) P(y) dx dy$, so that

$$E(x + y) = \int (x + y) P(x) P(y) dx dy$$
$$= \int x P(x) P(y) dx dy + \int y P(x) P(y) dx dy = E(x) + E(y)$$

Superconductivity in Nanowires, First Edition. Alexey Bezryadin.
© 2013 WILEY-VCH Verlag GmbH & Co. KGaA. Published 2013 by WILEY-VCH Verlag GmbH & Co. KGaA.

Similarly, one can derive within a few integration steps that for two independent variables x and y, the variance is

$$\text{Var}(x + y) = \int [x - E(x) + y - E(y)]^2 P(x) P(y) dx dy$$
$$= \text{Var}(x) + \text{Var}(y)$$

The same result can be obtained as

$$\text{Var}(x + y) = E(\Delta x + \Delta y)^2 = E(\Delta x^2 + \Delta y^2 + 2\Delta x \Delta y)$$

If x and y are independent, then $E(\Delta x \Delta y) = E(\Delta x) E(\Delta y) = 0$, and, in this case

$$\text{Var}(x + y) = \text{Var}(x) + \text{Var}(y)$$

The results above can be generalized for many independent variables as

$$\text{Var}(x_1 + x_2 + \cdots + x_N) = \text{Var}(x_1) + \text{Var}(x_2) + \cdots + \text{Var}(x_N) \quad (B.2)$$

Another well-known and useful formula is

$$\text{Var}(x) = E[(x - \langle x \rangle)^2] = E(x^2 + \langle x \rangle^2 - 2x\langle x \rangle)$$
$$= E(x^2) + \langle x \rangle^2 - 2\langle x \rangle E(x) = \langle x^2 \rangle - \langle x \rangle^2$$

where two equivalent notations, $\langle x \rangle$ and $E(x)$, have been used to denote the mean value.

Consider now a special case of normal or Gaussian distributions, which is defined by the following probability density

$$P(x) = \frac{1}{\sigma\sqrt{2\pi}} \exp\left[-\frac{(x-\mu)^2}{2\sigma^2}\right] \quad (B.3)$$

Such distributions are very common and sometimes called bell-shaped distributions. In such cases, the meaning of σ is easy to grasp. Suppose x is measured N times. Then, about $0.68 N$ of all measured data points fall within the interval $\langle x \rangle \pm \sigma$ and about 95% of all measurements fall in the 2σ interval, that is, from $\mu - 2\sigma$ to $\mu + 2\sigma$. One can show that only about $1.973 \times 10^{-7} N$ of all measurements would fall *outside* the 6σ interval, which is $\langle x \rangle \pm 6\sigma$. Of course, if N is not sufficiently large and this number is much smaller than unity, then none of the measurements would typically produce results outside the 6σ interval. To get an idea of how low this probability is, consider that it is roughly equal to the probability of an airplane crashing during one flight.

Note that the switching current distributions are not exactly Gaussian: They are not symmetric with respect to the center, as Figure 8.6 illustrates. Thus, the estimates given above are only approximately correct if applied to I_{sw} distributions. Also, note that even for a non-Gaussian random physical quantity x, the standard deviation is defined; it is defined for the switching current, for example.

The mean value and the dispersion of x can be estimated by performing N measurements and then computing the average $\overline{x_N} = N^{-1}\sum_{i=1}^{N} x_i$ and the standard deviation $\sigma_N^2 = N^{-1}\sum_{i=1}^{N}(x_i - \overline{x_N})^2$, where x_i is the result of the ith measurement of the random quantity x. If $N = \infty$, then $\overline{x_N} = \langle x \rangle$ and $\sigma_N^2 = \sigma^2$. In practice, it is not possible of course to perform an infinite number of measurements in any experiment. If N is large but not infinite, then $\overline{x_N} \approx \langle x \rangle$ and $\sigma_N^2 \approx \sigma^2$. In what follows, we will discuss how to estimate probable deviations between $\overline{x_N}$ and $\langle x \rangle$, and also between σ_N^2 and σ^2. Among other things, we will see that if the experiment involving N measurements is repeated an infinite number of times, then σ_N^2 is computed for each such experiment, and after that, the mean value of the thus obtained standard deviation is computed. Then, the result would be $E(\sigma_N^2) = [(N-1)/N]\sigma^2$. Thus, for a result of one experiment involving N measurements, one can also compute so-called adjusted standard deviation of the set of N data points, $\sigma_{Na}^2 = (N-1)^{-1}\sum_{i=1}^{N}(x_i - \overline{x_N})^2$. If N is large, the difference between the two is not significant. In the limit $N \to \infty$, both variance estimators approach the variance: $\sigma_N^2 \to \sigma^2$ and $\sigma_{Na}^2 \to \sigma^2$.

The precision of $\overline{x_N}$ and σ_N^2 compared to the corresponding exact values, μ and σ, can be estimated by repeating the N-point measurement many times and by observing by how much the mean value and the dispersion fluctuate from one experiment to the next one. Under an experiment, we understand an acquisition of N data points for the value of x. The variation of the mean value from one experiment to the next one can also be estimated. Such estimates are needed for the simple reason that experiments cannot be continued indefinitely. Usually, the experimentalist makes just one experiment of N measurements and has to estimate the expected error. Such error estimates are discussed below.

Suppose we measure the x value N times (under the same conditions) and thus obtain an array of data points $(x_1, x_2, \ldots, x_N)$. If such an experiment is repeated many times, we will find that each point in this array is a random number having the same mean value and dispersion. These points $(x_1, x_2, \ldots, x_N)$ are considered to be "iid" points, that is, independent and identically distributed random variables. In particular, it is true that $E(x_i) = \mu$ and $\text{Var}(x_i) = \sigma^2$, for any point number i, which, by definition, satisfies $1 \leq i \leq N$.

Let us find the dispersion or the probable error for the entire experiment. The probable error for the average value $\overline{x_N}$ obtained by performing one experiment involving N identical measurements of x can conveniently be defined as $\text{Err}_x(N) = \sqrt{\text{Var}(\overline{x_N})}$. Its meaning is that in 68% of experiments, the result will be such that $\mu - \text{Err}_x(N) \leq \overline{x_N} \leq \mu + \text{Err}_x(N)$, where μ is the true mean value of x, which is unknown in practice. In 95% of identical experiments, the result for the mean value estimator $\overline{x_N}$ will be such that $\mu - 2\text{Err}_x(N) \leq \overline{x_N} \leq \mu + 2\text{Err}_x(N)$. One can naturally expect that if $N \to \infty$, then $\text{Err}_x(N) \to 0$.

The error can be calculated as follows

$$\text{Err}_x^2(N) = \text{Var}(\overline{x_N}) = \text{Var}[N^{-1}(x_1 + x_2 + \cdots + x_N)]$$
$$= N^{-2}[\text{Var}(x_1) + \text{Var}(x_2) + \cdots + \text{Var}(x_N)]$$

where we have used (B.1), (B.2), and have assumed that the result of the second measurement is independent of the result of the first measurement, etc. Therefore, $x_1 \ldots x_N$ are iid variables. Now, remember that $\mathrm{Var}(x_i) = \sigma^2$. Therefore, $\mathrm{Err}_x^2(N) = N^{-2} N \sigma^2 = N^{-1} \sigma^2$. Thus, the estimate for the statistical error of the finite data set mean value $\overline{x_N}$ is $\mathrm{Err}_x(N) = \sigma/\sqrt{N}$. As expected, $\mathrm{Err}_x(N) \to 0$ if $N \to 0$.

Of course, in practice, the errors could be larger than this. The estimate above is the minimum possible error under ideal conditions. If the systematic errors of the setup is Err_{sys}, then for one experiment of N measurements the total error can be estimated as $E_{tot}^2 = \mathrm{Err}_{sys}^2 + \mathrm{Err}_x(N)^2$. We talk about estimating the errors because usually it is not possible to know the error with a precision better than roughly 20%. In practice, even larger errors can occur if the experiment is done over a long period of time. It is known that the noise Fourier components at frequency f_n increase as $1/f_n$. Such noise is sometimes called "flicker noise." This means that the properties of the sample drift with time or perform a "random walk." Such noise, also called "pink noise," cannot be eliminated by continuing the experiment longer in order to perform a larger number of measurements and then average them.

To conclude this section, let us keep assuming that the noise is Gaussian or close to Gaussian. Then, $E[(x-\mu)^4] = 3\sigma^4$. The variance for the variance estimator is

$$\mathrm{Var}\left(\sigma_N^2\right) = \left(\frac{N-1}{N}\right)^2 \frac{E[(x-\mu)^4]}{N} - \frac{(N-1)(N-3)}{N^3} \{E[(x-\mu)^2]\}^2$$

If $N \gg 1$, the formula above becomes

$$\mathrm{Var}(\sigma_N^2) = \frac{3\sigma^4}{N} - \frac{\sigma^4}{N} = \frac{2\sigma^4}{N}$$

Thus, the expected error for the finite data set variance is

$$\mathrm{Err}(\sigma_N^2) = \sqrt{\mathrm{Var}(\sigma_N^2)} = \sigma_N^2 \sqrt{\frac{2}{N}}$$

Thus, one can write

$$\sigma^2 = \sigma_N^2 \pm \sigma^2 \sqrt{\frac{2}{N}} \quad \text{or} \quad \sigma_N^2 = \sigma^2 \left(1 \pm \sqrt{\frac{2}{N}}\right)$$

By taking a square root, this can be further transformed into $\sigma_N = \sigma(1 \pm \sqrt{2/N})^{0.5}$, and assuming that N is large, one gets $\sigma_N = \sigma(1 \pm \sqrt{1/2N})$. Now, if we use an approximate formula for small x, $1/(1+x) \approx 1-x$, we get the desired estimate for the precision of σ_N as $\sigma \approx \sigma_N \pm \sigma_N/\sqrt{2N}$. One observes again that the uncertainty approaches zero as N approaches infinity.

If the distribution is not Gaussian, the deviation from Gaussian is characterized by so-called skewness Sk

$$\mathrm{Sk} = \frac{E[(x-\mu)^3]}{\{E[(x-\mu)^2]\}^{3/2}}$$

The skewness defines the degree of asymmetry with respect to the center of the distribution ($x = \mu$). The normal distribution is symmetric so that for normal distributions, one always gets Sk = 0.

The degree of deviation from the Gaussian form, without including asymmetry, is given by kurtosis Kr

$$\text{Kr} = \frac{E[(x-\mu)^4]}{\{E[(x-\mu)^2]\}^2}$$

For normal (Gaussian) distributions, one finds Kr = 3. Any different kurtosis obtained experimentally would indicate that the distribution is non-Gaussian. For example, the Kurkijärvi model generates Sk ≈ -1 and Kr ≈ 4.9. Experiments confirm these predictions. Interested readers can learn much more about the methods of statistical analysis from [198, 199].

Appendix C
Problems and Solutions

Problem 1

Use Heisenberg's uncertainty principle and estimate the size, ξ_0, of a superconducting (Cooper) pair of electrons in a clean superconductor. Assume that the critical temperature, T_c, and the Fermi velocity, v_F, are known.

Solution The electrons are bound in the BCS condensate, in which they pair up in the momentum space. Electrons forming a pair move in exactly opposite directions. The characteristic velocity of this motion is the Fermi velocity, v_F. The binding energy is twice the superconductor's energy gap, 2Δ. The time uncertainty, which can be identified with the lifetime of each pair, is $\tau_p \sim \hbar/2\Delta$, according to the Heisenberg's principle. The size of a pair is estimated as the distance over which two electrons move away from each other during the pair lifetime. Thus, the size is $\xi \sim \hbar(2v_F)/2\Delta$. According to the exact solution of the BCS theory, which is beyond the scope of this book, the exact formula is $\xi_0 = \hbar v_F/\pi\Delta$. Note that the energy gap can be obtained from the known critical temperature as $\Delta = 1.76 k_B T_c$, according to the BCS theory.

Problem 2

Use the Heisenberg's uncertainty principle and estimate the size, $\xi(0)$, of a superconducting pair of electrons (Cooper pair) in a dirty or disordered superconductor. Assume that the elastic mean free path for single electrons, l_e, the clean-limit coherence length, ξ_0, and the critical temperature in the clean-limit case (i.e., the same material, but without disorder), T_c, are known. Use the Anderson–Abrikosov–Gor'kov (AAG) theorem, which states that the critical temperature does not depend on the presence of disorder.

Solution The electrons are bound in the BCS condensate. The binding energy for one pair is twice the energy gap, 2Δ. The time uncertainty, which can be identified with the lifetime of the pair, according to Heisenberg's uncertainty principle, is $\tau_p \sim \hbar/2\Delta$. The size of the pair, $\xi(0)$, can be estimated by computing how far two paired electrons diffuse from one another during the lifetime of the pair. The diffusion distance is defined according to the elementary diffusion theory as $\xi(0) \sim$

Superconductivity in Nanowires, First Edition. Alexey Bezryadin.
© 2013 WILEY-VCH Verlag GmbH & Co. KGaA. Published 2013 by WILEY-VCH Verlag GmbH & Co. KGaA.

$\sqrt{D\tau_p} = \sqrt{\hbar D/\Delta}$. According to the diffusion theory, the diffusion constant is $D = l_e v_F/3$. Thus, $\xi(0) \sim \sqrt{\hbar l_e v_F/3\Delta}$. The coherence length of the clean limit is $\xi_0 = \hbar v_F/\pi\Delta$. Therefore, $v_F = \xi_0 \pi \Delta/\hbar$. Therefore, $\xi(0) \sim \sqrt{\pi l_e \xi_0/3} \approx \sqrt{l_e \xi_0}$. Note that the energy gap can be obtained from the known critical temperature as $\Delta = 1.76 k_B T_c$, according to the BCS theory. According to the AAG theorem, the critical temperature is independent of disorder, and therefore T_c is the same as in the clean-limit case, which corresponds to $l_e = \infty$.

Problem 3

Derive the expression for the supercurrent density in the case when the order parameter magnitude is constant in space and time, i.e., $|\psi| = $ const. Write the expression for the superfluid velocity. To solve the problem, use the general expression for the supercurrent density,

$$j_s = \frac{ie\hbar}{2m}(\psi \nabla \psi^* - \psi^* \nabla \psi) - \left(\frac{2e^2|\psi|^2}{mc}\right) A$$

(This equation was derived in the Section "Ginzburg–Landau equations.")

Solution To solve the problem, one needs to substitute the general form of the order parameter, $\psi = |\psi|\exp(i\phi)$, into the general equation for j_s and treat the amplitude, $|\psi|$, as a constant and the phase, ϕ, as a function of coordinates. Then, $\nabla \psi = |\psi|\exp(i\phi)(i\nabla \phi) = i\psi \nabla \phi$. Similarly, $\nabla \psi^* = -i\psi^* \nabla \phi$ (note that $|\psi|$ and ϕ are both real). We can now substitute these results into the general expression to get the expression for the supercurrent density as $j_s = 2e|\psi|^2(\hbar/2m)(\nabla\phi - \gamma_0 A)$, where $\gamma_0 \equiv 2e/c\hbar$ and m is the mass of a free electron. Remembering that the superfluid density (density of superpairs) equals, by definition, the square of the absolute value of the order parameter (i.e., $n_s = |\psi|^2$), we can rewrite the result as (in Gaussian units)

$$j_s = 2en_s \frac{\hbar}{2m}\left[\nabla\phi - \left(\frac{2e}{c\hbar}\right) A\right]$$

Note that the number of pairs, n_s, is related to a physical quantity which can be directly measured experimentally, namely, to the magnetic field penetration depth λ. The relationship, in Gaussian units, is

$$n_s = \frac{mc^2}{8\pi e^2 \lambda^2}$$

Since the current of any kind of charged particles of density n_s and of charge $2e$ can generally be written as $j_s = 2en_s v_s$, where v_s is the average velocity of the particles, we conclude that the superfluid velocity can be expressed as $v_s = (\hbar/2m)(\nabla\phi - \gamma_0 A)$. Note that this result is in agreement with (1.2), written for the case $A = 0$.

Problem 4

Derive the GL equation for the case when the amplitude of the order parameter is constant everywhere in the sample, $|\psi| = $ const. Start from the general form of the GL equation, which is

$$\left[\alpha + \beta n_s + \frac{\hbar^2}{4m}(-i\nabla - \gamma_0 A)^2\right]\psi = 0$$

where $n_s = |\psi|^2$ is the density of pairs of electrons in the condensate (superpairs) and $\gamma_0 \equiv 2e/c\hbar$. Solve the problem assuming that the gauge choice is such that the divergence of the vector potential is zero, $\nabla A = 0$.

Solution To solve the problem, we present the third operator as

$$(-i\nabla - \gamma_0 A)^2 \psi = (-i\nabla - \gamma_0 A)(-i\nabla - \gamma_0 A)\psi$$

Then, substitute the expression for the order parameter $\psi = |\psi|\exp(i\phi)$ and keep in mind that the amplitude $|\psi|$ is assumed constant while the phase ϕ varies in space. Thus, we obtain

$$(-i\nabla - \gamma_0 A)^2 \psi = |\psi|(-i\nabla - \gamma_0 A)\left[\exp(i\phi)(\nabla\phi - \gamma_0 A)\right]$$

We also need to remember that the rule of differentiation of a product for two functions a and b is $\nabla[a\nabla b] = \nabla a \nabla b + a\Delta b$, where $\nabla \cdot \nabla \equiv \Delta$. Therefore, for example,

$$-i\nabla\left[\exp(i\phi)(\nabla\phi - \gamma_0 A)\right] = \exp(i\phi)\left[(\nabla\phi)^2 - \gamma_0 A\nabla\phi - i\Delta\phi\right]$$

(In the last equation, it was taken into account that $\nabla A = 0$.) If we take into account the expressions derived above and divide each term of the equation by $|\psi|\exp(i\phi)$, we arrive at the GL equation (valid only if $|\psi| = $ const)

$$\alpha + \beta n_s + m\left(\frac{\hbar}{2m}\right)^2 (\nabla\phi - \gamma_0 A)^2 - i\frac{\hbar^2}{4m}\Delta\phi = 0$$

Since the left part is complex, both real and imaginary terms have to be equal to zero so the entire left part is zero. Therefore, the equation splits into two independent equations. If we put the imaginary part to zero, we get the Laplace equation

$$\Delta\phi = 0$$

in which the operator is $\Delta = \partial^2/\partial x^2 + \partial^2/\partial y^2 + \partial^2/\partial z^2$. Do not confuse this operator with the energy gap. Notice that the energy gap is not explicitly included in the GL equation.

Now, we put to zero the real part of the GL equation and get

$$\alpha + \beta n_s + mv_s^2 = 0$$

This equation implies that the superpair density is suppressed if the condensate is moving

$$n_s = -\frac{a}{\beta}\left(1 - \frac{mv_s^2}{a}\right) = n_{s0}\left(1 - \frac{mv_s^2}{|a|}\right)$$

The equations are written by taking into account the expression for the superfluid velocity $v_s = (\hbar/2m)(\nabla\phi - \gamma_0 A)$ (found in Problem 3). We also use $n_{s0} = -a/\beta$ to denote the density of the condensate in the case when $v_s = 0$. The suppression of n_s can be neglected if the superfluid speed is much smaller than the critical speed, $v_s \ll v_c = \sqrt{|a|/m} = \hbar/2m\xi(T)$.

Problem 5

Unit conversion rules (based on [1], Appendix 1). To convert any formula from Gaussian (also called cgs) units to SI units, make the following replacements in expressions related to electromagnetic quantities: speed of light $c \to \sqrt{1/\mu_0\epsilon_0}$; magnetic induction $B \to B\sqrt{4\pi/\mu_0}$; magnetic flux $\Phi \to \Phi\sqrt{4\pi/\mu_0}$; magnetic field $H \to H\sqrt{4\pi\mu_0}$; magnetization $M \to M\sqrt{\mu_0/4\pi}$; electric charge (as well as charge density) $Q \to (1/\sqrt{4\pi\epsilon_0})Q$; electric current (and also current density) $I \to (1/\sqrt{4\pi\epsilon_0})I$; electric capacitance $C \to C/4\pi\epsilon_0$; polarization $P \to (1/\sqrt{4\pi\epsilon_0})P$; voltage (or electric field, E, or electric potential) $V \to V\sqrt{4\pi\epsilon_0}$; Displacement $D \to \sqrt{4\pi/\epsilon_0}D$; conductivity $\sigma \to \sigma(4\pi\epsilon_0)^{-1}$; resistance $R \to 4\pi\epsilon_0 R$; magnetic inductance $L \to 4\pi\epsilon_0 L$; permeability $\mu \to \mu/\mu_0$; dielectric constant $\epsilon \to \epsilon/\epsilon_0$. Here, the permeability of free space is $\mu_0 = 4\pi \times 10^{-7}$ H/m and the vacuum permittivity is $\epsilon_0 = 8.85 \times 10^{-12}$ F/m. Using the rules listed above, answer the following questions:

a) If the expression for the flux quantum in Gaussian units is $\Phi_0 = hc/2e$, what is it in SI units?
b) Density of superpairs, when expressed in Gaussian units, is n_s. What is it in SI units?
c) The magnetic field penetration depth, in Gaussian units, is $\lambda = \sqrt{mc^2/8\pi e^2 n_s}$. What is the corresponding expression in SI units?

Solution (a) Make substitutions to the flux, the speed of light and the electric charge according to the unit conversion rules. The result is $\Phi_0\sqrt{4\pi/\mu_0} = h\sqrt{4\pi\epsilon_0}/2e\sqrt{\mu_0\epsilon_0}$. Thus, in SI units, the magnetic flux quantum is $\Phi_0 = h/2e$.
(b) The density of pairs is just the number of pairs per unit volume. Its notation does not change when the units are changed because n_s is not an electromagnetic quantity. Therefore, the answer is n_s. (c) The magnetic screening length can be written as $\lambda = (c/e)\sqrt{m/8\pi n_s}$. The expression under the square root does not change as we switch to SI units since it does not have electromagnetic quantities. The length, λ, does not change also. The ratio changes as $c/e \to (1/\sqrt{\mu_0\epsilon_0})(1/e)\sqrt{4\pi\epsilon_0} = (1/e)\sqrt{4\pi/\mu_0}$. By substituting into the formula

for the magnetic length, we finally get $\lambda = \sqrt{m/2n_s\mu_0 e^2}$. It follows also that the superpair density in SI units is $n_s = m/2\mu_0 e^2 \lambda^2$.

Problems 6, 7 and 8 pertain to the same situation

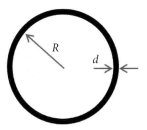

Problem 6

A thin nanowire of constant diameter is connected into a loop. The radius of this closed circular loop is $R = 10^{-7}$ m (so it can be called a nanoloop). The loop contains one superconducting vortex. Find the magnitude of the superfluid velocity v_s in the wire. No external magnetic field is applied. Assume the wire forming the loop is so thin that the magnetic field produced by the supercurrent is negligible. Comment on how v_s would change if the loop is deformed from the circular to some arbitrary shape, assuming that the length of the wire forming the loop remains unchanged. Electric capacitance of the loop is assumed negligibly small and the system is in local equilibrium.

Solution The presence of one vortex in the loop, by definition of a vortex, means that the integral of the phase gradient, $\nabla\phi$, taken over the entire loop must be equal 2π. In other words, the phase of the order parameter increases by 2π along the entire loop. Since the total current is conserved and the normal current is zero (assuming a steady state), then the supercurrent is also constant along the loop. The supercurrent is proportional to the phase gradient. Therefore, the phase gradient is constant along the loop. The length of the wire forming the loop is $2\pi R$. Therefore, the magnitude of the phase gradient is the total phase change around the loop divided by the loop length, i.e., $|\nabla\phi| = 2\pi/(2\pi R) = 1/R$. In the case when the magnetic field is negligible, the superfluid velocity is given by (1.2), so its magnitude is $v_s = (\hbar/2m)|\nabla\phi|$. Combining these expressions, we obtain the answer to this problem, namely, $v_s = \hbar/2mR \approx 570$ m/s.

If the loop is deformed but its length remains unchanged, all the arguments and the formulas presented above are still applicable. Therefore, the superfluid velocity remains fixed as the shape of the loop is changing.

Problem 7

The same circular loop as in Problem 6, having radius $R = 10^{-7}$ m and made of a superconducting wires of a constant diameter $d = 10$ nm, holds one vortex. Find the magnetic flux through the loop due to the vortex. Find the ratio of this flux to the flux quantum, $\Phi_0 = h/2e$. Does the considered vortex in the superconducting nanoloop carry one flux quantum or much less than that? Assume no external magnetic field is applied. Additional information: The presence of a vortex inside the loop means that there is a supercurrent in the loop and that the phase of the order parameter increases by 2π on a path around the loop. Assume $A = 0$ since there is no external magnetic field. The self-inductance of a circular loop (in SI units) is $L = \mu_0 R[\ln(16R/d) - 2]$, where $\mu_0 = 4\pi \times 10^{-7}$ H/m. In SI units, the magnetic field penetration depth, λ, in the pair density, n_s, are related as $n_s = m/2\mu_0 e^2 \lambda^2$. Assume that the material from which the wire loop is made has $\lambda = 400$ nm.

Solution If only one vortex is present inside the loop, the superfluid velocity is $v_s = \hbar/2mR$, as was determined in Problem 6. This expression is valid both in SI and Gaussian units since it does not contain electromagnetic quantities. Correspondingly, the supercurrent density is $j_s = 2en_s v_s = en_s\hbar/mR$. The total supercurrent in the loop is $I_s = (\pi d^2/4)j_s = \pi\hbar n_s d^2/4mR$. To find the current, one needs to know n_s. It is related to the magnetic penetration length as $n_s = m/2\mu_0 e^2 \lambda^2$. So, the current is (in SI units) $I_s = (h/2e)(d^2/8R\mu_0\lambda^2)$, where $h = 2\pi\hbar$.

The magnetic flux is therefore

$$\Phi = LI_s = \frac{h}{2e}\frac{d^2}{8\lambda^2}\left(\ln\frac{16R}{d} - 2\right) = \Phi_0 \frac{d^2}{8\lambda^2}\left(\ln\frac{16R}{d} - 2\right)$$

The ratio of the flux created by the coreless vortex sitting in the loop and the flux quantum is $\Phi/\Phi_0 = (d^2/8\lambda^2)[\ln(16R/d) - 2]$. We can put the numbers now and find $\Phi = 5 \times 10^{-19}$ Wb and $\Phi/\Phi_0 = 2.4 \times 10^{-4}$. Thus, a single coreless vortex present in the nanoloop creates a magnetic flux that is much smaller than the flux quantum.

Problem 8

Consider the same superconducting wire loop as in the previous problem. Estimate the maximum number, n_m, of vortices the loop can hold within its perimeter. The coherence length of the material from which the wire is made is $\xi(0) = 5$ nm. Assume that the wire diameter is sufficiently small, so the magnetic field produced by the supercurrent is negligible. The external magnetic field is zero. The temperature is sufficiently low so that $\xi(T) \approx \xi(0)$.

Solution Each vortex, contained inside the superconducting loop, produces a phase increase of 2π around the loop. If the number of vortices is n, the phase gradient is $|\nabla\phi| = 2\pi n/2\pi R = n/R$. The corresponding superfluid velocity is therefore $v_s = n\hbar/2mR$. The magnitude of the supercurrent density, according to (1.3), is $j_s = 2en_s v_s = nen_s\hbar/mR$. Note that here we make a plausible assumption that the vector quantities of interest, namely, the phase gradient, the superfluid velocity, and the supercurrent density are always directed parallel to the wire. Thus, we only use

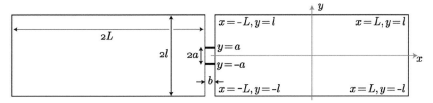

Figure C.1 Schematic presentation of a device with two nanowires connecting two thin-film electrodes. Such a structure was used to study the Little–Parks effect with nanowires [105, 110, 146]. The same device is shown on the cover of this book. It is analogous to Figure 2.1, but the notations are changed here as $X \to 2l$ and $Y \to 2a$. In this sample, two rectangular thin-film electrodes are connected by two nanowires. The nanowires are typically made using either DNA templates (see Figure 2.1 for the actual picture of the structure) or carbon nanotube templates. Here, $2l$ is the width of the rectangular thin-film electrodes and $2a$ is the distance between the nanowires. External magnetic field, B, is applied perpendicular to the electrodes, i.e., along z-axis. The length of each of the electrodes is $2L$, such that $L \gg l$. The length of each nanowire is b. The choice of coordinates is shown.

their absolute values (magnitudes). The value of the vorticity, n, corresponding to a given supercurrent density is therefore $n = mRj_s/en_s\hbar$. The maximum possible value of the supercurrent equals the critical current density j_c by the definition. The corresponding maximum vorticity then is $n_m = mRj_c/en_s\hbar$. As is explained in the Section "Kinetic Inductance and the CPR of a Thin Wire," the density of the superpairs, n_s, is related to the critical current density as $en_s = (3\sqrt{3}/4)j_c\sqrt{m/|\alpha|}$. The well known formula for the coefficient α is $\xi^2(T) = \hbar^2/4m|\alpha|$ (see [1] on the GL theory). If we combine these two equations, we get $en_s = 3\sqrt{3}j_c m\xi/2\hbar$. Therefore, the maximum number of vortices that can be trapped in the loop is $n_m = 2R/3\sqrt{3}\xi = 7.7$. Since the number of vortices is an integer, the correct answer is $n_m = 7$.

Problems 9 and 10 correspond to the same situation

Problem 9

For the device shown on the picture (Figure C.1) and for a given external magnetic field B, find the phase difference $\phi(B) = \phi_1 - \phi_2$ between the points which the wires are connected to. Assuming $a \ll l$, then $\phi_1 - \phi_2 = 2a\nabla\phi|_{(-L,0)}$. Note that the coordinates of the first connection point, having the phase ϕ_1, are $(-L, a)$ and the coordinates of the second connection point, having the phase ϕ_2, are $(-L, -a)$. Make the following additional assumptions: (a) the width of the electrodes, $2l$, is much smaller than the perpendicular magnetic field penetration depth, $\lambda_\perp$, so the magnetic field everywhere equals the external applied field; (b) the magnetic field is sufficiently weak, so that the order parameter amplitude in the sample is approximately constant and equals $|\psi|$; (c) the nanowires are sufficiently thin, so their effect on the phase gradients in the electrodes is negligible; (d) $L \to \infty$. Use the following Fourier analysis results $\int_{-l}^{l} \sin(k_n y)\sin(k_m y)dy = l\delta_{nm}$ and

$\int_{-l}^{l} y \sin(k_n y) dy = 2(-1)^n / k_n^2$. Here, $k_n = (\pi/l)(n+1/2)$, $n = 0, 1, 2, \ldots$ and $m = 0, 1, 2, \ldots$ are integer numbers, and $\delta_{nm} = 1$ if $n = m$ and $\delta_{nm} = 0$ if $n \neq m$. Also, use the Catalan number defined as $G = \sum_{n=0}^{\infty} (-1)^n/(2n+1)^2 \approx 0.916$. Assume also that the phase in the electrodes obeys the Laplace equation $\Delta \phi = 0$, if we choose $\nabla A = 0$.

Solution According to the analysis given in the solution of Problem 4, the pair density is $n_s = n_{s0}(1 - mv_s^2/|\alpha|)$. If the applied field is weak, the induced Meissner current is weak and the superfluid velocity is much lower than the critical velocity. Then, $n_s \approx n_{s0}$. As the amplitude of the order parameter is approximately constant, the phase varies according to the Laplace equation

$$\Delta \phi = 0$$

(see the solution of Problem 4 for details). If we solve this equation for the rectangular electrode (for the right electrode, for example), then we will be able to find the requested phase difference as

$$\phi = \phi_1 - \phi_2 = 2a \left. \frac{\partial \phi}{\partial y} \right|_{(x=-L, y=0)}$$

The vector potential has to be such that $\nabla A = 0$. To satisfy this condition, we choose $A = By e_x$, where e_x is a unit vector directed along the x-axis. In other words, the vector potential is parallel to the x-axis everywhere. It is then clear that the presence of the Meissner current, which circles around each electrode, requires some phase gradient to occur near those electrode's edges which are parallel to the y-axis. The edge of the electrode which is of interest to us is the line connecting the point $(-L,l)$ to the point $(-L,-l)$. This is the edge to which the wires are connect.

Since the current is $j_s \propto v_s \propto \nabla \phi - \gamma_0 A$ and it is parallel to the edges of the electrodes, the boundary condition for the Laplace equation is

$$\mathbf{n} \cdot (\nabla \phi - \gamma_0 A)|_C = 0$$

where $\mathbf{n}$ is the unit vector normal to the edges and C is the boundary of the electrode. The current which flows into nanowires is assumed negligible. Note that vector quantities, including $\mathbf{n}$, are assumed to be localized in two dimensions, namely, in the x and y plane. The only exception is the magnetic field, which is parallel to z-axis.

The solution to the Laplace equation, which can be checked by the direct substitution into the equation, is

$$\phi = \sum_k [A_k \exp(-kx) + B_k \exp(kx)] \sin(ky)$$

where k are some unknown "wave numbers." The coefficients A_k and B_k need to be chosen to satisfy the boundary condition of zero current flowing across the boundary. The solution with these boundary conditions is unique, up to adding a constant to the entire phase distribution.

Let us discuss the right electrode for certainty. Since the vector potential is parallel to the long edge of the electrode, the phase gradient in the direction perpendicular to the edge must be zero (otherwise, there would be a supercurrent flowing outside the electrode). Mathematically, this means that at the edges $y = \pm l$ (i.e., the long edges), we have

$$\partial \phi / \partial y |_{y=\pm l} = 0$$

This is satisfied if the wave numbers, defined above, satisfy $k = k_n \equiv \pi(n+1/2)/l$, where $n = 0, 1, 2, \ldots$ are integers.

The boundary condition at the edge $x = -L$ is $v_s e_x = 0$ which translates into $\partial \phi / \partial x |_{(-L,0)} = \gamma_0 A_x = \gamma_0 B y$, where $A_x = A e_x$ is the x-component of the vector potential. Taking into account the general solution for the phase, the boundary condition at the $x = -L$ edge is

$$\sum_n 2 B_n k_n \cosh(k_n L) \sin(k_n y) = \gamma_0 B y$$

where we set $x = -L$ (which is the position of the edge for which the boundary condition is written), $A_k|_{k=k_n} = A_n$, and $B_k|_{k=k_n} = B_n$. We also put $A_n = -B_n$. We will be able to find the solution with this restriction. Since the solution is unique, finding the solution proves that this choice is valid. Both sides can be multiplied now by $\sin(k_m y)$ and integrated over the interval $-l < y < l$. Here, m is an integer number. The equation then becomes

$$2 B_m k_m \cosh(k_n L) l = \frac{2 \gamma_0 B (-1)^m}{k_m^2}$$

where we have used the equalities $\int_{-l}^{l} \sin(k_n y) \sin(k_m y) dy = l \delta_{nm}$ and $\int_{-l}^{l} y \sin(k_m y) dy = 2(-1)^m / k_m^2$. The coefficient defining the phase are therefore

$$B_n = \frac{\gamma_0 B (-1)^n}{\cosh(k_n L) l k_n^3}$$

From this, the phase is

$$\phi = 2 \gamma_0 B \sum_n \frac{(-1)^n}{l k_n^3 \cosh(k_n L)} \sinh(k_n x) \sin(k_n y)$$

The y-derivative of the phase at the edge $x = -L$ at the point $y = 0$ is

$$\frac{\partial \phi}{\partial y} = 2 \gamma_0 B \sum_n \frac{(-1)^n}{l k_n^2} \tanh(k_n L)$$

In the limit $L = \infty$, the derivative is

$$\frac{\partial \phi}{\partial y}\bigg|_{-L,0} = \frac{8 \gamma_0 B l}{\pi^2} \sum_n \frac{(-1)^n}{(2n+1)^2} = \frac{8 G \gamma_0 B l}{\pi^2}$$

Thus, the phase difference between the tips of the wires is

$$\Delta\phi = 2a\left.\frac{\partial\phi}{\partial y}\right|_{-L,0} = \frac{16 a l\, G\, \gamma_0\, B}{\pi^2} = \frac{8G}{\pi}\frac{4 a l\, B}{\Phi_0}$$

where $\Phi_0 = hc/2e = 2\pi/\gamma_0$ is the magnetic flux quantum.

Problem 10

Find the magnetic field period, ΔB, of Little–Parks (LP) oscillations in this system with two parallel nanowires.

Solution The LP oscillations occur because whenever the phase imposed on a closed superconducting loop or counter is $2\pi n$ (where n is an integer, as before), it is possible to put n vortices (or antivortices) in the loop and compensate the externally imposed phase increase or supercurrent created by vector-potential to zero. A vortex sitting in a loop is, by definition, a quasistable state of the condensate characterized by a phase rotation of 2π around the loop. An antivortex is the same as the vortex, but the phase increases by -2π around the host loop. If the externally imposed phase is not $2\pi n$, then it is not possible to compensate the phase gradient to zero by introducing vortices (or antivortices) into the loop. Therefore, the kinetic energy of the condensate is zero when the imposed phase shift is $2\pi n$, and it is above zero if the imposed phase increase is different than $2\pi n$. Thus, the free energy of the condensate shows periodic oscillations with the applied field, with all consequences, such as T_c oscillations. Note, by the way, that an entrance of a vortex into the loop is equivalent to the occurrence of exactly one phase slip on one of the wires forming the loop.

In the example considered, the loop is formed by the two parallel wires and the two identical electrodes to which the wires are connected. Each electrode imposes a phase difference $\Delta\phi = (8G/\pi)(4 a l\, B/\Phi_0)$ on the loop. Thus, the total externally imposed phase difference is $2\Delta\phi$ (because there are two electrodes and each of them generates the same phase difference). When the imposed phase difference equals 2π, one period of the LP oscillation is complete. Thus, the period is defined by the condition $2\Delta\phi = 2\pi$, which leads to

$$\Delta B = \frac{\pi^2}{8G}\frac{\Phi_0}{4 a l}$$

In this respect, it is different from the LP effect that would occur in a loop made exclusively of thin wires. In such cases, as is well known, the period would be $\Delta B = \Phi_0/2ab$ (where $2ab$ is the area of the loop), and the superfluid velocity is defined by the vector potential. In the case considered here, involving nanowires *and* thin film electrodes, the electrodes effectively act as vector-potential antennas. It can be said that they sense the applied vector potential, convert it into the phase gradient and impose it on nanowires. The operation of such vector-potential antennas is based on the Meissner effect, which is a manifestation of the fact that the supercurrent is proportional to the vector potential.

Appendix C Problems and Solutions | 233

Problems 11–13 are mutually related In a short nanowire ($L < 3.5\xi$), biased with a constant current I, a phase slip can be modeled as an effective phase particle crossing a potential barrier in the Gibbs energy in the space of phase difference ϕ. The barrier is discussed in Section 8.7 titled "Universal 3/2 Power Law for Phase Slip Barrier." The height of the barrier was derived as

$$\Delta G = \frac{2\hbar\sqrt{2}}{3e} I_c^{3/2} (-Y_2)^{-1/2} \left(1 - \frac{I}{I_c}\right)^{3/2}$$

The constant Y_2 can be found if the current–phase relationship (CPR), $I_s = I_s(\phi)$, of the nanowire is known. The general form of the CPR near the critical current (the Taylor expansion in fact) is

$$I_s = I_c + \frac{Y_2}{2}(\phi - \phi_c)^2$$

As was argued in Section 2.12, titled "Kinetic Inductance and the CPR of a Thin Wire," the CPR for a short wire can be approximately written as the cubic CPR

$$I_s = \frac{3\sqrt{3}}{2} I_c \left[\frac{\xi\phi}{L} - \left(\frac{\xi\phi}{L}\right)^3 \right]$$

The constant Y_2 is defined by the equations $Y_2 \equiv (d^2 I_s/d\phi^2)|_{\phi=\phi_c}$, which follows from the general form of the CPR. Here, ϕ_c is the phase difference on the wire at which the supercurrent is the maximum and Y_2 is the curvature of the CPR function at $\phi = \phi_c$.

Problem 11

For the wires described by the cubic CPR (see above), estimate the height of the barrier for phase slips as a function of bias current. Assume that the critical current is known and equals I_c. Comment on how the expression for the barrier would change if the wire is long compared to the coherence length.

Solution The expression for the height of the Gibbs potential barrier was found in Section 8.7, "Universal 3/2 Power Law for Phase Slip Barrier." It is

$$\Delta G = \frac{4\sqrt{2}}{3} \frac{\hbar}{2e} I_c^{3/2} |Y_2|^{-1/2} \left(1 - \frac{I}{I_c}\right)^{3/2}$$

To complete the analysis, we need Y_2. The CPR for a short nanowire is

$$I_s = \frac{3\sqrt{3}}{2} I_c \left[\frac{\xi\phi}{L} - \left(\frac{\xi\phi}{L}\right)^3 \right]$$

The first derivative is

$$\frac{dI_s}{d\phi} = \frac{3\sqrt{3}}{2} I_c \left[\frac{\xi}{L} - 3\phi^2 \left(\frac{\xi}{L}\right)^3 \right]$$

Therefore, using the condition for a maximum, $dI_s/d\phi = 0$, we find the critical phase $\phi_c = L/\sqrt{3}\xi$. The second derivative of the CPR is

$$\frac{d^2 I_s}{d\phi^2} = -9\sqrt{3} I_c \left(\frac{\xi}{L}\right)^3 \phi$$

$$Y_2 \equiv \left.\frac{d^2 I_s}{d\phi^2}\right|_{\phi=\phi_c} = -9 I_c \left(\frac{\xi}{L}\right)^2$$

Therefore, the barrier for phase slips can be presented as

$$\Delta G = \frac{4\sqrt{2}}{9} \frac{\hbar}{2e} I_c \frac{L}{\xi} \left(1 - \frac{I}{I_c}\right)^{3/2}$$

This expression is valid if $L < 3.5\xi$, i.e., when the wire is sufficiently short to be considered as a junction [18]. Such limitation is due to the fact that the cubic CPR (sometimes called the Likharev CPR) is derived under the assumption that the order parameter is suppressed in the entire wire during each phase slip event.

Problem 12

For a short nanowire, described by the cubic CPR given above, estimate the width of the potential barrier as a function of the bias current I.

Solution The expression for the phase values corresponding to the minimum and the maximum of the Gibbs potential was derived in Section 8.7, "Universal 3/2 Power Law for Phase Slip Barrier." It is

$$\phi_{\pm} = \phi_c \pm \frac{\sqrt{2}(I_c - I)^{1/2}}{\sqrt{-Y_2}}$$

Assuming the barrier is approximately symmetric, the barrier width, $\Delta\phi$, can be estimated as twice the distance from the location (in the space of phase) of the minimum of the potential, ϕ_-, to the location of the maximum of the potential, ϕ_+. Thus, the barrier width is

$$\Delta\phi = 2(\phi_+ - \phi_-) = 4\frac{\sqrt{2}(I_c - I)^{1/2}}{\sqrt{-Y_2}}$$

In the solution to the previous problem, it was found that $Y_2 = -9 I_c(\xi/L)^2$. Therefore, the barrier width is

$$\Delta\phi = 4\frac{\sqrt{2}}{3}\frac{L}{\xi}\left(1 - \frac{I}{I_c}\right)^{1/2}$$

Problem 13

The situation is similar to the previous two problems, but now we consider a wire that is significantly longer than the coherence length, so it can not be treated simply as a junction but has to be treated as an extended weak link. In this case, the free energy barrier for a phase slip in zero bias current was estimated in [19]. It is $\Delta G(0) = \sqrt{6}\hbar I_c/2e$. Assume also that the CPR can still be presented as a Taylor series $I_s = I_c + (Y_2/2)(\phi - \phi_c)^2$, according to (8.26). Find the height and the width of the Gibbs energy barrier for a phase slip as functions of bias current I. The critical current of the wire is I_c.

Solution The free energy barrier at zero bias can be estimated as (see Section 8.7 titled "Universal 3/2 Power Law for Phase Slip Barrier")

$$\frac{\sqrt{6}\hbar I_c}{2e} = \Delta G \approx \frac{4\hbar\sqrt{2}}{6e} I_c^{3/2}(-Y_2)^{-1/2}\left(1 - \frac{I}{I_c}\right)^{3/2}\bigg|_{I=0}$$

From this, it follows that

$$Y_2 \approx -\frac{16 I_c}{27}$$

The expression for the height of the Gibbs potential barrier (found in Section 8.7 titled "Universal 3/2 Power Law for Phase Slip Barrier") is

$$\Delta G \approx \frac{4\sqrt{2}}{3} \frac{\hbar}{2e} I_c^{3/2}(-Y_2)^{-1/2}\left(1 - \frac{I}{I_c}\right)^{3/2}$$

Using the expression for Y_2, the barrier height can be written as

$$\Delta G \approx \sqrt{6}\frac{\hbar}{2e} I_c \left(1 - \frac{I}{I_c}\right)^{3/2}$$

The expression for the phase values corresponding to the minimum and the maximum of the Gibbs potential was derived in Section 8.7 titled "Universal 3/2 Power Law for Phase Slip Barrier." It is

$$\phi_\pm = \phi_c \pm \frac{\sqrt{2}(I_c - I)^{1/2}}{\sqrt{-Y_2}}$$

Assuming the barrier is approximately symmetric, the barrier width, $\Delta\phi$, can be estimated as twice the distance from the minimum of the potential to the maximum of the potential. Thus, the barrier width is

$$\Delta\phi = 2(\phi_+ - \phi_-) = 4\frac{\sqrt{2}(I_c - I)^{1/2}}{\sqrt{-Y_2}}$$

Substituting the expression for Y_2, we get

$$\Delta\phi \approx 3\sqrt{6}\left(1 - \frac{I}{I_c}\right)^{1/2}$$

The fact that at zero current the formula predicts the barrier width to be slightly larger than 2π is unphysical. This fact illustrates the approximate nature of the obtained expression, which is valid only near I_c. Yet, even at zero bias, the formula is not too different from the correct barrier width, which is 2π.

Problem 14

Use the Kurkijärvi–Garg (KG) model (Section 9.6) and the corresponding theorem (Section 9.7) to predict qualitatively how the standard deviation, σ, of the switching current changes as the bias-current sweep speed is increased. Assume $b > 1$ in the expression for the energy barrier for phase slips, which is $\Delta G = U_c \exp(1 - I/I_c)^b$.

Solution The theorem proven in Section 9.7 gives the following relation between σ and the slope of the logarithm of the switching rate versus current:

$$\sigma = \frac{\pi}{\sqrt{6}} \frac{1}{z} = 1.28 \left[\frac{d(\ln \Gamma)}{dI} \right]^{-1} \Bigg|_{I = \langle I_{sw} \rangle}$$

where the derivative is taken at $I = \langle I_{sw} \rangle$. The derivative can be explicitly written as

$$\frac{d(\ln \Gamma)}{dI} \Bigg|_{I=\langle I_{sw}\rangle} \approx \frac{U_c}{k_B T_{esc}} b \left(1 - \frac{\langle I_{sw} \rangle}{I_c}\right)^{b-1} \left(-\frac{1}{I_c}\right)$$

If the sweep speed is increased, the mean value of the switching current, $\langle I_{sw} \rangle$, becomes larger. This is because if the current is increased rapidly, the switching has a lower probability to happen at a low current. Thus, the term $(1 - \langle I_{sw} \rangle/I_c)^{b-1}$ becomes smaller as the sweep speed is made larger. Since σ is inversely proportional to $d(\ln \Gamma)/dI$, one concludes that σ becomes larger if the current sweeping rate is increased.

Problem 15

Use the Kurkijärvi–Garg (KG) model (Section 9.6) to determine the approximate functional dependence of the standard deviation, σ, as a function of the sweeping speed $v_I \equiv dI/dt$. Assume the phase slip barrier is $\Delta G = U_c \exp(1 - I/I_c)^b$, where $b = 3/2$.

Solution According to the KG model, the standard deviation is

$$\sigma = \frac{\pi I_c}{\sqrt{6b}} \left(\frac{k_B T_{esc}}{U_c}\right)^{1/b} \kappa^{(1-b)/b}$$

Here, the only factor which depends on v_I is $\kappa = \ln(\Omega \sigma/v_I)$. The attempt frequency Ω is assumed constant. The power is $(1 - b)/b = -(1/3)$. Thus, the dispersion depends on v_I as

$$\sigma(v_I) \propto \left[\ln\left(\frac{\Omega \sigma}{v_I}\right)\right]^{-1/3}$$

This definition is slightly circular. Apparently, the dependence of σ on the sweeping speed is very slow, namely, logarithmic. Thus, one can put into the logarithm any generic value of the standard deviation, σ_0. Thus, we find the following approximate function, describing how the dispersion depends on the speed at which the bias current is ramped up,

$$\sigma(v_I) \propto \left[\frac{1}{\ln(\Omega\,\sigma_0/v_I)}\right]^{1/3}$$

Here, σ_0 is the dispersion corresponding to some value of v_I withing the range of interest. The obtained result also means that

$$\sigma(v_I)^{-3} \propto \ln(\Omega\,\sigma_0) - \ln(v_I)$$

i.e., σ^{-3} is a linear function of $\ln(v_I)$. The result, as it stands, clarifies the shape of the function $\sigma(v_I)$, but it cannot be used to find the absolute value of σ.

Problem 16

The normal state resistance of a nanowire is $R_n = 9.4\,\mathrm{k\Omega}$. The nanowire is connected to a pair of macroscopic electrodes. According to the model of superconductor-insulator transitions in thin wires, described in Section 10.1, "Simple Model of SIT in Thin Wires," the wire has to be resistive (i.e. nonsuperconducting). Namely, it can either act as a normal resistor or as an insulator as the temperature approaches zero. Can this wire be converted into the superconducting regime (meaning zero resistance in the limit of zero temperature and zero bias current) if it is shunted with a nanoscale ohmic resistor of value $3000\,\Omega$? Explain your answer.

Solution The answer is "no." The reason is that the wire is already shunted with the macroscopic electrodes, which always have an impedance of the order of $100\,\Omega$. Thus, connecting an addition $3000\,\Omega$ resistor in parallel will not significantly change the impedance that shunts the wire. According to the model, the wire remains insulating not because it is not shunted strongly enough, but because its impedance, which is approximately equal to its normal state resistance, is larger than the quantum resistance $R_q = h/(2e)^2 \approx 6.5\,\mathrm{k\Omega}$. Thus, the wire is on the resistive side of the Schmid quantum transition.

Problem 17

A thin wire of length $L = 100\,\mathrm{nm}$, having critical current of $I_c = 300\,\mathrm{nA}$, is connected to a capacitor (made of a superconducting metal also) of value $C = 10^{-14}\,\mathrm{F}$. The coherence length for the wire is $\xi = 10\,\mathrm{nm}$. Estimate the root-mean-square (rms) value of the supercurrent in the wire, in the limit of zero temperature and assuming that the system is in its ground state.

Solution Superconducting wire acts as an inductor. When connected to a capacitor, an LC-circuit is formed, which can be considered, approximately, as a harmonic oscillator, in which the current through the wire and the charge on the capacitor are the two conjugate variables. Any harmonic oscillator undergoes zero-point motion or fluctuations due to Heisenberg uncertainty principle. The energy of zero-point fluctuations is know from elementary quantum mechanics and it is $E_{ZPE} = \hbar\omega_0/2$, where ω_0 is the frequency of free oscillations of the oscillator. For an LC-circuit, this frequency is $\omega_0 = 1/\sqrt{L_k C}$, where L_k is the kinetic inductance of the wire. To solve the problem, we need to find the zero-point energy first. The kinetic inductance of the wire is (see (2.80))

$$L_k = \frac{2}{3\sqrt{3}} \frac{L}{\xi} \frac{\hbar}{2e I_c} \approx 4\,\text{nH}$$

So the natural frequency of the device is $\omega_0 \approx 1.5 \times 10^{11}\,\text{s}^{-1}$. The zero-point energy is then $E_{ZP} \approx 8 \times 10^{-24}\,\text{J}$. To estimate the rms current, $I_{rms} \equiv (\langle I_s^2 \rangle)^{1/2}$, take into account that the energy stored in a harmonic oscillator is shared equally among various forms of energy. In the case considered, there are two forms, namely, the electrostatic charging energy of the capacitor and the kinetic energy of the condensate in the wire. Both are equal to $E_{ZPE}/2$. The kinetic energy of the condensate is $L_k I_{rms}^2/2$. Therefore, $L_k I_{rms}^2/2 = \hbar\omega_0/4$ and consequently $I_{rms} = (\hbar/2L_k\sqrt{L_k C})^{1/2} \approx 45\,\text{nA}$.

Problem 18

The standard deviation, σ, of the switching current, I_{sw}, measured on a thin superconducting wire, exhibits a saturation at temperature below a crossover temperature T^*. In other words, $d\sigma/dT = 0$ for $T < T^*$. The saturation signifies that the rate of quantum phase slips is much larger than the rate of thermally activated phase slips (TAPS), so TAPS can be completely neglected. Assume that each phase slip causes a switching event and find the temperature dependence of the mean switching current.

Solution According to the Kurkijärvi–Garg model, the mean switching current and its dispersion are related as (9.27)

$$\langle I_{sw} \rangle = I_c - \frac{\sqrt{6b\kappa}}{\pi}\sigma$$

where b and κ do not depend on temperature, at least within the quantum regime $T < T^*$. The temperature dependence occurs due to the critical current, $I_c = I_c(T)$, which follows the Bardeen formula (11.1)

$$I_c = I_c(0)\left(1 - \frac{T^2}{T_c^2}\right)^{3/2}$$

where $I_c(0)$ is the extrapolation of the critical current to zero temperature, which is a constant. Combining the equations we arrive at the following temperature dependence of the switching current

$$\langle I_{sw} \rangle = I_c(0) \left(1 - \frac{T^2}{T_c^2}\right)^{3/2} - \frac{\sqrt{6}b\kappa}{\pi}\sigma$$

Problem 19

The normal state resistance of a thin wire of length $L = 100$ nm is $R_n = 3000\,\Omega$ and the coherence length is $\xi(0) = 5$ nm. Estimate its critical current.

Solution According to (9.15), the critical current can be estimated as

$$I_c(0) = \frac{\pi L \Delta}{3\sqrt{3}e R_n \xi(0)} \approx 2.4\,\mu A$$

where $\Delta = 1.76 k_B T_c$ is the energy gap.

Problem 20

A thin MoGe nanowire is biased with a current $I = 10\,\mu A$, while its critical current is $I_c = 14\,\mu A$. The wire length is $L = 100$ nm and its diameter is 15 nm. The bath temperature is $T = 0.3$ K. Initially, the wire is in the superconducting state. Evaluate whether a single phase slip would switch the wire into the normal state or if it would not. The critical temperature of the wire at zero bias current is $T_c = 4$ K. The density of the wire is $\rho_{MoGe} = 9700$ kg/m^3. The electronic specific heat is given by the Sommerfeld formula $c_w = (2/3)\pi^2 N_0 k_B^2 T$. For the heat capacity estimate, choose $T = 1$ K. The density of states of MoGe is $N_0 \approx 4 \times 10^{46}\,\text{J}^{-1}\,\text{m}^{-3}$.

Solution The Joule heat released by one phase slip is $Q_J = I\Phi_0 \approx 2 \times 10^{-20}$ J. The specific heat of the electrons is $c_w = (2/3)\pi^2 N_0 k_B^2 T \approx 50$ J/K m^3. The specific heat of phonons can be neglected at low temperatures. The volume of the wire is $V_w = L(\pi d^2/4) \approx 1.8 \times 10^{-23}$ m^3. So the heat capacity of the entire wire is $C_w = c_w V_w \approx 9 \times 10^{-22}$ J/K. Since, generally speaking, the phase slips occur on a very short time scale compared to the time needed for the heat to flow from the wire to the electrodes, the temperature jump can be estimated as $\Delta T = Q_J/C_w \approx 23$ K. Let us also estimate the current-dependent critical temperature, $T_c(I)$. According to the Bardeen formula (11.1), we have

$$T_c(I)^2 = T_c^2 \left[1 - \left(\frac{I}{I_c}\right)^{2/3}\right]$$

Thus, $T_c(I) = 0.8$ K. Now, we conclude that the temperature jump associated with just one phase slip would definitely switch the nanowire into the normal state, because $T_c(I) < T + \Delta T$, where $T = 0.3$ K is the temperature of the wire before the phase slip. Note that the temperature after the occurrence of the phase slip is $T + \Delta T \approx 23.3$ K.

References

1 Tinkham, M. (1996) Introduction to Superconductivity, 2nd edn., McGraw-Hill, New York.
2 de Gennes, P.G. (1966) *Superconductivity of Metals and Alloys*, W.A. Benjamin, New York, Amsterdam.
3 Kamerlingh Onnes, H. (1911) Further experiments with liquid helium, B. On the change of resistance of pure metals at very low temperatures, etc. Communication No. 119 from the Physical Laboratory at Leiden, Koninklijke Akademie van Wetenschappen, Proceedings of the Section of Sciences, volume XII, pp. 1107–1113.
4 Bardeen, J., Cooper, L.N., and Schrieffer, J.R. (1957) *Phys. Rev.*, **108**, 1175.
5 Hoddeson, L. and Daitch, V. (2002) *True Genius: The Life and Science of John Bardeen*, Joseph Henry Press.
6 Bardeen, J. and Pines, D. (1955) *Phys. Rev.*, **99**, 1140.
7 Gor'kov, L.P. (1959) *Sov. Phys. JETP*, **9**, 1364.
8 Tuominen, M.T., Hergenrother, J.M., Tighe, T.S., and Tinkham, M. (1992) *Phys. Rev. Lett.*, **69**, 1997; also (1993) *Phys. Rev. B*, **47**, 11599.
9 Fulton, T.A. and Dolan, G.J. (1987) *Phys. Rev. Lett.*, **59**, 109.
10 Gor'kov, L.P. (1958) *Exp. Theor. Phys. (USSR)*, **34**, 735; (English transl.: (1958) *Sov. Phys. JETP*, **7**, 505.)
11 Anderson, P.W. and Dayem, A.H. (1964) *Phys. Rev. Lett.*, **13**, 195.
12 Mccumber, D.E. (1968) *J. Appl. Phys.*, **39**, 2503.
13 Joyez, P., Vion, D., Götz, M., Devoret, M., and Esteve, D. (1999) *J. Superconduct.*, **12**, 757.
14 Anderson, J.T., Carlson, R.V., and Goldman, A.M. (1972) *J. Low Temp. Phys*, **8**, 29.
15 Josephson, B.D. (1962) *Phys. Lett.*, **1**, 251.
16 Abrikosov, A.A. and Gor'kov, L.P. (1961) *Sov. Phys. JETP*, **12**, 1243.
17 Little, W.A. (1967) *Phys. Rev.*, **156**, 396.
18 Likharev, K.K. (1979) *Rev. Mod. Phys.*, **51**, 101.
19 Tinkhama, M. and Lau, C.N. (2002) *Appl. Phys. Lett.*, **80**, 2946.
20 Bardeen, J. (1962) *Rev. Mod. Phys.*, **34**, 667.
21 H.A. Kramers (1940) *Physica*, **7**, 284.
22 Lukens, J.E., Warburton, R.J., and Webb, W.W. (1970) *Phys. Rev. Lett.*, **25**, 1180.
23 Newbower, R.S., Beasley, M.R., and Tinkham, M. (1972) *Phys. Rev. B*, **5**, 864.
24 Tumulka, R. (2006) *Proc. R. Soc. A*, **462**, 1897.
25 Blanchard, P., Giulini, D., Joos, E., Kiefer, C., and Stamatescu, I.-O. (eds) (1996) *Decoherence: Theoretical, Experimental, and Conceptual Problems*, Springer, Berlin.
26 Schrödinger, E. (1935) *Naturwissenschaften*, **23**, 807.
27 Byrne, P. (2007) *The Many Worlds of Hugh Everett*, Scientific American.
28 Leggett, A.J. (1978) *J. Phys. Colloq.*, **39**, 1264.
29 Leggett, A.J. (1980) *Prog. Theor. Phys. Suppl.*, **69**, 80.
30 Caldeira, A.O. and Leggett, A.J. (1981) *Phys. Rev. Lett.*, **46**, 211.

31 Leggett, A.J. et al. (1987) *Rev. Mod. Phys.*, **59**, 1.
32 Caldeira, A.O. and Leggett, A.J. (1983) *Ann. Phys. (N.Y.)*, **149**, 374.
33 Martinis, J.M., Devoret, M.H., and Clarke, J. (1987) *Phys. Rev. B*, **35**, 4682.
34 Clarke, J., Cleland, A.N., Devoret, M.H., Esteve, D., and Martinis, J.M. (1988) *Science*, **239**, 992.
35 Wernsdorfer, W., Bonet Orozco, E., Hasselbach, K., Benoit, A., Mailly, D., Kubo, O., Nakano, H., and Barbara, B. (1997) *Phys. Rev. Lett.*, **79**, 4014.
36 Nakamura, Y., Pashkin, Y.A., and Tsai, J.S. (1999) *Nature*, **398**, 6730.
37 Friedman, J.R., Patel, V., Chen, W., Tolpygo, S.K., and Lukens, J.E. (2000) *Nature*, **406**, 43.
38 Chiorescu, I., Nakamura, Y., Harmans, C.J.P.M., and Mooij, J.E. (2003) *Science*, **299**, 5614.
39 Wallraff, A., Lisenfeld, J., Lukashenko, A., Kemp, A., Fistul, M., Koval, Y., and Ustinov, A.V. (2003) *Nature*, **425**, 155.
40 Shor, P.W. (1994) Proc. 35th Annual Symposium on Foundations of Computer Science, 124, IEEE Computer Society Press, p. 124.
41 Shor, P. (1997) *SIAM J. Comput.*, **26**, 1484.
42 Bouchiat, V., Vion, D., Joyez, P., Esteve, D., and Devoret, M.H. (1998) *Phys. Scr. T*, **76**, 165.
43 Wallraff, A., Schuster, D.I., Blais, A., Frunzio, L., Huang, R.-S. Majer, J., Kumar, S., Girvin, S.M., and Schoelkopf, R.J. (2004) *Nature*, **431**, 162.
44 Schoelkopf, R.J. and Girvin, S.M. (2008) *Nature*, **451**, 664.
45 Khlebnikov, S. (2003) arXiv:quant-ph/0210019v3.
46 Khlebnikov, S. and Pryadko, L.P. (2005) *Phys. Rev. Lett.*, **95**, 107007.
47 Khlebnikov, S. (2008) *Phys. Rev. B*, **77**, 014505.
48 Khlebnikov, S. (2008) *Phys. Rev. B*, **78**, 014512.
49 Kerman, A.J. (2012) Flux-charge duality and quantum phase fluctuations in one-dimensional superconductors. arXiv:1201.1859 [cond-mat.mes-hall].
50 Mooij, J.E. and Harmans, C.J.P.M. (2005) *New J. Phys.*, **7**, 219.
51 Mooij, J.E. and Nazarov, Y.V. (2006) *Nat. Phys.*, **2**, 169.
52 Van Run, A.J., Romijn, J., and Mooij, J.E. (1987) *Jpn J. Appl. Phys.*, **26**, 1765.
53 Giordano, N. (1988) *Phys. Rev. Lett.*, **61**, 2137.
54 Giordano, N. (1989) *Phys. Rev. Lett.*, **63**, 2417.
55 Giordano, N. (1994) *Physica B*, **203**, 460.
56 Giordano, N. (1990) *Phys. Rev. B*, **41**, 6350.
57 Cyrot, M. (1973) *Rep. Prog. Phys.*, **36**, 103.
58 Altomare, F., Chang, A.M., Melloch, M.R., Hong, Y., and Tu, C.W. (2006) *Phys. Rev. Lett.*, **97**, 017001.
59 Li, P., Wu, P.M., Bomze, Y., Borzenets, I.V., Finkelstein, G., and Chang, A.M. (2011) *Phys. Rev. Lett.*, **107**, 137004.
60 Tian, M., Kumar, N., Xu, S., Wang, J., Kurtz, J.S., and Chan, M.H.W. (2005) *Phys. Rev. Lett.*, **95**, 076802.
61 Fu, H.C., Seidel, A., Clarke, J., and Lee, D.-H. (2006) *Phys. Rev. Lett.*, **96**, 157005.
62 Bogoliubov, N.N. (1958) *Nuovo Cim.*, **7**, 794.
63 Bogoliubov, N.N. (1958) *Sov. Phys. JETP*, **7**, 41.
64 McCumber, D.E. (1968) *Phys. Rev.*, **172**, 427.
65 Meissner, H. (1959) *Phys. Rev. Lett.*, **2**, 458.
66 Holm, R. and Meissner, W. (1932) *Z. Phys.*, **74**, 715.
67 Holm, R. and Meissner, W. (1933) *Z. Phys.*, **86**, 787.
68 Meissner, H. (1958) *Phys. Rev.*, **109**, 686.
69 Thornton, S.T. and Marion, J.B. (2003) *Classical Dynamics of Particles and Systems*, 5th edn., Brooks Cole.
70 Ginzburg, V.L. and Landau, L.D. (1950) *Zh. Eksp. i Teor. Fiz.* **20**, 1064.
71 van der Marel, D., Leggett, A.J., Loram, J.W., and Kirtley, J.R. (2002) *Phys. Rev.B*, **66**, 140501(R).
72 Korn, G.A. and Korn, T.M. (2000) *Mathematical Handbook for Scientists and Engineers: Definitions, Theorems, and For-

mulas for Reference and Review, Dover Publications, Inc., Mineola, New York.

73 Volmer, M. and Weber, A. (1925) *Z. Phys. Chem.*, **119**, 277.

74 Farkas, L. (1927) *Z. Phys. Chem.*, **125**, 236.

75 Becker, R. and Döring, A. (1935) *Ann. Phys.*, **24**, 719.

76 Volmer, M. (1929) *Z. Elektrochem.*, **35**, 555.

77 Giaever, I. (1960) *Phys. Rev. Lett.*, **5**, 464.

78 Niemeyer, J. (1974) *PTB-Mitt.*, **84**, 251,.

79 Dolan, G.J. (1977) *Appl. Phys. Lett.*, **31**, 337,.

80 Jackel, L.D., Gordon, J.P., Hu, E.L., Howard, H.E., Fetter, L.A., Tennant, D.M., Epworth, H.W., and Kurkijärvi, J. (1981) *Phys. Rev. Lett.*, **47**, 697.

81 Voss, R.F. and Webb, R.A. (1981) *Phys. Rev. Lett.*, **47**, 265.

82 Chakravarty, S. (1982) *Phys. Rev. Lett.*, **49**, 681.

83 Schmid, A. (1983) *Phys. Rev. Lett.*, **51**, 1506.

84 Geigenmüller, U., and Ueda, M. (1994) *Phys. Rev. B*, **50**, 9369–9375.

85 Anderson, P.W. (1964) in *Lectures on Many Body Problem*, (ed. E. Caianello), Academic, New York.

86 Guinea, F. et al. (1985) *Phys. Rev. Lett.*, **54**, 263.

87 Fisher, M.P.A. and Zwerger, W. (1985) *Phys. Rev. B*, **32**, 6190.

88 Fröhlich, J. and Zegarlinski, B. (1991) *J. Stat. Phys.*, **63**, 455.

89 Schön, G. and Zaikin, A.D. (1990) *Phys. Rep.*, **198**, 237.

90 Bulgadaev, S.A. (1984) *JETP Lett.*, **39**, 315.

91 Bulgadaev, S.A. (1981) *Phys. Lett. A*, **86**, 213.

92 Werner, P. and Troyer, M. (2005) *Phys. Rev. Lett.*, **95**, 060201.

93 Penttilä, J.S., Parts, Ü., Hakonen, P.J., Paalanen, M.A., and Sonin, E.B. (1999) *Phys. Rev. Lett.*, **82**, 1004.

94 Miyazaki, H., Takahide, Y., Kanda, A., and Ootuka, Y. (2002) *Phys. Rev. Lett.*, **89**, 197001.

95 McCumber, D.E. (1968) *J. Appl. Phys.*, **39**, 3113.

96 Stewart, W.C. (1968) *Appl. Phys. Lett.*, **12**, 277.

97 Kautz, R.L. and Martinis, J.M. (1990) *Phys. Rev. B*, **42**, 9903.

98 Allen, M.P. and Tildesley, D.J. (1989) *Computer Simulation of Liquids*, Oxford University Press.

99 Broers, A.N. (1988) *IBM J. Res. Dev.*, **32**, 502.

100 Bezryadin, A., Lau, C.N., and Tinkham, M. (2000) *Nature*, **404**, 971.

101 Bezryadin, A. and Dekker, C. (1997) *J. Vac. Sci. Technol. B*, **15**, 793.

102 Bollinger, A.T., Rogachev, A., Remeika, M., and Bezryadin, A. (2004) *Phys. Rev. B – Rapid Commun.*, **69**, 180503(R).

103 Kelly, K.F. et al. (1999) *Chem. Phys. Lett.*, **313**, 445.

104 Bezryadin, A. (2008) *J. Phys. Condens. Matter*, **76**, 43202.

105 Hopkins, D. et al. (2005) *Science*, **308**, 1762.

106 Zhang, Y. and Dai, H. (2000) *Appl. Phys. Lett.*, **77**, 3015.

107 Bezryadin, A., Bollinger, A., Hopkins, D., Murphey, M., Remeika, M., and Rogachev, A. (2004) in *Dekker Encyclopedia of Nanoscience and Nanotechnology*, (eds J.A. Schwarz, C.I. Contescu, and K. Putyera), Marcel Dekker, Inc. New York, p. 3761.

108 Lau, C.N., Markovic, N., Bockrath, M., Bezryadin, A., and Tinkham, M. (2001) *Phys. Rev. Lett.*, **87**, 217003.

109 Lau, C.N. (2001) Ph.D. Thesis, Harvard University.

110 Hopkins, D.S., Pekker, D., Wei, T.-C., Goldbart, P.M., and Bezryadin, A. (2007) *Phys. Rev. B – Rapid Commun.*, **76**, 220506(R).

111 Hopkins, D.S. (2006) Ph.D. Thesis, University of Illinois at Urbana-Champaign.

112 Johansson, A., Sambandamurthy, G., Shahar, D., Jacobson, N., and Tenne, R. (2005) *Phys. Rev. Lett.*, **95**, 116805.

113 Liu, Y., Zadorozhny, Y., Rosario, M.M., Rock, B.Y., Carrigan, P.T., Wang, H. (2001) *Science*, **294**, 2332.

114 Aref, T. and Bezryadin, A. (2011) *Nanotechnology*, **22**, 395302.

115 Aref, T. (2010) Ph.D.Thesis, University of Illinois at Urbana-Champaign.

116 Aref, T., Levchenko, A., Vakaryuk, V., and Bezryadin, A. (2012) *Phys. Rev. B*, **86**, 024507.

117 Watson, J.D. and Crick, F.H.C. (1953) *Nature*, **171**, 737.
118 Rothemund, P.W. (2006) *Nature*, **440**, 297.
119 Andersen, E.S. et al. (2009) *Nature*, **459**, 73.
120 Aldaye, F.A., Palmer, A.L., Sleiman, H.F. (2008) *Science*, **321**, 1795.
121 Braun, E., Keren, K. (2004) *Adv. Phys.*, **53**, 441.
122 Watanabe, H., Manabe, C., Shigematsu, T., Shimotani, K., Shimizu, M. (2001) *Appl. Phys. Lett.*, **79**, 2462.
123 Seeman, N.C. (1998) *Angew. Chem. Int. Ed.*, **37**, 3220.
124 Braun, E., Eichen, Y., Sivan, U., Ben-Yoseph, G. (1998) *Nature*, **391**, 775.
125 Richter, J. et al. (2001) *Appl. Phys. Lett.*, **78**, 536.
126 Richter, J. et al. (2002) *Appl. Phys. A*, **74**, 725.
127 Mertig, M., Ciacchi, L.C., Seidel, R., Pompe, W. (2002) *Nano Lett.*, **2**, 841.
128 Averin, D.V. and Likharev, K.K. (1991) in *Mesoscopic Phenomena in Solids*, (eds B.L. Altshuler, P.A. Lee, and R.A. Webb), Elsevier, Amsterdam.
129 Grabert, H. and Devoret, M.H. (eds) (1992) *Single Charge Tunneling*, Plenum, New York.
130 Bollinger, A.T., Dinsmore, R.C., Rogachev, A., and Bezryadin, A. (2008) *Phys. Rev. Lett.*, **101**, 227003.
131 Sahu, M. et al. (2009) *Nat. Phys.*, **5**, 503.
132 Bollinger, A. (2005) PhD Thesis, University of Illinois at Urbana-Champaign.
133 Mael, D., Carter, W.L., Yoshizumi, S., and Geballe, T.H. (1985) *Phys. Rev. B*, **32**, 1476.
134 Kortright, J.B. and Bienenstock, A., *Phys. Rev. B* **37**, 2979 (1988).
135 Chu, S.L., Bollinger, A.T., and Bezryadin, A. (2004) *Phys. Rev. B*, **70**, 214506.
136 Bollinger, A.T., Rogachev, A., and Bezryadin, A. (2006) *Europhys. Lett.*, **76**, 505.
137 Halperin, B.I. and Nelson, D.R. (1979) *J. Low Temp. Phys.*, **36**, 599.
138 Hebard, A.F. and Fiory, A.T. (1983) *Phys. Rev. Lett.*, **50**, 1603.
139 Little, W.A. (1964) *Phys. Rev.*, **134**, A1416.
140 Ferrell, R.A. (1964) *Phys. Rev. Lett.*, **13**, 330.
141 Rice, T.M. (1965) *Phys. Rev.*, **140**, A1889.
142 McCumber, D.E. and Halperin, B.I. (1970) *Phys. Rev. B*, **1**, 1054.
143 Langer, J.S. and Ambegaokar, V. (1967) *Phys. Rev.*, **164**, 498.
144 Meidan, D., Oreg, Y., Refael, G. (2007) *Phys. Rev. Lett.*, **98**, 187001.
145 Zharov, A., Lopatin, A., Koshelev, A.E., and Vinokur, V.M. (2007) *Phys. Rev. Lett.*, **98**, 197005.
146 Pekker, D., Bezryadin, A., Hopkins, D.S., and Goldbart, P.M. (2005) *Phys. Rev. B*, **72**, 104517.
147 Krempasky, J.J. and Thompson, R.S. (1985) *Phys. Rev. B*, **32**, 2965.
148 Belkin, A., Brenner, M., Aref, T., Ku, J., and Bezryadin, A. (2011) *Appl. Phys. Lett.*, **98**, 242504.
149 Halperin, B.I., Refael, G., and Demler, E. (2010) *Int. J. Mod. Phys. B*, **24**, 4039.
150 Zhao, G. (2005) *Phys. Rev. B*, **71**, 113404.
151 Graybeal, J.M. and Beasley, M.R. (1984) *Phys. Rev. B*, **29**, 4167.
152 Graybeal, J.M. (1985) Ph.D. Thesis, Stanford University.
153 Graybeal, J.M. (1987) *Phys. Rev. Lett.*, **59**, 2697.
154 Ku, J., Manucharyan, V., and Bezryadin, A. (2010) *Phys. Rev. B*, **82**, 134518.
155 Mooij, J.E. and Schön, G. (1985) *Phys. Rev. Lett.*, **55**, 114.
156 Duan, J.M. (1995) *Phys. Rev. Lett.*, **74**, 5128.
157 Abrikosov, A.A., Gor'kov, L.P., and Dzyaloshinski, I.E. (1975) *Methods of Quantum Field Theory in Statistical Physics*, Dover, New York.
158 Kurkijärvi, J. (1980) in SQUID 80: Superconducting Quantum Interference Devices and their Applications, (eds H.D. Hahlbohm and H. Lubbig), de Gruyter, Berlin, p. 247.
159 Ivanchenko, Y.M. (1966) *Phys. Lett.*, **23**, 289.
160 Ivanchenko, Y.M. (1967) *Pis'ma Zh. Eksp. Teor. Phys.*, **6**, 879. For English translation see (1967) *JETP Lett.*, **6**, 313.
161 Dmitrenko, I.M., Yanson, I.K., and Yurchenko, I.I. (1967) Reports of

14th All-union Conference on Low-temperature Physics, Khar'kov.
162 Ivanchenko, Y.M. and Zilberman, L.A. (1968) *Zh. Eksp. Teor. Fiz.*, **55**, 2395. English translation: (1969) *Sov.Phys.-JETP*, **28**, 1272.
163 Larkin, A.I. and Ovchinnikov, Y.N. (1983) *Phys. Rev. B*, **28**, 6281.
164 Caldeira, A.O. and Leggett, A.J. (1983) *Ann. Phys.*, **149**, 374.
165 Mühlschlegel, B. (1959) *Z. Phys.*, **155**, 313.
166 Sheahen, T.P. (1966) *Phys. Rev.*, **149**, 368.
167 Thouless, D.J. (1960) *Phys. Rev.*, **117**, 1256.
168 Zaikin, A.D., Golubev, D.S., van Otterlo, A., and Zimányi, G.T. (1997) *Phys. Rev. Lett.*, **78**, 1552.
169 Zaikin, A.D., Golubev, D.S., van Otterlo, A., and Zimányi, G.T. (1998) *Phys. Usp.*, **41**, 226.
170 Arutyunov, K.Y., Golubev, D.S., and Zaikin, A.D. (2008) *Phys. Rep.*, **464**, 1.
171 Golubev, D.S. and Zaikin, A.D. (2008) *Phys. Rev. B*, **78**, 144502.
172 Golubev, D.S. and Zaikin, A.D. (2001) *Phys. Rev. B*, **64**, 014504.
173 Razavy, M. (2011) *Heisenberg's Quantum Mechanics*, World Scientific Publishing Co. Pte. Ltd, New Jersey.
174 Takagi, S. (2002) *Macroscopic Quantum Tunneling*, Cambridge University Press, Cambridge.
175 Arndt, M., Nairz, O., Voss-Andreae, J., Keller, C., Van der Zouw, G., and Zeilinger, A. (1999) *Nature*, **401**, 680.
176 Gerlich, S. et al. (2011) *Nat. Commun.*, **2**, 263.
177 Tinkham, M., Free, J.U., Lau, C.N., Markovic, N. (2003) *Phys. Rev. B*, **68**, 134515.
178 Shah, N., Pekker, D., and Goldbart, P.M. (2008) *Phys. Rev. Lett.*, **101**, 207001.
179 Kurkijärvi, J. (1972) *Phys. Rev. B*, **6**, 832.
180 Fulton, T.A. and Dunkleberger, L.N. (1974) *Phys. Rev. B*, **9**, 4760.
181 Garg, A. (1995) *Phys. Rev. B*, **51**, 15592.
182 van Otterlo, A., Golubev, D.A., Zaikin, A.D., and Blatter, G. (1999) *Eur. Phys. J. B*, **10**, 131.
183 Rogachev, A., Bollinger, A.T., and Bezryadin, A. (2005) *Phys. Rev. Lett.*, **94**, 017004.
184 Zgirski, M., Riikonen, K.-P., Touboltsev, V., and Arutyunov, K. (2005) *Nano Lett.*, **5**, 1029.
185 Altomare, F., Chang, A.M., Melloch, M.R., Hong, Y., and Tu, C.W. (2006) *Phys. Rev. Lett.*, **97**, 017001.
186 Altomare, F., Chang, A.M., Melloch, M.R., Hong, Y., and Tu, C.W. (2007) **98**, 169901(E).
187 Rogachev, A. and Bezryadin, A. (2003) *Appl. Phys. Lett.*, **83**, 512.
188 Rogachev, A., Wei, T.-C., Pekker, D., Bollinger, A.T., Goldbart, P.M., and Bezryadin, A. (2006) *Phys. Rev. Lett.*, **97**, 137001.
189 Xiong, P., Herzog, A.V., and Dynes, R.C. (1997) *Phys. Rev. Lett.*, **78**, 927.
190 Golubev, D.S. and Zaikin, A.D. (2001) *Phys. Rev. Lett.*, **86**, 4887.
191 Graybeal, J.M. and Beasley, M.R. (1984) *Phys. Rev. B*, **29**, 4167.
192 Brenner, M.W., Gopalakrishnan, S., Ku, J., McArdle, T.J., Eckstein, J.N., Shah, N., Goldbart, P.M., and Bezryadin, A. (2011) *Phys. Rev. B*, **83**, 184503.
193 Refael, G., Demler, E., and Oreg, Y. (2009) *Phys. Rev. B*, **79**, 094524.
194 Büchler, H.P., Geshkenbein, V.B., and Blatter, G. (2004) *Phys. Rev. Lett.*, **92**, 067007.
195 Coskun, U.C., Brenner, M., Hymel, T., Vakaryuk, V., Levchenko, A., and Bezryadin, A. (2012) *Phys. Rev. Lett.*, **108**, 097003.
196 Manucharyan, V.E., Koch, J., Glazman, L.I., and Devoret, M.H. (2009) *Science*, **326**, 113.
197 Brenner, M.W., Roy, D., Shah, N., Bezryadin, A. (2012) *Phys. Rev. B*, **85**, 224507.
198 Stuart, A. and Ord, K. (2009) Distribution theory, in *Kendall's Advanced Theory of Statistics*, Vol. 1, John Wiley & Sons.
199 James, F. (1988) *Statistical Methods in Experimental Physics*, 2nd edn., World Scientific Publishing Co. Pte. Ltd., Amsterdam.

Index

a

Abrikosov–Gor'kov (AAG) theorem 197, 223
AC measurements 97
aluminum-aluminum oxide-aluminum junction (Al-Al$_2$O$_3$-Al) 53
Ambegaokar–Baratoff formula 54
antiphase-slip 127
antiphase-slip (APS) 116
antiphase-slips 117, 119, 120, 125–127, 136, 180
antiproximity effect in nanowires 165
arbitrary infinitesimal variation 38
Arrhenius activation 177
 – exponent 106, 110
 – factor 68
 – law 8, 104, 109, 118, 164, 176, 189, 201, 202, 211
Arrhenius resistance 180
attempt frequency XIV, XXI, XXIII–XXV, 61, 67, 68, 72, 73, 107

b

Bardeen formula 136, 149, 157, 170, 171, 213, 238, 239
battery-operated preamplifier 98
bell-shaped distributions 218
black body radiation 94
Bloch wave 75
BNC cable 98
bogoliubons 29, 44, 53, 55, 56, 67, 114, 116, 121, 122, 155, 169
Bogoliubov quasiparticles 7, 121
Boltzmann constant 7
bosons 2
Brownian diffusion 124

c

calculus 11
 – of variations 40

Caldeira–Leggett
 – damping effect 180
 – macroscopic quantum tunneling theory 208
 – mechanism 173
 – quantum dissipation model 205
 – theory 73
carbon nanotube 82, 86
 – fluorinated 86
Carnot cycle 30
Catalan number 230
chip carrier 92
coefficient of viscosity 11
coherence length 47
condensation energy density 8, 18, 110
Cooper pair 2, 223
 – density 2
coplanar waveguide resonator 49
cotton insulation 94
Coulomb
 – blockade 87, 105, 202
 – energy 59, 60
 – island 3
 – repulsion 1, 87, 112
Crick–Watson double-helix 86
critical current 213
cryostat
 – cold finger 92, 93
cubic potential 53, 66, 71, 73
 – thermal escape 67
current-conservation equation 62
current–phase relationship 233
current–phase relationship (CPR) 50, 65
 – of the superconducting device 54

d

data acquisition (DAQ) cards 96
Dayem bridge 122, 123

Superconductivity in Nanowires, First Edition. Alexey Bezryadin.
© 2013 WILEY-VCH Verlag GmbH & Co. KGaA. Published 2013 by WILEY-VCH Verlag GmbH & Co. KGaA.

DC

- measurements 96
- transport setup 135

de Gennes solution 118, 120
deoxyribonucleic acid (DNA) 86
- as a structural template 86
- metal-coating 87
- molecules 86

depairing current 213
dichloroethane 81
- 1,2-dichloroethane 86

dielectric constant 113
digital analogue converter 98
Dirac delta function 145
dispersion 217
dissipative Schmid–Bulgadaev transition 11
distribution of the switching currents 139
DNA
- molecule 87

Drude formula 51
dry nitrogen gas 81
dynamic equilibrium condition 116

e

Eddy current 93
electric current 44
Electrodag-1415M 84
electrode 28, 35, 49, 66, 79, 101
- shunting effect 184

electromagnetic
- quantity 226

electromagnetic (EM)
- noise filter 93
- wave 94

electron 2, 28
- beam lithography 54, 79
- macromolecule 2
- uncondensed/unpaired 3

electron–electron
- attraction 1
- Coulomb repulsion 200
- repulsion 13

electron–phonon interactions 1
electronic transport measurement 95
electrostatic charging energy 238
elementary diffusion theory 223
equilibrium state 25

f

Fabry–Pérot interferometer 49
Faraday cage 29, 92
Fermi
- distribution energy 1
- energy 51
- velocity 51, 126, 223

fermion 2
flicker noise 220
fluorinated carbon nanotube 81, 86
fluorine atom 86
focused ion beam machine (FIB) 84
Fokker–Planck equation 68
Foucault current 93
free energy 17, 22, 24
- density 17
- minimization 38

function 38
functional 38

g

Gaussian
- noise 77
- normal distribution 70
- unit 112, 169, 224, 226

Geiger counter 135
Giaever junction 53, 54
Gibbs energy 37, 59, 143, 155, 233
- barrier 143, 145, 173
- density 35
- of superconducting wires 31

Gibbs energy barrier 235
Gibbs free energy 18, 21, 125
Gibbs functional 126
Gibbs potential 34, 156
- barrier 233

Ginzburg–Landau (GL)
- equation 30, 39, 46, 118
  - time-dependent 107
- free energy 17

Ginzburg–Landau–Abrikosov–Gor'kov (GLAG) theory 15
Ginzburt–Landau (GL)
- equation 225

Giordano model 149, 165, 169, 174, 210
- experimental tests 175

Golubev–Zaikin
- attempt frequency 128
- microscopic theory of TAPS 125, 130
- theory 173, 174, 183

Gor'kov–Josephson equation 5
Gor'kov phase equation 5, 28, 119, 128, 134
Gor'kov's phase evolution 54
Green's function 15

h

Hamiltonian operator 69
harmonic oscillator 11, 29, 69, 237

heat exchange gas   29
Heisenberg uncertainty   9, 70, 74, 163, 166, 223
helium   92
 – cryostat   92
 – thermal conductivity   92
Helmholtz
 – free energy   174
Helmholtz energy   17, 37, 65, 154
   – barrier   173
   – density   35
   – Taylor series   154
Helmholtz free energy   4, 18, 21, 27, 31, 40, 65
   – density   16
   – functional   117
Hilbert space   163
hyperbolic sine function   120

i

indium   92
insulating   187
integral differentiation rule   147
inverse Kurkijärvi–Fulton–Dunkleberger (KFD) transformation   152
ion   18
isopropanol   81

j

Johnson–Nyquist thermal current noise   76
Josephson
   – energy   157
   – equation   65, 68
   – junction   57, 59, 61, 76, 113, 190, 191, 210
      – macroscopic quantum tunneling (MQT)   165
Josephson junction   53, 54
   – kinetic inductance   61
Joule
   – heated normal state (JNS)   132
   – heating   27, 35, 48, 136
   – power   55

k

Keithley voltmeter   98
Khlebnikov theory (K-theory)   185, 186
kinetic energy   17, 58, 75, 136
kinetic inductance   49, 60, 208
Kirchhoff equation   76
Kosterlitz–Thouless (KT) transition   12, 103
Kramers escape rate theory   113
Kramers theory   67, 157, 160

Kurkijärvi
 – critical current law   161
 – power law   160, 162, 187, 190
 – temperature law   152, 161
 – theory   143, 157
Kurkijärvi–Fulton–Dunkleberger (KFD) transformation   143, 147, 148, 191
Kurkijärvi–Garg (KG model)   236
Kurkijärvi–Garg model (KG model)   189, 190
 – derivation   195
Kurkijärvi theory   131

l

LabView program   97, 140, 147
LAMH
   – model   107, 114, 115, 122, 168
   – resistance   121
Langer–Ambegaokar (LA) theory   160
Laplace equation   225, 230
law
   – of quantum mechanics   10, 62
   – of the energy conservation   74
Likharev current–phase relationship   234
linear resistance measurement   152
linear transport measurement   164
Little–Parks (LP)
   – effect   46
   – oscillation   49, 232
Little's
   – fit (LF)   108–110, 115, 122, 201, 206
   – Geiger counter   135
   – helix   163
   – phase slip (LPS)   7, 12, 23, 33, 40, 79, 95, 104, 105, 110, 166
   – phase slip (PS)   187
London penetration depth   3
Lorentz force   157
low pressure chemical vapor deposition (LPCVD) process   79

m

macromolecule   105
macroscopic
   – quantum tunneling (MQT)   9, 55, 131, 163, 164, 197
   – solenoid   210
macrostate   25
magnetic
   – field   40
   – flux   228
      – quantum   47, 162, 226, 232
   – nanoparticle   10
master equation   146
Maxwell equation   42, 46

McCumber
- barrier 158
- theory 173
Meissner
- current 7, 49, 230
- effect 3, 232
metallic nanowire 81
microscopic effective action formalism 183
microstate 25
microwave radiation 70
molecular templating 79
molecule 86
molybdenum-germanium (MoGe) 83, 86, 89, 215
- alloy 215
- film 82
- nanowires 172, 184
- optimized alloy 102
monotonic function 144
Mooij–Schön velocity 160, 185
mounting procedure 91
multiphase-slip switching 188
multiple phase slips 151

n

nanoloop 227, 228
nanometer-scale wire 101
nanoparticle 10
nanoscale superconducting device 61
nanowire 10, 37, 48, 65, 74, 101
- antiproximity effect 165
- fabrication 79
- homogeneity 104
- insulating 198
- kinetic inductance 51, 172, 199, 209
- made of superconducting materials 101
- normal 198
- resistance 169
- S-type 198
- superconducting
  - destruction of superconductivity 197
  - temperature-dependent coherence length 183
- truly superconducting 198
- uniformity 104
National Instruments digitizer card 140
negative-curvature Arrhenius-like activation 179
net fluctuation effect 119
net phase slippage rate 116, 149
Newton equation 35, 63, 69
Niemeyer–Dolan technique 54

nonsuperconducting wire 187
nucleation theory 23

o

Ohm's law 44, 98, 115, 202

p

parabolic law 175
parity effect 3, 6
perfect superconductor 117
phase-evolution equation 5
phase fluctuation 71
phase slip (PS) 180, 189
- barrier 153
- heating effects 129
- overheating effect 138
phase-slip (PS)
- barrier 156
- tunneling 186
phase slippage
- event 169
- LAMH model 115
- net rate 120
photolithography 81–83
photon 208
- detector 53
photoresist 82, 83
pink noise 220
Planck's constant 3, 125, 136, 163
Planck's formula 74
plasma
- frequency 56, 61, 62
- oscillation 62, 175
positive
- magnetic
  - flux 55
potential energy 58
- barrier 163
probability
- density 217
- distribution 144, 148
pseudo-four-probe geometry 102

q

quantum
- dissipation 11
- fluctuation 24, 71, 105, 141, 177
- mechanics theory 163
- phase slip (QPS) 9, 128, 141, 142, 151, 164, 167, 183
  - fugacity 180
  - Giordano model 12, 149, 165
  - saturation behavior 194
  - ZGOZ theory 199

– phase transition   74, 200
– resistance   205
– Schmid–Bulgadaev (SB) transition   175
– temperature   173
– theory   10
– tunneling   95, 151, 163, 167, 201, 210
   – quantum-coherent   106
– Zeno effect   11
quasiparticle   7, 53
qubit effect   55

r

reactive ion etching (RIE)   79
renormalization group model   181, 206
reproducibility   90
resistively and capacitively shunted junction (RCSJ) model   57
retrapping current   133
return current   133
Riemann zeta function   126

s

sample installation   91
scalar product   41
scanning electron microscope (SEM)   82, 84
Schmid–Bulgadaev (SB)
   – phase diagram   179, 180
   – transition   11, 55, 74, 175, 187, 205, 210
Schrödinger equation   5, 6, 16, 69
   – time-dependent   5, 6
   – time-independent   6
semiconductor   7
series resistor   96, 98
shunt capacitor   37
shunting
   – effect of the electrode   184
   – resistor   74, 213
silicon (Si) chip   84
silver paste   94
single-phase-slip switching   152
skewness   220
solenoid   17, 33, 60
Sommerfeld formula   239
sputtering   81, 88
standard resistor   98
Stewart–McCumber model   53, 57, 113, 205
   – macroscopic quantum phenomena   68
   – of normalized Variables   76
Stewart-McCumber model
   – mechanical analogy   62
stochastic
   – premature switching   131
   – quantity   217

streptavidin protein   87
superconducting
– condensate
   – free energy   15
   – usable energy   15
– electron   3
   – density   3
– energy gap   166
– flux quantum   159
– loop   17, 49
– macromolecule   105
– nanaloop   227
– nanowire   29, 204
– order parameter   1, 15, 105, 126
– parity effect   3, 4
– quantum interference devices (SQUID)   53
– qubit   53
– wavefunction   163
– wire   32, 103, 187
superconducting state (SS)   131
superconductivity   1, 215
– energy gap Δ 2
– suppression   112
superconductor   9
– free energy   38
– Helmholtz free energy   46
– density   16
superconductor-insulator-superconductor (SIS) junction   53, 190
superconductor–insulator transition (SIT)   197
superconductor-insulator transition (SIT)   11, 90, 177, 181, 200, 202, 210, 237
superconductor-normal quantum transition   177, 200
superconductor-resistor transition (SRT)   187
superconductor-to-insulator transition (SIT)   187
supercurrent   2, 3, 28, 227
– density   3, 43, 50
superfluid
– density   4, 43
– velocity   3, 36, 224, 230
superpair   2, 4, 18, 43, 225
switching
– current   133, 143, 188, 191
   – distribution   139, 218
   – measurement   140
– event   133
– probability
   – density   139

– function 165
– statistic 139

t

Taylor
  – expansion 56, 66, 233
  – theorem 71, 154, 162
Taylor expansion 121
TDGL equation 166
thermal
  – energy 24
  – fluctuation 23, 74, 106, 141, 145, 155
  – phase slip 129
thermal fluctuation 36
thermally activated phase slip (TAPS) 8, 106, 118, 164, 187, 201, 238
  – Arrhenius model 149
  – Golubev and Zaikin theory 125, 130
  – high currents 157
  – LAMH model 107
  – TAPS-only model 151
thermodynamic
  – critical field 158, 169
  – equilibrium 46
  – potential 39
  – system 18
thermodynamic relation 30
thermodynamics
  – definition of entropy 20
  – first law 19, 32
  – second law 20, 32, 33
thermometer 94
tilted washboard potential 64
time-dependent Ginzburg–Landau equation (TDGL) 119
time-evolution equation 5
total entropy 25
transmission electron microscope (TEM) 82, 84
true superconductor 200, 205, 211
tunneling 163
  – Caldeira–Leggett principle of dissipative suppression 184
  – event 180
typical fluctuation 110

u

usable energy 22, 23, 36, 58

v

van der Waals forces 87
variance 217
  – estimator 220
vector potential 41, 47
voltage 27
voltmeter 9
vortex-antivortex pair 20

w

washboard potential 60
wave number 231
wavefunction 2, 10, 15, 69
  – collapse 10
wavevector 4, 105
white spots 88
WKB expression 72

z

Zeno effect 11
zero-bias resistance 97
zero-point fluctuation 72, 163, 166, 193
zero-temperature coherence length 47
ZGOZ theory 199